AF478347

UNIVERSAL BIOLOGY

Living systems consist of diverse components and form a hierarchy, from molecules to cells to organisms, which adapt to external perturbations and reproduce stably. This book describes the statistical and physical principles governing cell growth and reproduction, as well as the mechanisms of adaptation through noise, kinetic memory, and robust cell differentiation via cell-to-cell interaction and epigenetics. The laws governing the rates, directions, and constraints of phenotypic evolution are derived through dimensional reduction. Throughout the book, the perspective of consistency is emphasized – between microscopic units (e.g., molecules) and macroscopic states (e.g., cells), and between distinct spatial and temporal scales. By integrating theoretical, computational, and experimental approaches, this book offers novel insights into biology from a physicist's perspective and provides a detailed picture of the universal characteristics of living systems. It is indispensable for students and researchers in physics, biology, and mathematics who are interested in understanding the nature of life and the physical principles on which it is based.

KUNIHIKO KANEKO has been Professor of the Graduate School of Arts and Sciences at the University of Tokyo for 27 years, teaching mathematical biology, biophysics, and complex systems, and he is currently at the Niels Bohr Institute. He was also Stanislaw Ulam Fellow at Los Alamos National Laboratory, a visiting professor at Osaka University (Graduate School of Frontier Biosciences), University of Lyon, Freiburg University, and part of the external faculty of the Santa Fe Institute, a member of the Institute for Advanced Study at Princeton, and is Founding Director of the Center for Complex Systems Biology and the Universal Biology Institute at the University of Tokyo.

UNIVERSAL BIOLOGY

The Physics of Life through the Macro-Micro Consistency Principle

KUNIHIKO KANEKO

Niels Bohr Institute

Shaftesbury Road, Cambridge CB2 8EA, United Kingdom

One Liberty Plaza, 20th Floor, New York, NY 10006, USA

477 Williamstown Road, Port Melbourne, VIC 3207, Australia

314–321, 3rd Floor, Plot 3, Splendor Forum, Jasola District Centre,
New Delhi – 110025, India

103 Penang Road, #05–06/07, Visioncrest Commercial, Singapore 238467

Cambridge University Press is part of Cambridge University Press & Assessment,
a department of the University of Cambridge.

We share the University's mission to contribute to society through the pursuit of
education, learning and research at the highest international levels of excellence.

www.cambridge.org
Information on this title: www.cambridge.org/9781009575669

DOI: 10.1017/9781009575690

First published 2025

A catalogue record for this publication is available from the British Library

A Cataloging-in-Publication data record for this book is available from the Library of Congress

ISBN 978-1-009-57566-9 Hardback

Cambridge University Press & Assessment has no responsibility for the persistence
or accuracy of URLs for external or third-party internet websites referred to in this
publication and does not guarantee that any content on such websites is, or will
remain, accurate or appropriate.

For EU product safety concerns, contact us at Calle de José Abascal, 56, 1°, 28003 Madrid, Spain,
or email eugpsr@cambridge.org

Contents

Preface

About 80 years ago, one of the pioneers of quantum mechanics, Schrödinger, published the seminal work *What Is Life?*, in which he predicted the properties of DNA as an information-carrying molecule and discussed the importance of nonequilibrium. This work was an attempt by a physicist to uncover the universal properties of life and played a significant role in the rise of molecular biology. With its advance, the properties of individual molecules within living organisms are revealed. However, the mere presence of specific molecules, such as nucleic acids or proteins, does not mean that something is alive. We need to understand how a collective of molecules creates a dynamic living state.

Let us rephrase the question *What is life?* to *Can we reveal the universal properties of a macroscopic life system of such a collective of molecules and formulate a systematic theory for it?*. Although this may still seem a difficult question to answer, mankind once succeeded in developing a macroscopic phenomenology for a system of collective of molecules, *thermodynamics*, by restricting our concern to stable equilibrium states. Now, instead of thermal equilibrium, let us focus on a robust living state. Can we extract universal properties from it, discover general laws in it, and formulate a macroscopic theory for it? This is the question addressed in this volume.

Living organisms generally form a hierarchy: Molecules assemble to form cells, cells assemble to form individual organisms, and individuals assemble to form an ecosystem. In this book, instead of examining each molecule or cell individually, we will consider the fundamental properties of reproduction, heredity, adaptation, memory, development, and evolution by emphasizing the principle of consistency across hierarchical levels, such as the collective of molecules versus a cell. First, we will reveal the nature of the nonequilibrium state that sustains living organisms and search for the laws that govern cell growth. Next, we will investigate how the seemingly contradictory properties of plasticity (the ability to change in response to environmental changes) and robustness (the ability to maintain internal stability) are compatible in a living system. We will explore adaptation, memory, and evolution by considering the consistency between phenomena occurring on different time scales. We will also uncover general principles of multicellular organisms, where interacting cells differentiate into different types to form a stable tissue or organism. Finally, by linking plasticity to fluctuations, we will unveil the direction and constraints in the evolution of a biological state (phenotype) that cannot be explained by random genetic mutation and selection alone.

Of course, the goal of this book – to understand the universal properties of life, comparable to thermodynamics, or to create a filed of "universal biology" – cannot be achieved overnight. Here I have used the term "universal biology," which may be unfamiliar to most readers. As far as I know, this field was first proposed in the science fiction novel *Who Will Inherit the Earth?* by Sakyo Komatsu. In this novel, universal biology is defined as a field that seeks to understand the universal properties of life, not only restricted to those that have evolved on Earth but also including those in the universe or in synthetic protocell systems as we are exploring today. In this novel, published in 1968, researchers in the twenty-first century explore the universal principles of possible patterns of life. How close have we come to this novel of half a century ago? This book may also be a report to the late Sakyo Komatsu.

Life is a unique state characterized by *having a variety of components, maintaining its state, and having the ability to reproduce.* Can we discover a state equation that represents this universal class? Can we uncover the principles of reproduction, adaptation (the compatibility of plasticity and robustness), memory (the creation of inherent time and history), differentiation (the tendency to diversify into different types, whether in cells or species), and evolution (the feasibility and constraints of phenotypic change)? I hope this book will serve as a first step in considering these questions.

In 2003, I published *What Is Life?* in Japanese, the English version of which was published in 2006. The Japanese version has sold over 10,000 copies. Encouraged by this support and motivated by the progress in the field over the past 20 years, I have written this volume. In this volume, readers will find that the names of genes or molecules are rarely mentioned. Although I discuss biological experiments, the focus is not on the molecular details but rather on how we can extract the universal properties of living systems from the experimental results. The theoretical discussions are based on mathematics and physics together, with the results of computer simulations. However, the focus is not on equations or specific models – one can skip them – but on exploration of the logic how a robust living state emerges in dynamic systems with a large number of components.

Chapters 1 and 2 provide an overview of the goals and methods of universal biology. I present five methodologies, emphasizing the interplay between microscopic entities (e.g., molecules) and robust macroscopic states (e.g., cellular states): (i) phenomenological theory for a robust macroscopic state (cells, organisms), (ii) exploration of universal statistical laws at the microscopic level (e.g., molecules for a cell, cells for an organism), (iii) extraction of characteristic properties as a result of consistency between the macroscopic states and constituent microscopic units that exist at different spatiotemporal scales, (iv) experimental approaches that pay attention to such macro–micro relationships, and (v) consequence of violation of macro–micro consistency leading to aberrant biological states. Following this general framework, the book explores topics such as cell reproduction and the origin of life (Chapter 3), the adaptation process of organisms balancing plastic change and internal robustness (Chapters 4 and 5), kinetic memory (Chapter 6), cell differentiation and development (Chapter 7), the relationship between fluctuations and evolution (Chapter 8), and the constraints and directionality in the evolution of biological state (phenotype) (Chapter 9). Chapter 10 discusses ongoing or future studies, including the origins of life, endosymbiosis and multicellular organisms, the resilience of ecosystems, possible correlations between developmental and evolutionary processes, and the dynamic memory and learning of the brain, all from the perspective of consistency across hierarchical levels. The chapter concludes with a discussion of the mathematical framework needed to address the issues presented in this book.

I have designed this volume so that, after reading Chapters 1 and 2, you can move on to any of Chapters 3 through 9, depending on your interests (though it is ideal to read Chapters 4 and 5 together, and the same for Chapters 8 and 9). Of course, reading the entire volume in that order is recommended. If parts of Chapters 1 and 2 seem too abstract or difficult, feel free to skip them for now and return to them after reading the later chapters.

While a full understanding of the research presented in this volume may require some knowledge of analysis in dynamical systems theory with mathematical equations, one does not necessarily need to follow the equations to grasp the essence. However, since some readers (especially those with a background in physics) may find the equations helpful, I have included them in some sections. But again, you can skip them and still understand the essence. I hope you can appreciate the book according to your own interests and background.

This book could not have been completed without the help of many people. First, I would like to express my deep gratitude to the Japanese people who have continuously supported the present research through research grants and support from the Center for Complex Systems Biology and the Universal Biology Institute at the University of Tokyo (2017–). The support from Novo Nordisk Foundation and Niles Bohr Institute from 2022 is also very much appreciated.

I am also deeply grateful to Professors Yuzuru Fushimi, Toshio Yanagida, and Akiyoshi Wada, and to the late Professor Fumio Osawa for their continuous encouragement. This book is the result of numerous collaborative research projects with many colleagues and students, including Tetsuhiro Hatakeyama, Yuichi Wakamoto, Saburo Tsuru, Masayo Inoue, Yusuke Himeoka, Nobuto Takeuchi, Atushi Kamimura, Yuuki Matsushita, Takahiro Kohsokabe, Tomoki Kurikawa, Jumpei Yamagishi and Riz Noronha. Special thanks are due to Professor Chikara Furusawa, who could almost be considered a coauthor, for the close, stimulating, and collaborative research over 25 years. Although I cannot mention everyone by name, I would also like to thank all the other collaborators and members of my research laboratory for their invaluable help. I am deeply grateful to all of them.

Most of this book is based on lectures and seminars on Universal Biology that I have given to date. Along with the establishment of the Universal Biology Institute at the University of Tokyo, I started a course on Universal Biology at the Graduate School and have been giving this course for eight years to students in physics, biology, bioinformatics, and so on, while giving some lectures at the Graduate School of Frontier Biosciences at Osaka University, Hokkaido University, Kyoto University, and Ochanomizu University, as well as seminars for high school students and public lectures. In addition, I have presented the contents of the preset volume in a two-week lecture course at the International Center for Theoretical Physics in Trieste, as well as at schools or conferences in Italy, Denmark, Germany, France, Spain, Ireland, Armenia, Argentina, the United States, India, China, and so forth. The stimulating feedback I received from students and participants was a tremendous help in completing this book.

Some parts of this book were developed during my visits to various institutions. I am particularly grateful for the hospitality of the Institute for Advanced Study in Princeton. Of course, the continued support and stimulating discussions with members of Universal Biology Institute at the University of Tokyo and the Biocomplexity Group at the Niels Bohr Institute have been of great help in completing this book, and I am truly grateful to all involved.

Most of the present volume is based on my book *Universal Biology* (Fuhen seibutsugaku in Japanese). I am grateful to the University of Tokyo Press for the kind permission. For the cover of this book, I chose to use "Bird and Animal Screens" by Ito Jakuchu 1716–1800. I am grateful to Shizuoka-prefectural museum for allowing the use of the picture. The picture vividly represents dynamics and diversity of life with interaction among species. Universal biology seeks to uncover the universal laws and properties that apply for diverse life. With the picture, I hope one can sense the stirring of the new field we are developing in the volume.

1

The Potential of Universal Biology

Is There a Universal Nature of Life?

1.1 Introduction

The time might now be ripe for "universal biology." To the best of my knowledge, the notion of universal biology was first proposed in the science fiction novel "Who will take over?" by the Japanese author, Sakyo Komatsu. In the novel published in 1968, universal biology is a field that seeks to understand the universal properties of organisms, not only those that happen to have evolved on Earth, but also those that might exist in the universe beyond Earth and those artificially constructed in the laboratory. According to this science fiction novel, universal biology was described as:

*Universal Biology has expanded rapidly at the end of the last century(*) to explore the universal patterns and possible variations of life phenomena in the universe. Although this new field so far remained mostly at theoretical studies, it has opened up a fresh approach to life science, based on topological geometry[1] and has made rapid progress, to the point where the advent of groundbreaking theory in the field of biology is expected in the near future. (*Note: This novel was set in the twenty-first century; so this refers to the end of the twentieth century.)*

Half a century has passed since this novel was written. Is a breakthrough theory just around the corner? This volume is a report on the answer to this question at the current stage.

Here the question of whether an object is alive or not cannot be reduced to existence of specific molecules. Even if proteins or nucleic acids (DNA or RNA) are important for life on Earth, their existence does not necessarily mean that they are alive. Certain aggregates with certain composition of a set of molecules and their dynamic processes provide life. Accordingly, the universal pattern of life that we

---

[1] This definition is not entirely clear, but given that this novel was written around the time when René Thoms catastrophe theory (Thom, 1976) came out, this term would probably imply "dynamical systems theory" in the present terminology. In fact, in the novel, there is a description that young graduate students carry out computer simulations and find that the trajectory of state variables converges to a few possible states, as in the process of reaching attractors.

seek to understand could not be reduced just to the nature of individual molecules. It is also possible that organisms outside Earth may use different molecules instead of a nucleic acid or protein. (The inability to answer *What is life?* as a property of individual molecules has already been discussed in *Life: An Introduction to Complex Systems Biology* [Kaneko 2006].)

Rather, should we think of life as a universal state with the characteristic dynamic processes of an ensemble of diverse molecules. In other words, one must ask how a group of diverse composite molecules maintains and reproduces itself independently. Understanding the general nature of these systems is the goal of universal biology.

1.2 Universal Nature of Life

Can we then think that there should be some universal nature of life? What are the universal properties of life?

To discuss such an issue, one may seek for the definition of life. Unfortunately, it will be difficult to reach a definition that everybody agrees with. Generally, one would think of life as a system that can sustain itself and reproduce. However, on its own, a growing crystal would also be life. Then, some might intend to add Darwinian evolution for the requirement of life. In fact, it is the view by NASA. However, this is a different category of condition, or in other words, Darwinian evolution could be a more of consequence.

So what is it that stands out as a universal property of the life state we see? What we immediately notice is that it is composed of unusually diverse components. Even a simple cell has at least several thousand different molecules species. Despite such a diversity of components, they can be maintained, can reproduce itself rather faithfully. Hence, in answer to the question "What is life?," most people would agree with the statement that **Life involves diverse components, which are maintained and can be reproduced**. Such a state, consisting of diverse components, differs from what is usually observed in physics. This diversity is what makes it different from simple crystal growth that can grow and reproduce itself.

Surprisingly, such diverse components are assembled and maintained. There are as many as 10,000 different proteins in most cells, and many other molecular species coexist. The amount of each component is generally not constant but can be synthesized under certain conditions, and these different components are replicated to maintain their abundances, when a cell grows and divides. The composition is not completely identical before cell division but is maintained within a certain range. Thus, one can postulate the following property for a life (cell) system:

(0) Sustainment of diverse components and capacity for reproduction: The capability to reproduce a set of diverse components. All abundances increase to reproduce the set. After reproduction of a set (say, a cell), its composition

is not significantly different. In contrast to the growth and reproduction of a nonliving system, a life system maintains a huge variety of components and reproduces the entire system with approximately the same composition. The maintenance and reproduction of these diverse components may serve as a starting point for discussing the universal properties of living systems.

From this postulate, then, for a living state the following characteristics are generally expected:

(1) Adaptation: The capacity of a living system to respond to the outside world (i.e., environment) and maintain its internal state so that it can survive and grow under novel environmental conditions; as the system includes diverse components, this adaptation is not necessarily a property of a few molecule species, but a property of a system sustaining diverse components. This implies "homeostasis"[2] – the compositions of most components within are sustained without much change in the concentrations. This adaptation is a consequence of a system with diverse components that maintains itself.

(2) Memory: The above response to external change generally depends on the past history and, as a result, maintains the previous experience of a cell (individual) in an internal state; however, this memory is not perfectly static and can vary continuously over time. With diverse components in a cell, various processes with different timescales coexist and mutually interfere, resulting in changes that persist over long timescales.

(3) Evolution: As there are many diverse components, replication of exactly the identical individual should be difficult. Reproduction is not perfect, and some variation appears in each individual (cell) at the event of reproduction. Here, the speed for reproduction can vary depending on each variant. If this variation has some memory, a state with higher growth is transferred to the offspring, which has a higher chance of survival and reproduction. Thus, the variation and selection follow and evolution arrives.

(4) Differentiation/diversification: Taking into account of internal diversity in components, cells (or species) tend to diversify. Still, the condition that many components can be sustained will provide a constraint on the range of variation in their possible compositions. According to the constraint on reproducibility, the diversified states that are allowed for survival are restricted within a certain range, which probably form several discrete types. This generation of discrete types is known as cell differentiation or speciation, in biology. In a high-dimensional phenotypic state space (which are, for instance, given by the concentrations of each component), there are several types between which no

---

[2] Originally it means "similar to stasis." Currently it means self-regulating process by which biological systems tend to maintain stability by adjusting to external conditions that can vary.

intermediate states can exist: Such intermediate states cannot be stably reproduced. In short, the states are not just broadly distributed continuously but are structurally organized.

At this time, some components that are well inherited over reproduction, change slowly in time, and control others are separated. This component is responsible for genetic information. In contrast, those components (or properties) that are affected by genetic information are called phenotypes; depending on the phenotype, the capacity of survival and the populations of offspring are determined. As will be discussed, the separation between genotype and phenotype is a necessary course of a reproductive system consisting of chemicals mutually catalyze each other. Furthermore, the relationship between genotypes and phenotypes can change through evolution, which leads to constraints and direction in evolution, as will be discussed in the following chapters.[3]

Properties (1)–(4) for a life system could be a consequence of a system with the **maintenance and reproduction of diverse components** in (0). Although logical *derivation* of the properties (1)–(4) from (0) is not available as yet, it is natural to assume that the properties (1)–(4) are general properties of a life system satisfying (0).

Note that these properties are not represented simply by the features of a specific molecule but rather by properties of a system consisting of diverse molecules (chemicals). Indeed, the smallest unit of life, the cell, consists of a collection of many different types of molecules. If we reduce this diversity of components, there is a risk of losing the essence of life.[4] All these properties are dynamic processes, and in this sense, the universal properties of life should be considered dynamic complex systems.

Furthermore, considering the universal properties of living organisms, it will be useful to note the following three properties.

(a) Activity: The cell grows, divides, and survives by taking in nutrients from the outside and maintaining a nonequilibrium state. Sometimes the flow of nutrients to cellular components is referred to as the activity of the cell. However, even though biologists have intuitively grasped the concept of activity, so far, it remains at a qualitative level, and a quantitative definition is still missing.

(b) Robustness: Even though biological states are perturbed by several environmental disturbances or internal noise, they are not destroyed so easily but

---

[3] The basis for the formation of such a genotype–phenotype structure (as will be discussed later) is the difference in the fast/slow time scales mentioned in (2), in which the slow component, which changes on a much longer scale than the generational, is responsible for inheritance and provides heredity.

[4] There is an allegory by the ancient Chinese philosopher Zhuangzi (about second century BC).

remain within a certain range. Even if perfect replication of a cell is not possible, it can produce "almost the same" two cells due to the process of growth and division, even under the presence of environmental fluctuations or other sources of noise. In addition, the cell state is maintained to a certain degree, even under external or internal disturbances which can be relatively large. Life systems thus exhibit robustness. The quantitative characterization and understanding of robustness are the main themes of the present volume.

(c) Plasticity: Although robustness implies unchangeability, biological systems or some parts of them also change flexibly against environmental changes in order to achieve robustness. Some parts of biological systems change to a relatively large degree during adaptation to environmental changes, whereas others remain unchanged to achieve robustness. During evolution, biological states change over a longer term to survive in a new environment. This changeability is called "plasticity." At first glance, robustness and plasticity seem to be in the opposite direction. Yet, somehow life systems manage to reconcile them. To avoid changes of most part, some part of the system must be changed. Understanding how robustness and plasticity are compatible is also a core theme of the present volume.

In any case, our rule of thumb is that there is a universal state that sustains activity and can reproduce even under external and internal changes, having robustness and plasticity. If the universality of such a state can be formulated mathematically, one can make an important step toward answering the question, "What is life?".

1.3 Characteristic Features of Life

1.3.1 Comparison with Machines

Given these characteristics, it would make sense to compare life systems to human-made machines. Machines are also influenced by the external world and function while changing or maintaining their internal state, storing the external inputs the system experiences, responding accordingly, and influencing the external world. A machine's function is robust to a certain degree against disturbances. The machine cannot replicate or evolve independently, but it does have some activity, robustness, and plasticity. In this sense, we can note some similarities between machines and life.

For instance, consider a computer. An input is applied to it, which changes its internal state. However, the computer returns to its original state after the computation is completed. For computers to operate, they generally consume power. Thus, robustness, plasticity, and activity are important also in machines. However, there are important differences between machine and life, which are essential in considering the nature of the life system.

1.3.2 Activity, Robustness, and Plasticity

Activity: In the case of machines, energy is supplied to perform a specific function, but the activity of a living system does not necessarily have a specific purpose. Of course, it requires material and energy to survive and reproduce, and reproduction may be regarded as a purpose of such activity. However, a life system maintains and proliferates a wide variety of components that seem unnecessary by themselves, and a single purpose of the activity is not necessarily identified. Rather than a purpose, an activity can be regarded as a property of the living state itself. In addition, while energy is supplied to machines through external manipulation, life systems generate activity through energy produced internally with interaction with the environment, rather than using energy for its activity as its purpose.

Robustness: The machine must be stable to function. Usually, it adopts a strategy to reduce external or internal noise such that the disturbances do not affect the internal state. However, such a strategy is not always taken in life systems. The abundances of many components in a cell are not necessarily so large; thus, there is a large fluctuation in each reaction process. Indeed, if the number of molecules is N, there are typically concentration fluctuations of the order $\sqrt{N}/N = 1/\sqrt{N}$, as per standard statistics. If this number is less than 100, a fluctuation of more than 10 percent is expected. Cells are sustained with such relatively large fluctuations.

Furthermore, genes that serve as blueprints in life systems may sometimes be varied by genetic mutations, in contrast to finely designed and accurate blueprints for machine. As cells are difficult to implement a machinery to eliminate noise or genetic variation, they need to acquire robustness to noise or genetic variations. In this sense, high-level robustness of a system is postulated for life systems beyond the standard machines.

Plasticity: Machines can also adapt to external disturbances to a certain extent. However, they are often designed to work under given conditions. Therefore, we sometimes need to tune the system externally under different conditions so that the machine does not misfunction. Living things, even though they may not survive too extreme conditions, can often survive under a rather large degree of environmental perturbations, by changing their internal state through adaptation. While machines are usually preconditioned for operation, life systems can adapt to unexpected environmental changes.

1.3.3 Autonomy

Another difference between machines and life systems lies in the degree of "autonomy" (i.e., the capacity to control itself, or make its own rules for the state change). Life is a self-evolving system that does not require external designers. To a certain degree, life systems have evolved to have certain functions or behaviors under certain conditions. However, even in such cases, there is no external designer, and they are generated from autonomous changes. Autonomy exists both for changes by adapting to the environment and longer-term evolution.

As mentioned, genes are often considered blueprints. They determine the rules of the possible reactions. However, there remains some variety for possible states in individuals given the same genes. Because the fluctuations in the concentration of intracellular or environmental variations are relatively large, genes cannot accurately determine cellular behavior. There remains a great deal of freedom regarding which of these broad possibilities can be realized.[5] In this sense, what the genes prescribe is more of a "recipe," rather than a blueprint (Kaneko & Tsuda, 2000).

Moreover, the genes themselves have evolved autonomously based on the individual organism's phenotype. There is no external designer with a specific purpose. Hence, a life system is inherently different from that of machines with external designers. The consequences of this essential difference in evolution compared to designed machines are discussed in Chapters 8 and 9.[6]

We generally regard that there is higher autonomy when the behavior of the system cannot be predicted or controlled from the outside. How is such unpredictable behavior generated in living system? Note that, there are dynamics governed by many degrees of freedom that cannot be observed from the outside. Since all degrees of freedom can neither be observed nor controlled, the system has autonomy.

Recall that the abundances of each component need to be amplified for the cell (system) to grow and divide. Hence, the noise in the reaction process may also be amplified. This represents a source of uncertainty that is not rigidly determined by the rule of genes. Accurate determination by the blueprint is not possible in the life systems, unlike machines.

1.3.4 Interdependence among Diverse Components

As previously mentioned, life systems generally consist of diverse components that interact strongly with each other. For example, in a bacterial cell, there are usually more than a few thousand protein species. Furthermore, there are many other metabolic species. These components mutually influence one another. This "complex system" with a large degree of freedom is characteristic of life, and this huge diversity will be related to robustness and plasticity.

---

[5] This is called epigenetics in a broad sense. However, the current use of the term *epigenetics* is sometimes restricted to the molecular modification process that changes the degree of gene expression. One must be careful when using the term.

[6] Currently, drastic changes in evolutionary studies are ongoing, to the direction that seriously consider the direction and constraint in evolution of living state (phenotype), which goes beyond the simplest view based on simple one-to-one correspondence between genotype and phenotype.

Of course, as machines become more complicated, the number of parts to be combined also increases. In this case, to design a functional machine, we would generally try to avoid too much interdependence. We usually design a machine by first creating a part with a few degrees of freedom to perform its function and then combining it. When combining them, mutual relationships among parts are needed, which should be minimized. As the demand for machine functions increases, blueprints become more complicated. Still, we adopt the strategy of a simple combination of parts.

Is it then possible to understand a life system from such a simple combination of components? We have discussed this question before (Kaneko, 2006). In a system that can grow and reproduce, such as life, it is often difficult to view each part as separate and independent functional modules because strong interactions with each other are inevitable.

In a life system, many interrelated components form a network. Research that cuts out some module parts with a function from the total network has attracted much attention and has been intensively studied. This extraction of modules is sometimes valid, and the appropriate function for each module is identified. However, each part of a biological network is generally interdependent. The function that is considered in isolation is not necessarily the same as in the whole system. For example, each component of the modules in the gene expression network extensively influences the others (as also seen in Chapter 5). It is also often found that when a gene that carries out a certain function is knocked out, others replace the function to compensate for the loss. Thus, we need some other logic beyond the simple view of a combination of modules to understand how the life system works with the mutual interaction of diverse components, even if under certain conditions, some functional modules may predominate.[7]

1.3.5 Diversity

One may still wonder why life systems consist of so diverse components, considering that maintaining and reproducing such diversity systems may be difficult or costly. Why are life systems not reduced to mere combinations of simple systems with fewer components by eliminating redundant components? If one simply wants to construct a system with reproduction, an autocatalytic system with just a few catalysts, in which catalysts that need to reproduce themselves are made from resource chemicals, could be sufficient, and further, such systems can grow and replicate faster. We return to this problem in Chapter 3, but let us briefly examine it here.

---

[7] In studies of statistical physics over this half-century, it has been shown that systems with strong interactions among many elements exhibit inherently different behaviors from those comprising an ensemble of independent components or slight perturbations from them.

1.3.5.1 Why Diversity?

(a) Originally diverse?

Consider the reproduction of molecules before an accurate replication system based on the genetic information is established. The replication of individual molecules is not so precise. Polymers with different monomer combinations than the original template may often be synthesized. Therefore, it is plausible that many similar polymers coexist.

For example, consider polymers consisting of four monomers, some of which have catalytic activity and catalyze polymerization reactions. Generally, macromolecules cannot be synthesized without catalytic reaction steps. Therefore, it is unlikely that only the necessary macromolecules will be synthesized. Under standard thermodynamic processes, some polymers with sequences different from those of the necessary polymers for catalysis are also produced. As a result, it is natural to assume that a particular set of polymers is not necessary for the synthesis of catalytic polymers. In addition to specific polymers, a set of these polymers was maintained. As long as the macromolecular replication process is error-prone and the number of polymer sequences is large, it is inevitable that a large variety of molecules will coexist. A simple replication system consisting of only a few molecules is not easily achieved. Rather, it is more likely that a collection of polymers with different sequences will be maintained and are reproduced without accuracy (e.g., see Dyson [1985] and Kaneko [2006]).

(a') Parasite problems (parasitic molecules) inherent in a system with growth and reproduction

Polymerization can hardly proceed without catalysts because such reactions are too slow. Even if the process occurs by chance, degradation dominates over replication, and macromolecules cannot be sustained. In this sense, a cell is a machinery to encapsulate catalysts (enzymes) within it so that they are sustained without flowing away, thereby promoting the reaction of macromolecules inside (see also Chapter 3). The resulting macromolecules will not always be catalytically active; therefore, there will be molecules that are synthesized with the aid of other catalysts but do not aid in the synthesis of other molecules, called as parasites (or parasitic molecules). There can be much greater diversity in such parasites, and there can be other parasites for the synthesis reaction of one parasite, leading to the formation of a chain of parasitic molecules, resulting in an unnecessarily diverse set of molecules (Eigen & Schuster, 1979; Eigen, 1992; Dyson, 1985).

From this perspective, one can expect that when the reproduction process is ongoing, there is diversity with many molecules that are not necessary or even harmful.

(b) Limitation in available nutrient abundances

It is often assumed that sufficient nutrients are supplied for the replication of molecules, even if their number increases. Then a simple replication system would be advantageous, as it can generally have a higher growth rate than a complex reproduction system with diverse chemicals requiring more reaction steps to replicate a set of all components. Once such a replication system increases, however, nutrients are quickly depleted as the number of molecules increases. At this time, it is necessary to note that there is still a large variety of nutrient species, even though they are sparse and may be less efficient for the synthesis of necessary macromolecules. As the "good" nutrients are depleted, it would be advantageous to utilize all these less efficient nutrients. Indeed, it has been theoretically and numerically shown that a replication system with diverse components can survive and grow as nutrient abundances are limited (see Kamimura & Kaneko [2015, 2016] and also Chapter 3). As the replication of molecules progresses, they are inevitably surrounded by the same (or similar) molecules; therefore, if the system is simple and uses only certain nutrients, the competition among replicators will be tight. In contrast, complex systems that use diverse resources may relax such competition.

On the other hand, macromolecules with complex synthetic processes may be advantageous for containing constituent small molecules, because each step in the synthesis of a complex molecule contains nutrients. Complex polymers also break down into molecules that can later be used as nutrients. In addition, the complete degradation of such complex polymers takes a long time. By contrast, simple molecules are easily degraded. In this sense, simple replication systems tend to break down first when they run out of nutrients. The reproduction system that consists of complex macromolecules can therefore be maintained under nutrient-poor conditions.

Furthermore, let us consider a situation in which many replicators (reproductive systems) compete under an environment with a variety of nutrients that are not available in sufficient quantities. One possible strategy for each reproduction system is to withhold nutrients so that they are not used by others. If each reproduction system is independent of the others and competes only for speed, then it would be an effective strategy to use only the most efficient nutrients and convert them for reproduction. On the other hand, if many individuals share a common source of nutrients, the strategy of taking in and keeping them before they are taken by others may be relevant.[8]

---

[8] The strategy of maintaining diversity in molecules may be analogous to the K strategy instead of the r strategy in evolution theory. In the r strategy, organisms try to produce more offspring per generation, of which only a small fraction survives, whereas in the K strategy, only a small number of offspring are

(b') Advantages of diverse replication systems in varying environments
A system with diverse molecular species is often thought to be more advantageous when the environment changes over time. Such a system consisting of diverse molecules will naturally deal with different nutrient sources and can survive in fluctuating environments. However, one must be cautious in adopting this argument because the system cannot predict future changes in the environment, and evolution would select a system depending on the fitness at each moment.

On the other hand, one may wonder if a life system with diverse components, when placed in a certain fixed environment, might evolve ("deteriorate") into a much simpler system. For example, by putting bacteria (*Escherichia coli*) in a constant nutrient environment over many generations, the answer to this question would be obtained. Lenski et al. carried out long-term evolution experiments. A slight reduction in the number of genes was observed, but no noticeable simplification was detected (Elena & Lenski, 2003).

Why has a remarkable simplification not been observed? There are several possible explanations for this finding. One is the problem of parasitic molecules inevitable in any reproductive system. In a system with diverse components, the increase in some parasitic molecules will be suppressed by other parasitic molecules. With this sequence of parasitic molecule components, the system is sustained, so that they remain even under a fixed, well-nourished environment.

In addition, once a variety of molecules coexist, they may not be easily eliminated individually, even if many of them are unnecessary. It is possible that fitness (growth rate) can be increased by eliminating a set of molecules simultaneously. For example, consider a case in which components A and B are harmful to the system, but this effect is mutually canceled by their coexistence. Even though in this case both A and B are unnecessary, if only one of the two components is removed, the balance is lost and there will be damage to the system. The system can only increase its fitness if both components are removed at the same time.

This is the simplest example. As the situation becomes more complicated and intertwined, it becomes more difficult to simplify it, even if all such diverse components are unnecessary. This suggests that once a complex system is created, it is difficult to simplify it without breaking it entirely.[9]

---

produced and nurtured so that a large proportion can survive (MacArthur & Wilson, 1967). The term r and K comer from the equation $dN/dt = rN(1 - N/K)$.

[9] This situation is seen not only in living systems, but also in social systems. One renowned example is Parkinson's Law, which states that once bureaucracy is established, the number of bureaucrats does not decrease. For example, departments that monitor other departments are introduced to prevent other departments from running out of control. These departments cannot be reduced unless they are all simultaneously reduced (by "revolution"?).

1.3.5.2 Summary

When considering reproduction processes in an environment with diverse external components with limited abundances, the components of each reproduction system (e.g., cells) are diversified, even if their growth speed may not be optimal. Even if a system with lower diversity can grow faster, it cannot be sustained.

1.4 Macroscopic Theory for a Living State

1.4.1 Seeking the Possibility That We Can Understand the Living System

Although we do not yet have an absolute answer to the origin of diversity, the life systems that we observe always consist of diverse reactions that contain many components. Hence, it will be a sound attitude to characterize a biological system by assuming such a diversity of components. Then the goal is to explore the universal nature of such a system with diverse components, that is, with a high-dimensional state space.

Is it possible, then, to formulate a theory for a life system as a whole, without going into too much detail about thousands of components in each cell? If this kind of theoretical formulation were not possible at all, it would be impossible for a human being to understand what life is. In such a case, all that would remain would be the mere enumeration of a large number of individual components and their modification by external conditions, which is far beyond the understanding of a human being.

For instance, assume that you take a medical checkup and have a huge list of data from all blood tests with a huge number of components. Machine learning, in future, may provide you a diagnosis for your health and may provide a cure. If you have a list of the concentrations of all the components, one may guess the behavior of the cells. Prediction of the behaviors of organisms and cells might be possible even without understanding what life is. We might go in such a direction, following the current trend of machine learning with a huge database. One might give up the hope of understanding universal laws as adopted in science up to the twentieth century. This could be the case.

However, even if a large database is available, one must know how to search for relevant information that describes the living state. Which items we should pay attention to understand what is life will not be deduced from the database. One must know the nature of life before referring to the database. Then, a priori understanding of life by thinking our brain is required. Of course, it might be possible that AI can deduce it from huge data through machine learning. However, at this stage, the understanding by AI may not be understandable by humans. At present, the relationship between understanding by our brain and by machine learning is not clearly established, so that we cannot solely rely on AI for our understanding of life.

In either case, we do not take such a *pessimistic(?)* standpoint that our brain cannot understand what life is, and our classic science will be replaced by "Database + Search + Machine Learning".[10] It is a natural desire of humankind to understand what kind of system and what type of a state life is. Pessimism may be inevitable after one has thoroughly examined whether it is really impossible to understand what life is. It is not a scientific attitude to abandon the possibility of understanding life with reduced degrees of freedom, at this stage.

On the other hand, the optimism that we can understand the life system and that we can find universal laws may not be well grounded. One may wonder if we can understand systems that involve large degrees of freedom. However, our science has one success story. It is thermodynamics, which is a kind of *systems physics* that is universally applicable to a system with a huge number of molecules. In Section 1.4.2, we will briefly review the structure of this *thermodynamics as systems physics* and discuss whether universal biology can be considered by learning from its successes. (Of course, this could be an endeavor to catch lightning in a bottle twice. Readers can judge whether this is the case throughout the present volume).

1.4.2 Idiosyncratic Review of Thermodynamics

The target system of thermodynamics generally involves many large molecules (although this was not known at the time when thermodynamics were developed) that move around and interact with each other. Nevertheless (or because?) the discoverers of thermodynamics did not know that fact), one can describe the system using a few variables, such as temperature, pressure, and entropy. The theory of thermodynamics does not depend on specific molecules or individual systems. A theory with a high universality was established. It has the following structure[11]:

(i) Restriction to **equilibrium states**

The system, without external operation, is assumed to reach an equilibrium state in which the macroscopic change ceases. This could be an idealized state, strictly speaking. Thermodynamics is based on the empirical observation that the

---

[10] Only classical scientists with a traditional standpoint might feel "pessimistic."

[11] In the argument in this section, I assume that readers have learned standard thermodynamics and present a viewpoint that may not be necessarily discussed in the standard textbook. Readers who are not familiar with thermodynamics can skip this subsection, whereas reading a standard textbook of thermodynamics is also highly recommended.

macroscopic state approaches such equilibrium state. Thus, we restrict our concern only to the equilibrium states and the transition processes between them.[12]

Under such restrictions, it is possible to formulate a theory for states at a macro level using only a few variables such as temperature and pressure, without resorting to the details of the positions and velocities of individual molecules.

(ii) Stability and irreversibility

A postulate for the equilibrium state is that the system returns to it after perturbations to the system. This implies the stability of the equilibrium state and also the existence of an irreversible process. Without such stability, it would be meaningless to construct a theory of the system by restricting it to the equilibrium state. In fact, if a theory is restricted to a state from which it freely deviates, unless we thoroughly control it, such a theory would be almost inapplicable. On the other hand, we empirically know the stability of an equilibrium state and the irreversible approach to it.

For example, consider subsystems at different temperatures, such as $T_1 \neq T_2 \neq T_3, ...$, and put them into contact. Then the system will eventually fall into an equilibrium state with a homogeneous temperature T. Temperature is not spontaneously differentiated from place to place, without special operations from the outside. Alternatively, if the temperature is raised locally in some regions, it will go down, and later the temperature will be uniform again. This is known empirically.

Indeed, this stability of the equilibrium state is the basis of why we can describe an equilibrium system by using a few variables. In contrast, suppose that the temperature spontaneously differentiated by spatial positions. In such case, we would have to define different temperatures at each location. As the inhomogenization progresses, we need more and more temperature variables (at each position), to characterize the system. Eventually, the assumption that we can describe the system by using a few state-variables would collapse. Thus, stability and irreversibility underlie the thermodynamic premise that a few numbers can characterize the equilibrium state.

---

[12] The extension of thermodynamics to a nonequilibrium state has been pursued for a long time. Linear nonequilibrium thermodynamics is well established, in which the deviation from equilibrium needs to be sufficiently small, where the theory is formulated to be consistent with equilibrium. Attempts have been made to generalize it to far-from-equilibrium systems, but a theory for any nonequilibrium system would be too general to be reasonable. Then one possible generalization would be to restrict our concern to steady state out of equilibrium. In this case, to maintain the system out of equilibrium, some external flow is required. Now the difficulty arises, because there exists a large degree of freedom for such flow. For instance, we need to provide appropriate boundary conditions to define the flow. Still, for it, there remains a large freedom. Therefore, whether a description by few variables as in thermodynamic formalization, is possible remains elusive. Of course, it may be possible to limit this to only one class of boundary conditions. However, such an extension might then be too restrictive to assume it to be a universal theory for macroscopic steady states.

(iii) Description by a few degrees of freedom

As described in (i) and (ii), the macroscopic behavior can be described with a small number of variables in the equilibrium state (such as temperature, pressure, chemical potential, and so forth). Because the temperature is homogeneous, there is no need to specify the temperature for each location and a description using only a small number of macro variables is then possible. On the other hand, we now know that even at thermal equilibrium, the velocity of constituent molecules is distributed. If we intend to describe the state of each molecule, we need as many degrees of freedom as Avogadro's number. In other words, the system is described by a small number of variables only when our interest is concerned on a macroscopic scale, which is coarse-grained compared to the microscopic level. The possibility of such a coarse-grained description is not self-evident. Assuming that the equilibrium state is stable, this is possible, as mentioned above, and a description with a few degrees of freedom is allowed.

(iv) The ordering of irreversibility as described by entropy

In an adiabatic process in a closed system, there is an order in the transition between equilibrium states. When there is an order between states such that A can adiabatically (i.e., without the exchange of heat with the outside of the system) reach B, but B cannot reach A, and B can reach C but C cannot reach B, then the following ordering exists: $A \rightarrow B \rightarrow C$... For example, in order to cool down a room, it is necessary to dispose of heat outside by an air conditioner, so that the room with lower temperature cannot be reached adiabatically. On the other hand, even if one cannot exchange the heat with the outside, one can just rub a board with a stick and heat it with friction to warm up a room. Hence, adiabatic ordering exists in the equilibrium states.[13] If all the states we are considering can be connected by transitions, one can define a quantity representing the ordering reached by the adiabatic process. Then, the quantity S is introduced as $S(A) < S(B) < S(C)$,[14] which is "entropy."

(v) Distinguishing heat, a form of energy transfer, from the energy that is a conserved quantity

Thermodynamics concerns heat. While heat has the same dimension as energy, it is defined only as a form of transfer and is not a quantity of the state of the equilibrium system itself. Although energy is conserved in closed systems, heat could be generated within. This is because the irreversibility mentioned above

---

[13] An attempt to construct thermodynamics by postulating the axiom that adiabatically unattainable states exist close to every state, as was carried out by Carathéodory (1976).

[14] In this way, by comparing different equilibrium states in terms of adiabatic attainment, one can introduce a quantity of entropy as that describes the ordering of the adiabatically achievable states (Lieb & Yngvason, 1999).

increases the entropy in the process toward equilibrium. Thus, entropy, which is an indicator of irreversibility, is related to heat. If we consider the ideal process in which the state changes quasi-statically by maintaining the state closest to the equilibrium state at each moment, the internal heat generation will be minimized. In this case, the change in entropy is resulted from the heat transfer divided by the temperature (of the heat bath). Thus, when trying to consume energy to do a certain amount of work, the energy of the system is dissipated as heat to some degree. Thus, the energy available (for a given temperature) is not the energy itself that the system has, but the energy minus the entropy times the temperature.

(vi) Thermodynamic potential defined by energy and entropy

Given the stability of the system, it is natural to have a landscape picture in which a ball falls down to a valley. However, since the energy E is conserved, such landscape cannot be represented by using the energy itself. Instead, because the available energy mentioned above involves a subtraction of entropy (multiplied by temperature), it can be introduced as a quantity minimized at equilibrium and used as an index of the landscape, that is, as a thermodynamic potential. This is given by $F = E - TS$, where F is the free energy.

(vii) Equation of state

At equilibrium, there is a relationship among thermodynamic variables. For instance, there is an equation of state relating the pressure p, volume V, and temperature T. For an ideal gas, $pV = NRT$ (N is the number of moles and R is a constant) is satisfied. Although the relationship itself can depend on each material, the existence of the relationship itself is general.

(viii) Le Chatelier's principle

Let us change the thermodynamic conditions of a system externally, for example, by increasing the temperature, and explore the resultant response of the thermodynamic system. In general, this response, in the process of returning to the new equilibrium, occurs in the direction that reduces the externally induced change. Such "law of resistance" is called Le Chatelier's principle. For example, if heat is applied to a gas, to raise the temperature, then the volume is expanded, resulting in the decrease in the raised temperature; if the temperature is increased in a system with chemical reaction, the degree of an exothermic reaction decreases, and the resulting temperature decreases. This is a consequence of the stability of the original thermodynamic state of the system; therefore, "thermodynamic force" acts in the direction toward returning to the original state. It should be noted that the principle appears to involve time-dependent process, since we discuss the subsequent change after the environmental changes that compensate applied changes. One might find this a bit tricky as thermodynamics is concerned with equilibrium states (without dynamics). Still, thermodynamics can discuss the ordering of the transitions between thermodynamic states, so it is possible to discuss the time's arrow.

(ix) Fluctuation–response relationship

The thermodynamic systems are now known to consist of many microscopic elements (molecules) whose changes (motions) exist in the equilibrium state, which are rather random. For example, if we observe molecules in a small box within the system, their number or local energy in the small box is not constant but fluctuates. Such fluctuations typically manifest as Brownian motion. Brown found that small particles in a pollen move randomly in water, which was later shown to be produced by the collisions of water molecules on the particle. In other words, this fluctuation is a manifestation of motion of microscopic constituents (molecules) of a thermodynamic system.

Now let us apply some macroscopic change in the system by external manipulation (applying force, heat, etc.). When such an external force is applied, the system changes; accordingly, a response is produced. This is a response to the force at a macroscopic level, so one might assume that it has nothing to do with microscopic fluctuations. Einstein proposed that there should be consistency between such micro and macroscale phenomena so long as the equilibrium system is stable. He demonstrated proportionality between microscale fluctuation and macroscopic response (Einstein, 1905, 1906) – the proportional relationship between the variance of a given quantity (say, velocity) without external force and the response rate of the quantity under the application of an external force (i.e., change in the quantity divided by the force). This is generally called the fluctuation-response relationship (Kubo et al., 1985). Such a relationship generally holds for fluctuations around the equilibrium state and the responses at the equilibrium state.

(x) Introduction of a simple idealized model relevant to the construction of thermodynamics

Since thermodynamics is a universal discipline, it is by no means model-dependent. However, the introduction of an idealized model and the consideration of its behavior is of great importance in the construction of thermodynamics. Examples include the ideal gas and Carnot's idealized heat engine (Carnot cycle), which played an important role in the foundation of thermodynamics. The results obtained from such idealized models provided a guide to construct thermodynamics. Such models worked as scaffolds. Once thermodynamics was established, they could be abandoned, but they were useful for establishing thermodynamics at its infancy.

So far, we have given a rather idiosyncratic overview of thermodynamics. Considering the discussion above, we will now present a plan for building a phenomenology of the living state.[15] To limit our discussion, we focus on cells as a living system. Examples of the practice in making such macroscopic theories are seen in Chapters 3–9.

1.4.3 Possibility of a Phenomenology of Biological Systems by Referring to the Success of Thermodynamics

(I) Restriction to a steady growth or stationary state
Instead of a formulation limited to thermodynamic equilibrium states, the theory of life phenomena can be formulated by limiting it to the states with a steady growth or stationary state. It is an empirical fact that life systems include diverse internal components, but they maintain their components under growth. Therefore, we first restrict ourselves to a stable state in which the abundances of these various components steadily increase (or approximately constant in the zero-growth limit). We restrict our concern to such steady growth states and the transition process between them. When the environment changes, the state of the cell changes in which compositions of each component changes for a while, but after some transient time, the system reaches a new state in which each component can grow steadily. We focus on these steady states, and for the moment we do not consider the transient process to reach such a steady state.

In addition, there is a *sleeping* state in the living organisms in which growth has nearly ceased. This state is different from the thermal equilibrium state, but it is sustained at a (nearly) stationary state. For the time being, this state can be included in the above steady-state growth state by regarding it as the limit state where the growth rate is +0. (However, it may have different characteristics from the growing state, and we will study this issue in Chapter 3). In any case, we first restrict ourselves to a steady state of growth or maintenance.

(II) Stability (robustness) and irreversibility
We already mentioned that robustness is essential to a living state. A cell is under various external and internal disturbances, so that its state is perturbed. Still, the cell (as long as it is living) returns (or comes back closer) to its original state,

---

[15] Some physicists regard phenomenology as a primitive level of a theory. However, this negative attitude is totally misplaced. Thermodynamics, which is a phenomenological theory, has the strongest universality in physics. In addition, in the history of physics, novel progress has often been made by relying upon thermodynamics when the physics at that time came to a stage that needed revolution. For example, in quantum mechanics, consistency with thermodynamics is key to revising the mechanics. To establish quantum gravity, one will probably have to rely on the thermodynamics of the black hole. We also note that it is still impossible to derive macroscopic thermodynamics completely from statistical physics of microscopic elements; choosing relevant probability measure in statistical mechanics needs appropriate initial or boundary conditions for the microscopic mechanics. It is incorrect to view phenomenology as a primitive theory before the establishment of a further advanced theory.

as is sometimes called homeostasis. Empirically, biologists know that an adapted cellular state has such robustness.

As already discussed for thermodynamics, stability and irreversibility are two sides of the same coin. First, when the replication rate of each component deviates from the steady-state growth state, it returns to the steady-state growth state after some time. This irreversibility is inherent in the system. Furthermore, it is possible that a certain steady state may be more reachable than the other when the system is left alone. This is discussed in (IV).

(III) Description by using a few variables

As already stressed, there are many components in a cell. They include a large number of protein species, mRNAs, and metabolic components. Now, let us state that there are N types of such components. Then, N can easily exceed several thousand even if we consider only the protein species. The number will be much higher if we also consider the metabolic components. Even if we disregard the spatial arrangement of molecules in the cell and consider only their abundances, it seems that a large number of degrees of freedom are needed to describe this system.

Now, let us recall that we focused only on steady-state growth. Then, by the time the cell divides into two in one generation, each N-component should have been synthesized and doubled in number. Because each component i is replicated, each abundances increase approximately $dN_i/dt = \mu_i N_i$ with time t, so that $N_i \approx exp(\mu_i t)$. Of course, there can be deviation from this average increase for a short time scale; however, if we assume a steady growth state, we can roughly expect this growth form. For this growth to be steady, the growth rate of each component μ_i must be equal. Otherwise, as they grow, the increase rate of abundances of some components would be larger or smaller than others, so that the composition of chemicals would be deviated from the original. With the growth-division process of cells, the deviation would increase (such that the components either dominate or become extinct), and steady growth keeping the compositions of each would be violated. In other words, for the $N - 1$ condition, $\mu_1 = \mu_2 = \ldots = \mu_N$ is required. This common growth rate for each component gives a macroscopic quantity, which is the growth rate of the total cell. In other words, even though there are many components, if we restrict ourselves to the macroscopic state to maintain steady growth of cells, a macroscopic variable (cell growth rate) constrains the dynamics of all the components. This gives the possibility of description of a cell system by few variables, as will be explained in Chapters 2 and 9.

Generally, biologists often describe organisms or cells only by few terms, such as plasticity, robustness, and activity. The existence of description by such few terms has often been implicitly assumed, but they have not yet been expressed in terms of concrete "quantities." In thermodynamics, the feeling of being "hot" is expressed quantitatively by defining heat or temperature. In contrast, our current

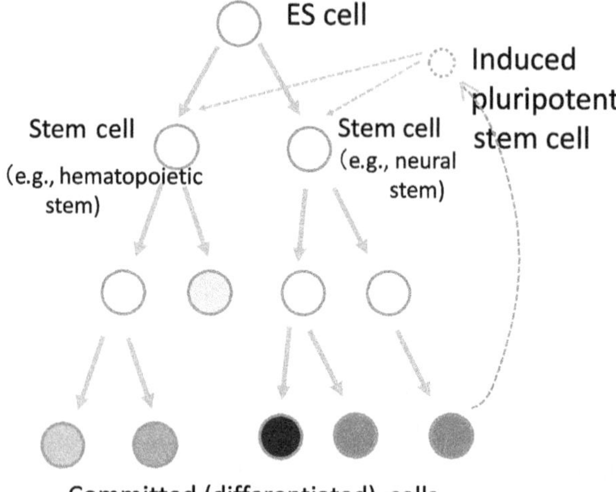

Figure 1.1 Irreversible differentiation process to lose potency from an embryonic stem (ES) cell, and reprogramming to generate an induced pluripotent stem cell as an operation to regain multipotency (schematic diagram).

understanding of the living state still remains in its infancy, that is, at the stage in which temperature was not quantitatively defined distinct from our feeling "hot or cold." One of the directions in this volume is an endeavor to overcome this issue.

(IV) Ordering of irreversibility

Even among cells that are in a steady (growth) state, there is an ordering in the state transition from one type to another is more feasible. In multicellular organisms, fertilized eggs, and pluripotent cells (ES cells) can differentiate into all other types of cells. From the pluripotent cells, cell differentiation progresses sequentially to committed cells under certain conditions (Fig. 1.1). This differentiation process cannot be reversed without special external manipulations. Thus, quantities can be introduced along one line of differentiation to describe the differentiation order.

On the other hand, by taking differentiated cells and applying external manipulation (such as over-expression of some gene expression level externally as adopted in making induced Pluripotent stem cells), the cells can be reverted to an undifferentiated state and then be induced to differentiate into other lines, as recent experiments have demonstrated (Takahashi and Yamanaka 2006). We could say that without such external operation, there is irreversible direction in cell differentiation, but external manipulation can revert it. (One may recall that in thermodynamics, there is irreversibility in the direction to increase of entropy for a *closed* system, but by the ejection of heat to external systems, the entropy can be decreased). Now by comparing the process of differentiation and the reverse process (called as reprogramming) of different cell types, it would be possible to characterize the ordering of cell states, and would be possible to introduce a quantity that characterizes an irreversible differentiation (see Chapter 7). Possible extraction of such "entropy-like" quantities from experiments on differentiation induction and their link to protein expression levels in thousands of dimensions will be discussed in Chapter 7.

(V) Activity and growth rate

When we consider cells as a system that takes in nutrients from the environment and grows steadily, we note that not all the nutrient abundances that are absorbed are used for cell growth. Some are used in metabolic activities to maintain the cellular state, which is responsible for the activity level of the cell. Some may be used for cell proliferation, while in other cases, most of them are used only for maintenance, not for growth. The efficiency of nutrient conversion to growth varies depending on the state of the cell. Therefore, in addition to the growth rate, we can also consider the amount of nutrient uptake, and/or the activity level of the cell as a macroscopic variable.

(VI) Activity and potential picture

As mentioned above, only a portion of the energy and nutrients that the cell receives from the outside is converted into growth. In other words, growth = nutrient uptake minus X, where X is the portion consumed in various reaction paths that do not directly lead to growth, and is the cost of maintaining the cell, which is related to the activity level of the cell. This structure is somewhat analogous to the relationship between free energy (that can be used for work) and internal energy. This issue will be discussed in Chapter 3.

(VII) Equation of state

At present, there is no definitive equation of state that expresses the cellular state in terms of a few macroscopic quantities. Still, many attempts have been made to determine the relationship between quantities that characterize the cell, such as the growth rate, stress, and nutrient consumption, by focusing on the steady-state growth state.

On the other hand, in thermodynamics, the equations of state for pressure, volume, and temperature can be used to discuss the phase changes between the solid, liquid, and gas phases. Similarly, in protist cells, three "phases" are universally observed.

(i) **Exponential growth phase**: the number of cells increases exponentially given a good nutrition condition.[16]

(ii) **Stationary phase**: When the cells become crowded and nutrients are limited, the cells cannot grow effectively, and the number of cells remains almost

---

[16] This is often referred to as the "logarithmic growth phase" or "log-phase," as the data are plotted in logarithmic in number. Still, it would be misleading to call the exponential-growth as logarithmic, we use the term "exponential" in this book.

constant. As a state of cells, it can be called the **dormant state**, where cells lose the activity and stop growing, and remain in a "sleeping" state. Here, cell growth is not necessarily stopped, completely, but the growth is much slower than exponential. In this state, upon appropriate supply of nutrients over sufficient time span, the cell can recover the exponential growth

(iii) **"Death phase"**: the cells stop growing (and eventually their structure will be collapsed), and cannot return to the state with growth. Thus, the transition to this phase is irreversible. This is in contrast to phase (ii), where the growth is recovered after the nutrient is supplied over a certain period.

These "phases" can be distinguishable by the differences in nutrient consumption, growth state, intracellular metabolism, and gene expression. Is there a way to describe the differences between phases and their transitions between them[17]? This issue is discussed in Chapter 3.

(VIII) Le Chatelier's principle
Life systems tend to return to the original state after external environmental changes, as is called by "homeostasis" by Cannon (1932). This is similar to Le Chatelier's principle in thermodynamics, which states that when there is an external change in conditions (temperature, pressure, etc.), there is a response in the system that reduces the imposed change. Similarly, the changes in the internal state of a cell (the abundances of individual proteins) caused by environmental changes, tend to be reduced by adaptation (Chapter 5) and evolution (Chapter 9).

(IX) Fluctuation and response
Recent studies have shown that the state of organisms or cells exhibit large fluctuations. For instance, recent measurements have demonstrated that the abundances of a particular protein in isogenic (cloned) cells vary from cell to cell. This is not surprising because the reactions in the cell are originally caused by molecular collisions, following random motion of the molecules. This fluctuation could be reduced if there is an internal feedback process to reduce it. However, as will be discussed, the reduction of fluctuation is limited because the growth of the cell needs amplification (see also Kaneko [2006]).

Here, external influences (say, environmental changes) can also change the state of the cell. The degree to which the internal state is susceptible to external environmental conditions can be quantified as the degree of "response." Recently, correlation (or proportional relationship) between the degree of response and magnitude of the fluctuation without external perturbation has also been found in biological phenomena. The relationship between this response and fluctuations

---

[17] In physics, the question is whether we can find macroscopic order variables that describe each phase.

is biologically significant because the response discussed here provides plasticity: the ability to adapt and evolve under external changes. In Chapters 3–5, we will discuss the relationship between this fluctuation and adaptation. Chapter 8 formulates the relationship between this fluctuation and the evolutionary response by extending the fluctuation–response relationship of thermodynamics.

(X) Ideal cell model

As thermodynamics can be developed by introducing ideal gas and referring to it, introducing an ideal cell model will be useful for formulating a macroscopic theory of life phenomena. As an ideal gas is not necessarily identical to a real gas, an ideal cell model is not the same as a real cell. Rather, it is important to strip away the details of reality and idealize them to make analysis easier. In a similar manner as thermodynamics is generally valid, not just for ideal gas, the ideal cell model will be a scaffold for constructing a macroscopic theory of a cell state. In an ideal cell model, for example, only the minimum conditions that a cell must meet are considered, such as external nutrients being taken up and various components being produced from them by catalytic reactions, while the catalyst (enzyme) is also produced as a result of the reactions; the cell grows and divides as a result, and the composition of the individual components in the cell is almost maintained. The aim of this book is to understand the properties of (I)–(IX) through [thought + computer] experiments and to derive universal laws from them.

1.4.4 Difference between Thermodynamics and Macroscopic Universal Biology

In the above, we discussed the similarities between thermodynamics and the macroscopic phenomenological theory for living states that we are planning to formulate. So far we have not established a theory comparable to thermodynamics, nor have we confirmed its possibility. I hope that by reading the present volume, readers will find that this possibility may not be just a dream or fantasy. Still, one may find that this will not be easy by considering the following important differences between the possible theory of life phenomena and thermodynamics.

- In contrast to a large number of molecules, we have a large number of (chemical) species. Statistical mechanics, which links thermodynamics to microscopic molecular dynamics, is based on a large number of molecules.[18] On the other hand, in a cell, the number of molecules for each chemical species sometimes decreases by something on the order of 1–10, whereas the number of species is large, at least 10^4. Thermodynamics itself is applied to any number of molecular

---

[18] In reality, we assume the order of Avogadro's number $\sim 10^{23}$. Still, according to the results of numerical simulations, even 10–100 is often sufficient.

species (kinds), but usually the number of species is not so large. Rather, it is based on the limit in which the number of each molecular species is large. In the theory of life phenomena, such a "many-kind" limit (i.e., diversity limit) is required, rather than a "large-number" limit. The theory that deals with the statistical properties in this limit is still underdeveloped.

- Multi-level hierarchy: There is a hierarchy with multiple (often more than two) levels, and micro and macro levels are sometimes not well separated.

Of course, there is a hierarchy between microscopic (atoms and molecules) and macroscopic (the whole system) scales, even in a system with which thermodynamics is concerned. In the case of thermodynamics, the hierarchy is relatively well separated.[19] On the other hand, in the case of living systems, it is not always easy to separate the micro and macro levels.

There, in contrast, mutual relationships between the levels are intrinsically important. There is a hierarchy of molecules, cells, individuals (tissues), and ecosystems (Fig. 1.2). Moreover, the numbers of elements (say molecules or cells) in each level in the hierarchy are not constant, and they can increase or decrease through the birth-and-death process (except for the entire ecosystem). As the hierarchy continues, each element (except for the lowest element, which is a molecule),[20] has internal degrees of freedom. For example, the internal composition of a cell can alter its state. This internal state is influenced by what elements are distributed around it, or, in other words, by its upper hierarchy (in the case of a cell, the state depends on what kind of tissue it belongs to). In other words, there is an influence from the higher to the lower levels. This situation is inherent to complex systems (Kaneko & Tsuda, 1994, 2000), where the state of a part changes depending on the nature of the whole consisting of the parts. This makes it difficult to construct a phenomenology of living states. However, by restricting ourselves to a state with steady growth or a steady state, as mentioned in Section 1.4.3 (III), "macro-micro consistency" should hold so that the upper and lower elements grow at the same rate (i.e., the number of each molecule doubles, when a cell doubles). Hence, we may expect a phenomenological theory that takes hierarchies into account.

- The internal state is not constant in time, but can oscillate or change over time.

As we will see in the examples in later chapters, the amount of each component in a cell increases steadily and can also oscillate in time. However, if the oscillation itself is stationary,[21] we can consider the phenomenology as a

---

[19] However, in some cases, we need to assume the existence of molecules, so that the separation of scales is not necessarily complete.

[20] Still, biomolecules are typically polymers that consist of monomers and thus have lower levels. For example, proteins change their state depending on external conditions.

[21] That is, if we can consider it as an attractor in a dynamical system (see Chapter 2).

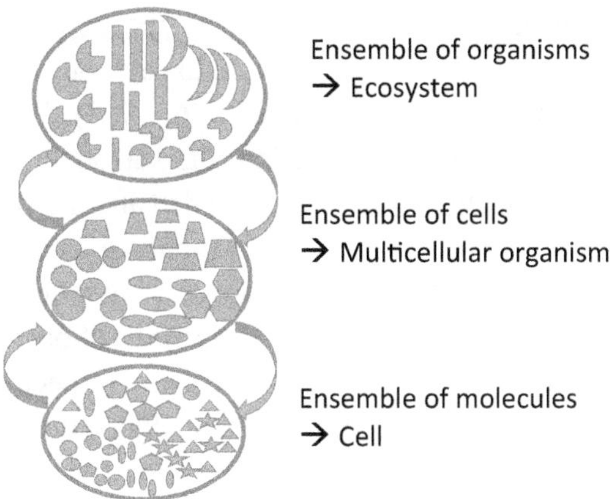

Figure 1.2 Hierarchy of molecules, cells, individuals, and ecosystems.

stationary growth state if the state is averaged over a longer time span than the time scale of the oscillation.

Another important point in life systems is that they usually consist of dynamics consisting of many orders of magnitude different time scales (one consequence of it will be the generation of memory). We will discuss this issue in later chapters.

Although it will be too optimistic to expect the success of formulating a theory of living systems in analogy to equilibrium thermodynamics and statistical mechanics, one may still hope that the macroscopic theory of life systems will be constructed by suitably noting the above differences. It is important to be aware of the connections between the micro level with many degrees of freedom and the macro level with few degrees of freedom, and to apply the idea of complex systems that seriously consider their mutual dynamics across levels. This will be discussed in Section 1.5.

1.5 View of Biocomplex Systems: Dynamic Consistency between Macro-Micro Levels

1.5.1 Hierarchical System

In Section 1.4, the importance of hierarchies in living systems such as molecules, cells, tissues, individuals, and ecosystems[22] is discussed. This hierarchy is empirically known for biological systems. Is this hierarchy necessary for a system that has diverse components and reproduces while keeping some degree of robustness?

---

[22] If we consider human society, there are hierarchies such as individuals, families, communities, and nations.

A system of replicating molecules requires the reproduction of catalytic molecules. Therefore, a set of molecules that help each other produce themselves must emerge and be maintained (Kauffman, 1986, 1993). However, as the number of molecules increases due to replication, the surrounding area will soon be filled up with them. The molecules are then crowded and interfere with each other. Therefore, the catalytic reaction processes and replication can be maintained only when the molecules form moderate aggregates. As mentioned earlier, parasitic molecules that lose their catalytic ability and replicate without supporting the replication of others can usually emerge therein. Parasitic molecules can replicate by being catalyzed by others. As they do not catalyze others, they are able to replicate while other catalysts help others, so that the fraction of parasites increases. If the replication of parasites continues, the system will soon be dominated by parasitic molecules. Then the replications of molecules will soon be stopped, which is a "parasite" problem (see also Chapter 3).

Suppose that a population of molecules forms a spatially compartmentalized structure, a protocell, that divides as the number of molecules increases and the protocell grows. In this case, the protocell with less parasites (more catalysts) will grow faster, whereas a protocell filled with parasitic molecules cannot grow and survive. As a result, the number of parasitic molecules will be suppressed in the survived protocells. The formation of a hierarchy to make a compartment (protocell) level, which consists of a population of molecules, is thus essential for reproduction to continue (see Chapter 3 and Eigen [1992]; Szathmary & Demeter [1987]; Boerlijst & Hogeweg [1991]; and Hogeweg [1994]). In other words, for the molecular replication process to continue, a cellular level of reproduction inevitably arises in molecular replication systems.

This leads to the first step in a **spatial** hierarchy. In general, in a replicating system, higher-level unit for reproduction is generated, to keep the growth sustained. On the other hand, as discussed in Kaneko (2006), in a system of mutually catalytic replication, the molecule species with a slower replication rate and having smaller numbers is shown to control the system, and are better conserved to the next generation. If the number of molecules of a given species is large, the change to such molecules would be averaged out and do not make much impact on the replicating system. In contrast, changes ("mutations") to a scarce component have much stronger influence on the system, as the averaging by the large number does not work. (Consider the extreme case that the number is just one, for instance). As a result, a molecular species with a small number and slower replication will play the role of "genes," that is, carrying heredity and controlling the behavior of a protocell. In other words, the existence of a temporal hierarchy allows for the separation of genotypes and phenotypes, making evolution more likely to occur. In this respect, a system with time-scale separation can evolve, as will be discussed in Chapters 3 and 9. Thus, not only the spatial hierarchy, but also the **temporal** hierarchy, is essential for a life system to survive.

As will be discussed in Chapter 3, the separation of the components carrying genetic information from those that carry function is a universal property of replication systems with a molecule–cell hierarchy. In this case, a small number of components that replicate slowly and change only slowly will be carriers of genetic information (as in DNA). In contrast, components that can change on a much faster time scale serve catalytic functions (e.g., proteins). This theme of separation of timescales and hierarchy is a common thread of the present book. In later chapters, we will explore how slow components can control fast components to enable adaptation and evolution.

There is a similar question regarding how an ensemble of diverse cells can maintain replication in the hierarchy between cells and multicellular organisms (consisting of aggregated cells). First, when a cell population forms a multicellular system, there may be multiple possibilities. Cell types with different genotypes can coexist and form a colony, as in bacterial biofilms. There is also a reversible change between multicellular and unicellular organisms, such as cellular slime molds, in which unicellular organisms aggregate into a multicellular system under nutrient-poor conditions, whereas they survive as unicellular organisms when nutrients are available. Most present multicellular organisms, in contrast, begin with a small number of germ cells and undergo a stable cell differentiation process to produce an individual. In this case, a small number of cells (i.e., germ cells) are transmitted (inherited) to the next generation. This differentiation between germ and somatic cells is probably important for multicellular organisms to avoid catastrophe by parasitic cells (e.g., cancer cells) and to ensure evolvability. This conceptualization of the stability and evolution of hierarchical replicating systems is common to molecule–cell and cell-multicellular-organisms.[23]

To sum up, understanding the common properties of hierarchical systems is a fundamental issue in universal biology. The fact that hierarchical systems can be maintained and grow stably requires that relationships for consistency be established between the levels.

1.5.2 Macro-Micro Consistency Principle across Levels in a Hierarchy

Let us call the lower level in the hierarchy "micro" and the higher level as "macro." Examples of each topic at the hierarchical level are listed in Table 1.1.

---

[23] The organism-ecosystem hierarchy has similar situation but could be somewhat different, because the ability of the ecosystem to replicate itself is uncertain.

Table 1.1 *Microscopic processes and macroscopic phenomena in each theme in universal biology.*

Theme	Micro	Macro
reproduction	molecule replication	cell reproduction
adaptation	expression of each gene	cell growth
memory	epigenetic modification	long-term change in cellular state
development	cell growth and differentiation	multicellular development
evolution	genetic change	phenotypic change

Here, the macro side results from (an ensemble of) micro sides, whereas the macro side provides the conditions (field) for the micro side to maintain, grow, and reproduce. Hence, instead of a one-way relationship in which the micro determines the macro, both sides determine each other.[24]

Let us consider the case where both the micro and the macro sides grow, keeping the microscopic constituents of the macroscopic state. Thus, the growth rate of each element on the micro side must be the same. Otherwise, the proportion of a particular component will decrease or increase. Moreover, this growth rate is equal to the growth rate on the macroscopic side (for example, the growth rate of a cell). In other words, the micro and macro sides must grow synchronously. This is just an example of **macro-micro consistency**, which is a guiding principle in constructing the phenomenology of complex living systems.

This consistency is achieved when the cell is in a state of steady maintenance or growth. When the environment change the cell adapts to the new environment. During this adaptation process, such macro-micro consistency is not necessarily sustained. The growth rate of each component is not identical, so as to allow each concentration to change for the adaptation. Therefore, after extracting the universal properties based on the consistency, the next step is to investigate how it collapses when the environment change and how the transition to a new state occurs in which the consistency is restored. From the perspective of consistency, the following points must be considered.

- How is consistency achieved? The micro level is smaller in spatial size and faster on the time scale than at the macro level. How can the phenomena at these distinct scales remain consistent? In addition, there are usually different factors on the micro and macro sides associated with the respective scales. The processes of different factors, such as physical processes, chemical

---

[24] In fact, even the relationship between thermodynamics and statistical mechanics is not unidirectional, but in the case of living systems, this factor truly comes into play because of their dynamics with growth and reproduction.

processes, cellular processes, population dynamics, and genetic changes, remain consistent. How then can the phenomena at these different scales be consistent?

It is also important to consider how phenomena on different timescales can be reconciled. In the process of cellular adaptation, there are fast and slow changes in intracellular chemical dynamics. Naturally, fast changes occur first, and are then transferred to slow changes. Often, slower changes compensate for faster changes to ensure stability, thus linking phenomena on different time scales (see Chapters 5 and 9). For instance, epigenetic modifications of DNA work at a much slower time scale than metabolic changes, and then, genetic evolution occurs at the slowest timescale.

- Universal laws as a consequence of consistency. To achieve consistency across scales, constraints are generally imposed on multiple components at the microscopic level. As mentioned, to achieve consistency with the cell growth, the growth rates of each component must be the same. This leads to a statistical law for the abundances of each component on the microscopic side. On the other hand, the macroscopic side may be described by a few variables owing to the constraints of the microscopic side, even though the microscopic side requires many degrees of freedom.

  Then, biologically relevant characteristics, such as plasticity, stability, and activity may be described by few degrees of freedom. If macroscopic variables and microscopic statistical laws can be linked in this manner, it will lead to an understanding of the universality of living systems following the spirit of thermodynamics.

- Collapse of consistency and its recovery. Of course, living systems do not always maintain consistency across the hierarchy and maintain a steady (growth) state. It would also be worthwhile to study how such a consistent state collapses. For example, in bacteria, as already mentioned, there is a phase in which steady exponential growth collapses, known as the dormant state. When strong stress is imposed, cellular states change over time as the consistency between molecular replication and cellular reproduction collapses. At the level of a multicellular-organism, cancer is regarded as a state in which the consistency between cells and tissues is disrupted.

When the consistency is not sustained, disordered states with temporal variations and large fluctuations appear even at a macroscopic level because the microscopic elements can no longer maintain coherence. When cells are not yet adapted to the new environment, larger fluctuations emerge over cells, leading to an increase in plasticity, and cells change in time before adaptation is recovered. After increasing the direction of change and plasticity, a novel, adapted phenotype is acquired, where the fluctuations decrease and the consistency between cellular growth and molecular replication is recovered (see Chapter 3). Therefore, it is important to investigate the temporal dynamics of many microscopic elements (e.g., concentrations of many components and gene expression levels) and their statistical distributions during this recovery process, and establish a possible connection between these changes and macroscopic dynamics (e.g., recovery of the cell growth rate). With such picture in mind, we intend to establish a fresh viewpoint on adaptation, differentiation, and evolution.

1.6 Outline

This book aims to explore the principles of living systems, noting inter-hierarchical consistency. Chapter 2 describes the research methodology and presents an example of a law that can be derived from the consistency principle of the steady-state growth theory. From Chapter 3, we look for universal laws for each subject, based on the methodology introduced in Chapter 2, which covers reproduction, adaptation, memory, differentiation, development, and evolution.

2

Methodology in Universal Biology

2.1 Introduction

In Chapter 1, macro-micro consistency is proposed as a guiding principle to formulate universal biology. In order to achieve the consistency between layers, a stable state must be generated both on the micro (molecular) and macro (cellular) sides. Of course, generation of such a state is not always achieved. However, when cells keep on reproducing themselves or maintain their states, this kind of consistency between micro and macro sides will be achieved. Then, can we describe such states quantitatively and uncover some laws therein, as a consequence of this consistency principle?

In order to conduct research from the perspective of consistency between such hierarchies, it is necessary to investigate universal properties from both the macroscopic (A) and microscopic (B) sides, and to explore the consequence of consistency (C) between them from both theoretical and experimental perspectives. It is also essential to consider how phenomena on different time scales are linked (D) in order to connect the micro and macro scales. On the other hand, it is also necessary to discuss the biological significance of the breakdown of consistency. For each of these methodologies, we will first take the position of (A)–(D) on the theoretical side, and then proceed to experimental investigations (E) based on the standpoint of (A)–(D). In addition, it is necessary to discuss the biological meaning of the collapse of consistency (F).

In this chapter, I will sketch the methodology of universal biology with a few examples for the above (A)–(C). Specific studies on the basic problems, based on (A)–(F) in the universal biology, will be presented in Chapter 3 and beyond, which cover reproduction, adaptation, memory, differentiation, and evolution.

2.2 Brief Description of Dynamical Systems Approach

First of all, we briefly discuss a concept of state space and dynamical systems, as most of the theoretical studies we develop in the present volume adopts the picture of dynamical systems. Even though the mathematics therein are not needed to understand the present volume, it is necessary to understand its viewpoint.

State space: Dynamical systems consist of the time, a set of state variables, the rule for their temporal change, and initial conditions of the states. The state is represented by a set of k variables. This k is called the number of "degrees of freedom." Then, the state at each time is represented by a point in the k-dimensional space, called the state space (or phase space).

For example, consider a cell consisting of k chemical species, whose concentrations are given by $x_1, x_2, \ldots, x_k$. These variables represent metabolite or protein abundances in a cell, or the gene expression levels. With recent advances in technology, they can be measured specifically. In general, k is rather large, so that the state space is high dimensional.

Dynamical systems: In dynamical systems, it is assumed that the temporal change of these state variables is determined by these state variables at a given time. In terms of an ordinary differential equation, the rule is given by

$$dx_i/dt = f_i(x_1, x_2, \ldots, x_k) \qquad\qquad (i = 1, 2, \ldots, k), \qquad (2.1)$$

The important point here is not a mathematical form or solution of this equation, but rather that this gives a k-dimensional flow in the state space. For each point in the state space, the rule of dynamical systems gives an arrow for its change (vector), with which the temporal change occurs. The temporal change of state variables is uniquely determined by each point in the state space (i.e., the set of variables at a given time). When an initial point is given, the temporal evolution of the state is provided as an orbit in the k-dimensional space, which follows the arrows.

Attractor: In the dynamical systems (to be precise, dissipative systems), an orbit cannot come back to the neighborhood of some of the initial points. With time, the region that an orbit visits and stays is restricted. The region within which the orbit recurrently returns to its neighborhood is called an attractor. Roughly speaking, the attractor is the region to which the orbit is attracted as time goes to infinity. The regime before the orbit reaches an attractor is called the transient.

The simplest and one of the most typical attractors is the fixed point. In this case the attractor is a point in the state space, where x_i does not change any more with the rule of dynamical systems. The flows around the fixed-point attractor are directed toward it. Another case is the cycle, in which the attractor is a closed curve. This explains the periodic cycle, that can explain, for instance, rhythmic behavior.

Stability against noise: Note that the description by k variables will not be often sufficient. The influence of some other unknown sources can often be regarded as noise in the system. The noise causes fluctuation around the orbit in dynamical systems. Depending on how attraction to the orbit is strong, stability of the system against the "noise term" can differ. If the magnitude of the flow to an attractor is large, then the attractor has higher stability against noise.

Basin of attraction: Starting from an initial condition, the state is attracted to an attractor. From a region of initial states, the orbits are directed to an attractor. In a simple case, the states are attracted to a single attractor starting from any initial conditions. In some dynamical systems, there are multiple attractors, and depending on initial conditions, the state is attracted to one of them. The region of initial states that are attracted to one attractor is termed as the *basin of attraction* to the attractor.

Bifurcation: In dynamical systems, there are some parameters that govern its rule. Such parameters do not change in time, once the system is given. For instance, the reaction rates are parameters that govern the dynamical systems of the concentration change of corresponding chemicals. Here, one can externally change the parameter (say, by changing the temperature). Then the behavior of the dynamical systems changes accordingly. The location of the attractor will change continuously by changing the parameter value continuously (say, the concentration of some chemical at steady state increases or decreases following the change in the reaction rate). Now, at a certain value of parameters, the attractor may lose its stability, because the direction of flow (arrow) changes, and the arrows that were originally directed to the attractor may now depart from them. Accordingly, the original state, (say a fixed-point) is no longer an attractor. Accordingly, the state is directed to a different attractor, which may have already existed or may be newly created. Such qualitative change of attractor according to the parameter change is called the *bifurcation*.

The above is a very brief introduction to dynamical systems. Please refer to standard textbooks for further understanding of dynamical systems (Strogatz, 2001; Hirsch, Smale & Devaney, 2003; Takagi et al., 2025).

2.3 Formation of Types as a Consequence of Consistency

2.3.1 State with Consistency between Macro and Micro Levels

We now discuss how a consequence of consistency is represented in the state space in dynamical systems. Recall that there is a strong constraint on each element in a hierarchical biological system, due to the consistency across levels. Then, such a consistency may not be satisfied globally within the total state space and the consistent state exists only within a limited region of it. When perturbed from outside, the state is attracted back to its original state (within a certain range), but if it is perturbed significantly, the state may move into a new consistent state that lies far from the original region. As a result, we expect the consistent state to exist as a discrete type only within a certain range (see Fig. 2.2(a)).

2.3.2 Types as Attractors

The attractor of the dynamical system, as discussed in Section 2.3.1, provides a coherent explanation of the above view of discrete types. The attractor is the region in the state space, to which the state is attracted with the time evolution (Fig. 2.1). The initial states within a certain range are attracted to the same attractor. The state can be attracted to different attractors from different regions of the initial state. If there are noise or environmental disturbances, the state fluctuates around the attractor. Fluctuations broaden the region in which these states can exist, but the states cannot stay beyond a certain range around the attractor. Different attractors may exist farther away, and there is a region that seperates each of them. Hence, multiple discrete states that fluctuate around the attractors exist like clouds in the state space. Thus, the concept of attractors provides a basis for understanding how living systems are diversified and divided into discrete types (see Fig. 2.2(a)).

Different cell types in multicellular organisms are examples of such discrete types that exist as clouds in the state space. In Chapter 7, we will see how such discrete cell types emerge as consistency between hierarchies of macro-micro levels (i.e., tissue-cell). In this case, both the state of each cell type (micro side) and the ratio of the number of cell types (macro side) are maintained within a certain range to achieve consistency. This stabilizes both the cell state and the number ratio of cells of each type.

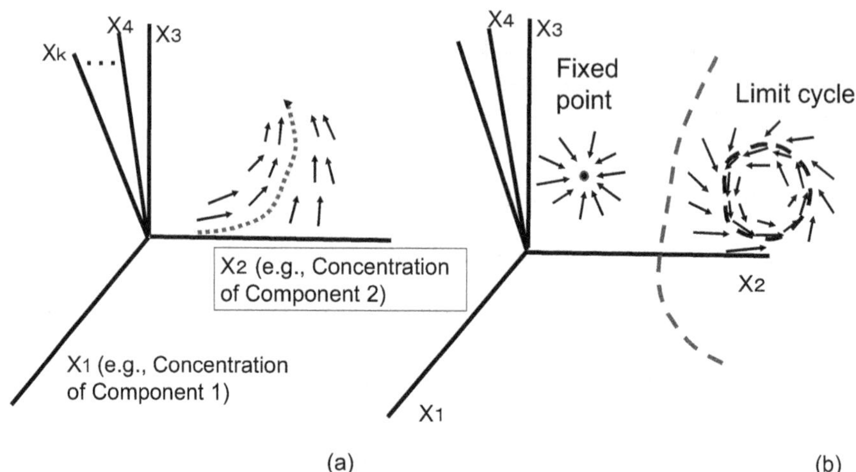

Figure 2.1 (a) In dynamical systems, the state change in time is represented by a flow in the state space. At every point of state space, the flow, represented by arrows, is provided, and the temporal orbit along these arrows (schematic diagram). (b) Attraction to a fixed-point attractor and a limit-cycle attractor.

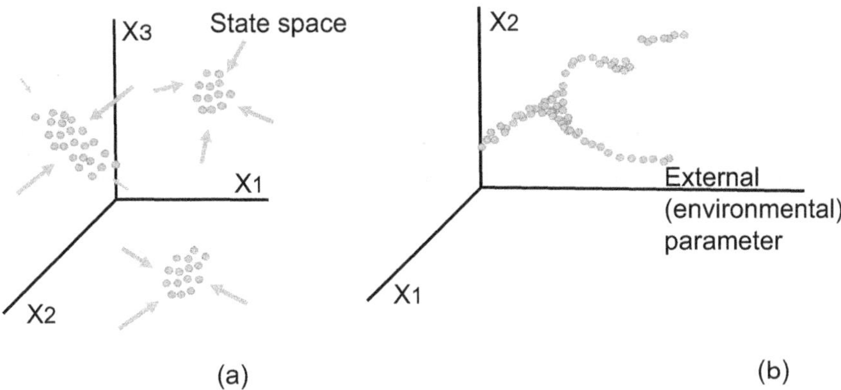

Figure 2.2 (a) Cellular states robust to perturbation exist as separated clusters in the state space. In the state space, they are understood as attractors (b) Upon environmental changes, the states change continuously up to a certain level, and then beyond it, bifurcation of the states or discontinuous changes can happen. (Schematic diagram).

2.3.3 Type Formation by Bifurcation

The attractor picture can explain the existence of discrete types of states that a cell (an individual) with the same genes reaches through evolution under a given environment. On the other hand, it is often observed that when external parameters such as environmental conditions change, the state changes gradually up to a certain range, beyond which drastic changes with different characteristics occur. Then, even in response to external changes, the state transitions into discrete types.

In other words, up to a certain point, the state changes gradually, maintaining an equilibrium that preserves macro-micro consistency. However, when external changes exceed a certain threshold, the constraints for consistency requirements can no longer be met. When this occurs, the state changes drastically, and the variables that characterize the micro and macro sides also change qualitatively. As a result, a new state is reached that satisfies consistency. In other words, in response to environmental changes, the state changes gradually up to a certain point, beyond which such continuous change can no longer maintain consistency, and the system adopts another state with a drastic qualitative change.

The concept of "bifurcation" in dynamical systems is important to understand the above qualitative change of attractors at a certain point of the parameters. The interval over which no bifurcation occurs against the external parameter changes defines the domain of each type, whereas the bifurcation provides a move to a different type. This view provides a way to consider the divergence of states into discrete types in response to environmental change (Fig. 2.2(b)).

This typification does not directly explain the existence of discrete species in evolution. Rather, as discussed in Chapters 8 and 9, the state (phenotype) remains stable and largely unchanged up to a certain extent in response to genetic changes. Furthermore, when a major environmental change alters the state of the cell (e.g., protein composition) and adaptive genetic evolution occurs, the state tends to return to its original state. In other words, the state of each type is quite stable and remains within a certain range, against both genetic and environmental changes. The original state is destabilized only when a major environmental or genetic change occurs beyond a certain level.

2.3.4 Transitions between Types

In Section 2.3.3, we discussed typification by consistency regarding the adaptation to environmental changes. Such changes in states are not only against external perturbation but are also seen as the result of internal changes generated by the living system itself. The response against the gradual increases in the number of cells during development and the changes in enzyme activity due to genetic changes are examples. In response to these changes, fast changes occur first, and then slow changes occur in line with them. Because microscopic changes are generally faster, it is natural that microscopic changes occur first, which are then transmitted to slow macroscopic changes, and then both the macroscopic and microscopic processes stabilize each other to stop their changes.

Let us now consider how this typification has evolved over time. We assume that the temporal changes consist of slow changes in the steady state that maintain the type at each level, and rare, drastic changes that alter the type. The development of multicellular organisms, for instance, consists of several epochs in which changes are gradual, which are interrupted by a transitional regime of drastic change. Such developmental processes are frequently observed, including cell differentiation, as described in Chapters 7 and 10.

Applying this view of fixation from fast-scale to slow-scale processes to evolution, we expect that the phenotypic states, such as the amount of protein or stress responses, are later stabilized over several generations by genetic changes. In other words, the phenotypes that change first are genetically consolidated. This view was proposed by Conrad Waddington, as is termed *genetic assimilation* (Waddington, 1954). This is to be understood as a correspondence between faster phenotypic changes and slower genetic changes, where fast changes are embedded in slower changes. We discuss the validity of this view both theoretically and experimentally in Chapters 8 and 9. Then, the direction in which phenotypic evolution is likely to occur will be predicted.[1]

2.4 A: Macroscopic Phenomenology

Macro-phenomenology is based on the premise that a biological steady state, at a macroscopic level, can be described by a few variables. This is a way of looking at a phenomenon and grasping its essence, just like the introduction of temperature and entropy in thermodynamics. An example of (partial) use of this style of research in biology can already be seen in the pioneering studies of Waddington's epigenetic landscape (Waddington, 1957).

Waddington was concerned with cell differentiation in multicellular organisms. Initially, there are cells with the potential to differentiate to many other cell types (pluripotency). After differentiation to each cell type during development, cells are gradually fixed to each state. Each cell type is stable, and even if perturbed, the cell returns to its original state. This process is represented by the change in the landscape as shown in the figure. This landscape represents the process of cells losing their pluripotency, and to differentiate into distinct cell types sequentially, and stabilizing themselves over developmental time, as branching valleys (Fig. 2.3). This landscape gives intuitive illustration of robust differentiation process. Then, what

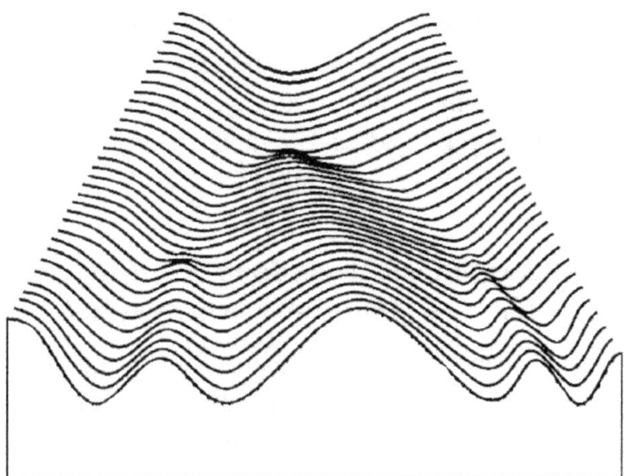

Figure 2.3 Epigenetic landscape by Waddington. Schematic figure based on Waddington (1957).

---

[1] Newman put forward the view of consolidation in the developmental process, in which generic processes in physics are later fixed by genes (Newman, 1994).

is the height axis of this landscape? In physics, we often depict potential landscape where the (free) energy changes with the state. However, the height of Waddington's landscape does not represent the (free) energy, and it is unsure what the axis represents. However, the landscape picture has high descriptive power for a macroscopic phenomenology of development. Can we define this height operationally, express it quantitatively, and describe the possibility of the switch from one valley to another? This will be one of the questions that macroscopic phenomenology aims to answer.[2] In addition, the depth axis of this figure can be considered to represent the developmental time of occurrence. The horizontal axis will represent the cellular state, but how can the cellular state with many gene expressions be represented by a single scalar quantity? How is the height represented as a macroscopic quantity to represent the robustness or plasticity of a cell? We need to answer these questions to understand the landscape of developmental process.

The above is an example of phenomenological approach to understand life, aimed at development. In general, it is essential for our understanding of life to develop such appropriate phenomenological description based on the intuition on a living system that is envisioned by excellent biologists. They, for instance, have an image of the activity of a cell, which itself is composed of a large number of *microscopic* components. The phenomenology that extracts the *macroscopic* cellular activity from microscopic components is needed, which may provide a relationship with the activity, cellular growth rate, and nutrient uptake rate. Such *macroscopic* representation may be given from a potential landscape as in thermodynamics. From the depth and width of the valley in the landscape, one may get information on the robustness and plasticity of the cellular state. Furthermore, Waddington also adopted this landscape picture to understand the direction of evolution, which will be discussed in detail in Chapter 9.

2.5 B: Statistical Laws of Consistency from Microscopic Side–Statistical Law in a Multi-component System

Although macroscopic phenomenology is important, its establishment is often difficult even in physics. It will then be more difficult for complex systems such as living organisms, where the relationships between the layers are bidirectional. Therefore, research from the microscopic side is also necessary.

When we consider a cell, the microscopic side consists of a collection of many molecules, and the macroscopic side is a cell, and both of their abundances are

---

[2] As mentioned in Chapter 1, we should not make negative evaluation on phenomenology, following the success of thermodynamics and the history of physics.

increasing. Since the amount of each component in the cell changes due to reactions, the micro side is described by a dynamical system with many degrees of freedom. Because reactions occur stochastically, it is necessary to view them as stochastic processes. In addition, it is important to deal with the statistical distribution of the quantities of the microscopic degrees of freedom. Simulation of ideal cell models with many degrees of freedom, analysis from dynamical systems and stochastic processes, and statistical physics theory of fluctuations are appropriate research methods from the microscopic side.

In the case of stationary growth with a large number of components, is there a universal statistical law among the microscopic components? For example, in the equilibrium state, statistical laws such as the Maxwell distribution and the Boltzmann distribution are well known. On the other hand, in the case of the cell in question, each component increases its number to achieve steady growth of a cell. Is there a common statistical law among the multiple components? Indeed, we previously found such laws theoretically and experimentally (Kaneko, 2006). Let us briefly review them.

2.5.1 Statistical Law Concerning the Abundances of Components

Let us first discuss reproduction of a cell, which consists of several replicating molecule species that catalyze the synthesis of new molecules. As a result of the replication of molecules, a cell grows until it divides to produce two cells with similar chemical compositions. These replications of molecules need to be somehow correlated; otherwise, the chemical composition of a cell cannot be maintained, and reproduction of cells with similar compositions will not continue. At the very least, a membrane that partly separates a cell from the outside has to be synthesized, and this synthesis must keep some degree of synchronization with the replication of other internal chemicals. How can such recursive production and chemical diversity be maintained at the same time? Is there some statistical law for a system that sustains such reproduction?

To investigate universal properties in the intracellular dynamics of replicating cells, we studied several cell models, consisting of intracellular catalytic reaction networks that transform nutrient chemicals into other chemical species. A large number of chemical species (say 10^4) in this model cell mutually catalyze reactions. Nutrient chemical(s) transported from outside of the cell are transformed to other molecules through the catalytic reactions. Successive transformations of chemicals can progress through these catalytic reactions starting from nutrients. When the number of specific (or total) molecule species goes beyond some threshold, the cell is assumed to divide into two. We studied a variety of models within this class, by adopting a stochastic simulation for reaction.

When a certain condition is met, we have discovered that the cell continues reproduction, approximately maintaining the compositions of chemicals, where the growth is optimized (Furusawa & Kaneko, 2003). Reproduction of a cell with diversity in chemicals is generally possible, even in this simple setup with mutual catalytic reactions. Universal statistical characteristics of such a reproduction state has been explored.

First, we studied the statistics of the abundance of chemicals, for cells that keep reproduction and chemical compositions. We measured the rank-ordered number distributions of chemical species, and found that the number of molecules N_k as a function of its rank k follows $N_k \propto k^{-1}$, that is, the distribution displays a power-law with an exponent of -1. Indeed, this kind of power-law was first studied in linguistics by Zipf (1949), measuring the frequency of word appearances, as is called Zipf's law.

In our model, this power-law of abundances is maintained by a hierarchical organization of catalytic reactions. Major chemical species are synthesized, cat-alyzed by chemicals with a slightly smaller amount of abundances. The latter chemicals are synthesized by chemicals with much less abundance, and so forth. This hierarchy of catalytic reactions continues until it reaches the minor chemi-cal species. (Indeed, with the aid of mean-field analysis in statistical physics, the appearance of power-law distribution with the exponent -1 is explained). Also, the universality of Zipf's law is confirmed by taking a variety of cell models with different parameters, reaction networks, and conditions for nutrient transport and cell division. Furthermore, this power-law is also confirmed by measuring gene expressions (i.e., by measuring the abundances of a huge variety of mRNAs). For all cell types we checked over a hundred, we confirmed this power-law with the exponent -1. This law is a consequence of consistency between replication of cells and faithful reproduction of cells, as briefly discussed below.

For a cell to grow to increase the abundance of its components, the com-position needs to be somehow organized so that it is kept after growth and division process. This growth process implies positive feedback process to amplify each of the components. Then, the more components tend to become more abundant (the rich get richer). However, since changes in each compo-nent proceed by reactions aided by catalysts, (enzymes), if one component wins out too much, there will be no more catalytic components to help the repro-duction of that component. As each component cannot be produced by itself alone, there must be other component(s) that catalyze that component for the steady growth. Then such component(s) also need other component(s) for their reproduction. To repeat this process, there must be a set of components that catalyze each other. In this situation, one (or few) components cannot over-dominate in the abundances. The components that catalyze each other need to be synthesized by utilizing nutrients, which will be maintained by keeping appropriate fractions.

Here, nutrient components branch out to create other components by different reaction paths. To maintain the components in a steady state under the conservation law, components that are lost in reactions are ensured to be synthesized. If the reaction paths in the network branch out with an average of K, component amounts decrease by $1/K$ per reaction while branch points increase by K. Consequently, the quantity of the component ranked K drops to about $1/K$. This illustrates Zipf's law, where quantity and rank are inversely related. While this explanation is somewhat simplified, for a deeper analysis, refer to (Kaneko, 2006; Chapter 6; Furusawa & Kaneko; 2003). Here, it should be noted that this discussion assumes the system is optimized for growth. Assessing whether the system can adapt to such optimal state under varying environmental conditions requires further examination, as discussed in Chapter 5.

In a system consisting of elements that can grow by simply catalyzing each other by taking nutrients, the structure to catalyze each other cannot be maintained, if the growth is over-optimized to leave only a few dominant components. Even though the growth rate increases with nutrient influx up to a certain point, the system collapses and cannot grow (or "die"), if it exceeds that point. In other words, a system with many degrees of freedom, which is maintained mutually, the system dies irreversibly, when its structure is destroyed. This point is suggestive in considering a living system consisting of many components.[3] The cell needs some adaptation process to avoid such death, which will be discussed in Chapter 5.

2.5.2 Log-Normal Distribution on Abundance Fluctuation

So far, we have discussed the average abundance of each chemical. Because the chemical reaction process is stochastic, the number of each molecule differs between cells. We then studied the distribution of each molecule number, sampled over cells, to find that the number distribution of each molecule number n_i is fitted reasonably well by the log-normal distribution that is,

$$P(n_i) \propto 1/n_i exp\left(-\frac{(logn_i - log\overline{n_i})^2}{2\sigma}\right),\qquad(2.2)$$

where $\overline{n_i}$ indicates the average of n_i over cells (Furusawa et al., 2005). In other words, the distribution of the chemical abundances is fitted by the normal (Gaussian) distribution when the logarithm of the abundances is taken. This means that

---

[3] In the story of chaos in the Chinese philosopher Zhuangzi (Second century BC) mentioned above, reducing the degree of freedom too much implies the death of the system.

the distribution has a rather long tail over the side of the larger amount. In a simple cell model consisting of mutual catalytic reactions, this log-normal distribution holds for the abundances of all chemicals that are reproduced within a cell.

Why does the log-normal distribution law generally hold, in spite of the trend expected from the central limit theorem caused by the addition of several fluctuation terms? In fact, the Central Limit Theorem, a fundamental theorem of probability theory, states that when uncorrelated random variables are added up, the average of them approaches a Gaussian distribution as the number of variables increases. On the other hand, the observed protein-concentration distribution in cells has a longer tail on the side with a larger value, as is closer to a lognormal distribution.

In general, when successive catalytic reactions for recursive production exist in a biochemical reaction network, fluctuations are multiplied successively through the catalytic reaction cascade. Then, by taking logarithms of concentrations (i.e., $logn_i$), these successive multiplications are transformed into successive additions, and the problem is reduced to the addition of random variable. According to the central limit theorem, the distribution of $logn_i$ is expected to converge to the Gaussian distribution. Hence, the log-normal distribution of n_i is derived. This log-normal distribution is also experimentally verified by measuring some fluorescent proteins introduced into cells with the help of flow-cytometry. Note the log-normal distribution here is also a consequence of the reproduction of a cell, and is a result of *consistency between replication of molecules and replication of a cell*. This also suggests that when a cell is located in a condition that cannot maintain its state for reproduction, it may exhibit deviation from these statistics.

The concentration fluctuation is caused by the noise of the chemical reaction through the random collision of molecules, but the magnitude of the noise depends on if the fluctuation in the abundances is amplified or reduced through other reactions. Fluctuations can be reduced if there is a "negative feedback," such that if an amount is too large, its synthesis reduced. On the other hand, a cell that is growing steadily will need some positive feedback process to produce itself, which will amplify the fluctuations.

The concentration distribution of the components under such a positive amplification process generally takes a form different from the usual Gaussian (normal) distribution: As mentioned above, successive catalytic processes lead to log-normal distribution. Furthermore, dilution by growth, as described in Section 2.6, is also a multiplication process that occurs in proportion to the concentration of each component. In this case, if the growth rate fluctuates, multiplicative noise process will be introduced, so that the distribution will be close to lognormal (see Chapter 3).

On the other hand, statistical theories that we will use in later chapters are often based on Gaussian distribution. Therefore, when applying such a theory, it is effective to use the logarithm of the abundance (concentration) as a variable that represents the cellular state, rather than the abundance (concentration) itself. The use of logarithm is not only for the distribution, but is adopted in the response of the (average) quantity to environmental changes, as will be seen in Section 2.6. In this case, too, the multiplicative process in which the quantity increases in proportion to its own is fundamental, so that the analysis using logarithms is effective.

Remark: The log-normal distribution is a simple and generic example for statistical variables with successive multiplications. In real cells, there can be some deviation, due to some factors. If the successive steps of multiplication are few, the so-called gamma distribution follows. When the concentration of some chemical is low and its number goes close to zero. There exists a deviation because the number is discrete (0,1,2,..) and cannot be less than zero. Some examples of deviation will be briefly discussed in Chapter 3.

2.6 C: Macro-Micro Consistency

2.6.1 Significance of Consistency

Of course, the macro and micro descriptions must be well connected, by the consistency principle. For example, in a steady-state growth condition, the fact that the microscopic components grow at the same rate constrains their statistical distribution on the micro side, which in turn constrains the state change on the macroscopic level. In this mode of research, we explore possible universal laws that link the two levels through the macro-micro consistency.

In the case of statistical mechanics, this macro-micro consistency has been most effective in the fluctuation–response relationship mentioned earlier. Here, fluctuations are derived from microscopic random motions at the level of atoms and molecules, while the response is the changes of macroscopic states. The consistency of these two scales led Einstein to derive the general relationship between fluctuations and responses (Einstein, 1905, 1906). In Section 2.6.2, we will extend this fluctuation–response relationship to life phenomena in the form of the relationship between fluctuations in the molecular concentrations in cells and the adaptability of cells.

2.6.2 Response Ratio as a Measure of Plasticity

The phenotypic plasticity in biology represents the degree of change in the phenotype in response to environmental variation. As a quantitative measure for it, we consider the ratio of change in a phenotype against environmental change. For example, consider the concentration (x) of a specific protein that depends upon the concentration (s) of an external signal molecule. Then, the response ratio, R, could be computed as the change in the protein concentration divided by the change in the signal concentration ($R = \Delta x/\Delta s$).[4]

Now the relationship between response ratio and the variances of phenotype is explored here. In physics, the changeability of a system is studied as the response of a system against external force. As mentioned, a precise relationship exists that relates the response of a system to its fluctuations in statistical physics (Kubo et al., 1985), as was pioneered by Einstein's Brownian motion theory. There, he proposed the proportionality between fluctuation and response, by noting that the same random forces that cause the erratic Brownian motion of a particle also underlie the resistance to the macroscopic motion of that particle when a force is applied (Einstein, 1905). This insight can be generalized so that the response of a system's variable to perturbation should be proportional to the fluctuation of that system in the absence of an applied force. Very roughly speaking, the more something varies, the more it will respond to perturbation, irrespective of the precise molecular details (see Fig. 2.4).

To apply the fluctuation–response relationship in biology, consider a system characterized by a parameter a and a state variable x, and discuss the change in x against the change in the parameter value a to $a \rightarrow a + \delta a$. Then, the proposed fluctuation–response relationship (Sato et al., 2003; Kaneko, 2006) is given by

$$\frac{<x>_{a+\Delta a} - <x>_a}{\Delta a} \propto <(\delta x)^2>, \tag{2.3}$$

where $<x>_a$ and $<(\delta x)^2> = <(x - <x>)^2>$ are the average and variance of the variable x for a given parameter value a, respectively.

The above relationship is derived by assuming that the distribution $P(x; a)$ is approximately Gaussian and that the effect of the change in a on the distribution is represented by a bilinear coupling between x and a. With this assumption, the distribution is written by

$$P(x; a) = N_0 exp\left(-\frac{(x - X_0)^2}{2\alpha} + v(x, a)\right), \tag{2.4}$$

with N_0 a normalization constant so that $\int P(x : a)dx = 1$.

Here, X_0 is the peak value of the variable at $a = a_0$, where the term $v(x, a)$ gives a deviation from the distribution at $a = a_0$, so that $v(x, a)$ can be expanded as

---

[4] In biological systems, the response against a signal often follows Weber's law, that is, the dependence of the response upon the signal concentration is on a logarithmic scale see Chapter 5. In this case, the response ratio should be chosen as $R = \Delta log(x)/\Delta log(s)$, which, indeed, is often adopted in cell biology.

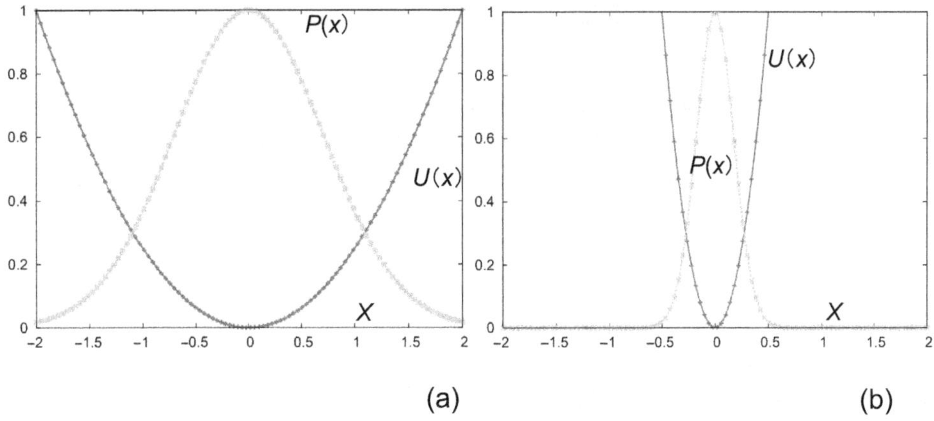

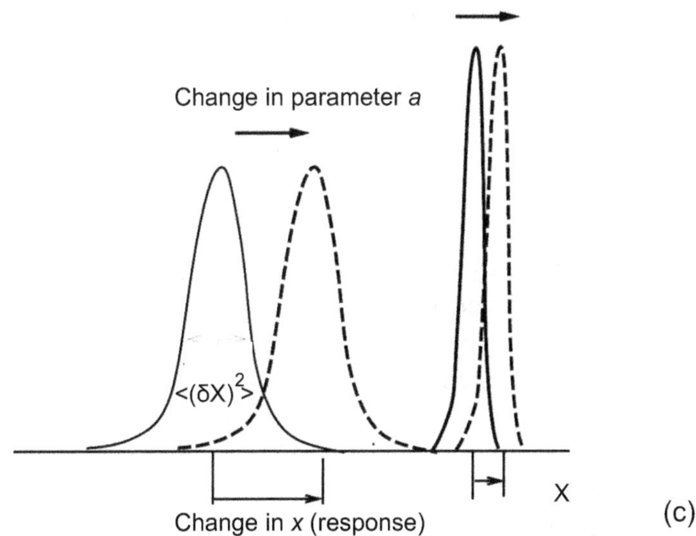

Figure 2.4 Schematic diagram on fluctuation–response relationship. (a) Probability distribution $P(x)$ is broader in a broader potential valley, and (b) sharper in a narrower potential valley. (c) The distribution function $P(x; a)$ is shifted by the change of the parameter a. If the variance is larger, the shift is larger.

$v(x, a) = C(a - a_0)(x - X_0) + \ldots$, with C as a constant, where $\ldots$ is a higher order term in $(a - a_0)$ and $(x - X_0)$, which will be neglected in the following analysis.

By assuming this distribution form, the mean $<x>_{a+\Delta a}$ is computed from $\int x\, exp\left(-\frac{(x-X_0-C\alpha\Delta a)^2}{2\alpha}\right) dx = X_0 + C\alpha\Delta a$, so that the change in the average value $<x>$ following the change of the parameter from a_0 to $a_0 + \Delta a$ follows

$$\frac{<x>_{a=a_0+\Delta a} - <x>_{a=a_0}}{\Delta a} = C\alpha. \tag{2.5}$$

Noting that $\alpha = <(\delta x)^2>$ Eq. (2.1) with the proportion coefficient C is obtained. Here we neglected dependence of α on a, which is a higher-order in Δa.

This relationship holds for Gaussian-like distributions, and for linear coupling between the variable and parameter, which brings about a shift of the average of the corresponding variable.

If a is assigned as a parameter to characterize the environmental condition, and x as a variable characterizing a phenotypic state, then the relationship means that response of the average phenotype due to the environmental change is proportional to $<(\delta x)^2>$, variance of the phenotypic distribution at a given fixed environmental condition. In other words, proportionality between the plasticity and the fluctuation is suggested, as that between the responsiveness and the variance.

If a is a parameter that specifies the genotype, then the relationship means that the proportionality between the evolutional change in the phenotype due to the genetic variation and the fluctuation in the phenotype by noise, that is, without genetic change, as will be discussed in Chapter 8.

2.6.3 Consequence of Steady Growth between Micro and Macro Scales

As mentioned already, thermodynamics was established by restricting our discussion to the equilibrium state, which is a state that can be reached if the system is left alone where it is macroscopically stationary. Of course, a cell is not in an equilibrium state. Cells can grow and multiply. Now, let us consider a state in which cells adapt to the outside world and grow steadily. In other words, instead of an equilibrium state, consider the state of a steadily proliferating cell that maintains the same composition even as it grows and divides (Kaneko et al., 2015; Kaneko & Furusawa, 2018). In this case, a constraint is imposed that the amount of all components in the cell (e.g., the amount of protein) increases at approximately the same rate, because if the cell is stable, the composition should not change significantly with each division. In other words, even if there are M kinds of components, there is a strong restriction that each component almost doubles by the time it divides: There are $M-1$ constraints that the growth rates of components $1, 2, ..., M$ are equal. In this case, even if the state of the cell is represented by the quantity of M components, we expect that in the adaptive state, these M quantities cannot change independently, and only one degree of freedom remains in the end because of the $M-1$ constraints.

2.6.4 Constraint in a Steady Growth System: Global Proportionality Law

Now we formulate the discussion of Section 2.6.3 quantitatively. To describe changes in the cellular state in response to environmental changes, we introduce a simple theory by assuming that cells undergo steady growth. Consider a cell consisting of M chemical components. In the cellular state under steady-growth conditions, the cell number increases exponentially over time, as does the cell volume V, as given by $dV/dt = \mu V$. In a steady-growth cell, the abundance of all components increases at the same rate, preserving the concentration of each component.

To formulate the constraint for steady growth, let us denote the concentration $x_i (> 0)$ for each component $i = 1, \cdots, M$. The cellular state is represented as a point in an M-dimensional state spaces. Here, each component i is synthesized or decomposed, depending on to other components, at a rate $f_i(\{x_j\})$, for instance, by the rate-equation in chemical kinetics. Additionally, all concentrations are diluted by the rate $(1/V)(dV/dt) = \mu$, so that the time-change of a concentration is given by

$$dx_i/dt = f_i(\{x_j\}) - \mu x_i. \tag{2.6}$$

For convenience, let us denote $X_i = \log x_i$, and $f_i = x_i F_i$. Then, Eq. (2.1) can be written as

$$dX_i/dt = F_i(\{X_j\}) - \mu, \tag{2.7}$$

where we assume that $x_i \neq 0$, that is, all components exist. Based on this, the stationary state is given by the fixed-point solution

$$F_i(\{X_j^*\}) = \mu, \tag{2.8}$$

for all i.

In response to environmental changes, the term $F_i(\{X_j\})$ and growth rate μ change, as does each concentration x_i^*; however, the $M - 1$ conditions $F_1 = F_2 = \cdots = F_M$ must be satisfied. Thus, a cell must follow a 1-dimensional curve in the M-dimensional space (see Fig. 2.5) under a given change in the environmental conditions (e.g., against changes in stress strength). Now, consider intracellular changes in response to environmental changes as represented by a set of continuous parameters, E^a, which denote environmental changes under the stress condition a. Here, each environmental change is parameterized by a single continuous parameter E^a (such as the temperature or nutrient limitation). Using this parameterization E^a, the steady growth condition leads to $F_i(\{X_j^*(E^a)\}, E^a) = \mu(E^a)$.

We consider the parameter change from $E^a = E_0$ to E, where each X_j^* changes from X_j^* at E_0, to $X_j^* + \delta X_j$, which is accompanied by a change from μ to $\mu + \delta\mu$.

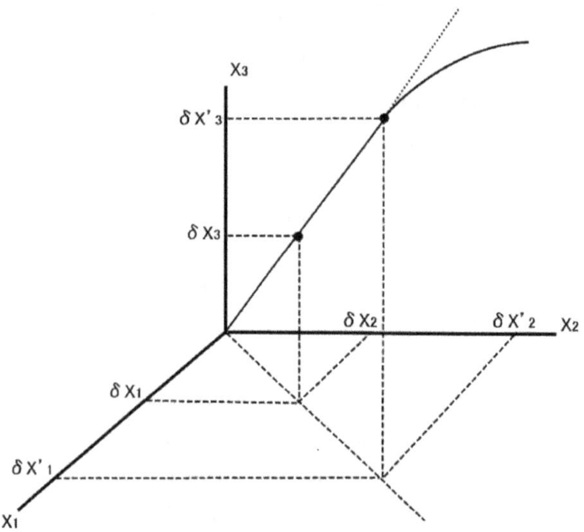

Figure 2.5 Schematic representation of our theoretical analysis: Changes in gene expression in a high-dimensional state space under different perturbations upon δE with increasing its strength are presented. By increasing the strength of a given environmental stress, the phenotypic changes in component concentrations follow a curve satisfying the constraint showing that the growth rates of all components are identical, that is, an iso-μ line $F_1 = F_2 = \cdots = F_M$, in an M-dimensional state space.

Assuming a gradual change in the dynamics x_j, we introduce a partial derivative of $F_i(\{X_j^*(E)\})$ by X_j at $E = E_0$, which gives the Jacobi matrix J_{ij}. Assuming that the environmental change is small and that phenotypic changes are sufficiently small and follow only the linear term in δX_j, we obtain

$$\sum_j J_{ij}\delta X_j(E) + \gamma_i \delta E = \delta\mu(E), \tag{2.9}$$

with $\gamma_i \equiv \frac{\partial F_i}{\partial E}$. Under our linear conditions, $\delta\mu \propto \delta E$, so that $\delta\mu = \alpha\delta E$ holds for a constant α. Accordingly, we obtain

$$\sum_j J_{ij}\delta X_j(E) = \delta\mu(E)(1 - \gamma_i/\alpha), \tag{2.10}$$

Hence, $\dfrac{\sum_j J_{ij}\delta X_j(E)}{\delta\mu(E)} = \dfrac{\sum_j J_{ij}\delta X_j(E')}{\delta\mu(E')}$ so that

$$\frac{\delta X_j(E)}{\delta X_j(E')} = \frac{\delta\mu(E)}{\delta\mu(E')}, \tag{2.11}$$

is obtained over all j.

The formula can be tested experimentally. Note that the formula can be applied to any component. For example, one can use the concentration of either mRNA or protein, depending on the available experimental data.

2.6.5 E: Experimental Confirmation of Global Proportionality

To explore the relationship between changes in global gene expression and growth rate, we analyzed transcriptome data of *Escherichia coli* obtained under three environmental conditions: osmotic stress, starvation, and heat stress, as presented in (Matsumoto et al., 2013). In the experiments, the cells were initially cultured under minimal medium at 37°C. After the transient response to the introduction of a given stress, the cells were cultured up to the time when the growth rate reached a constant value. The data were taken only after an exponentially growing steady state was reached. For each stress condition, three levels of stresses (s = high, medium, low) were used, so that the absolute expression levels, represented by x_j for j-th gene, were measured over 3×3 conditions.

Through transcriptome analysis under different environmental conditions, we calculated the change in gene expression levels between the original state and for a system experiencing environmental stress. We investigated the difference in gene expression using a log-scale ($X_j = \log x_j$), that is $\delta X_j(E) = X_j(E) - X_j^O$ (i.e., $\log(x_j(E)/x_j^O)$) for genes j, where E represents a given environmental condition, and X_j^O represents the log-transformed gene expression level under the original condition (Kaneko et al., 2015).

To examine the validity of the theory for global changes in expression induced by the environmental stresses, we plotted the relationship between the differences in expression ($\delta X_j(E_{s_1}^a), \delta X_j(E_{s_2}^a)$) in Fig. 2.6(a–c) for $s_1 =$ low and $s_2 =$ medium, where a is either osmotic, heat, or starvation stress. A common proportionality was observed in concentration changes across most mRNA species, which is consistent with the theory. Furthermore, the comparison with the response under stronger stress ($s_3 =$ strong) also supported this common proportionality (Kaneko et al., 2015).

According to our theory, the slope in the expression level changes should agree with the growth rate. Here, for each condition, the change in the growth rate $\delta\mu(E_s^a)$ was also measured (a is either osmotic, heat, or starvation stress). As shown in Fig. 2.6, we also plotted the slope computed from the growth rate change. As predicted by the theory, this result agrees with the common ratio $\delta X_j(E_{s_1}^a)/\delta X_j(E_{s_2}^a)$.

In this respect, the theory based on the steady growth state and linearization of changes in stress works well for analyzing transcriptome changes in bacteria. However, the global proportionality is satisfied even under a stress condition that

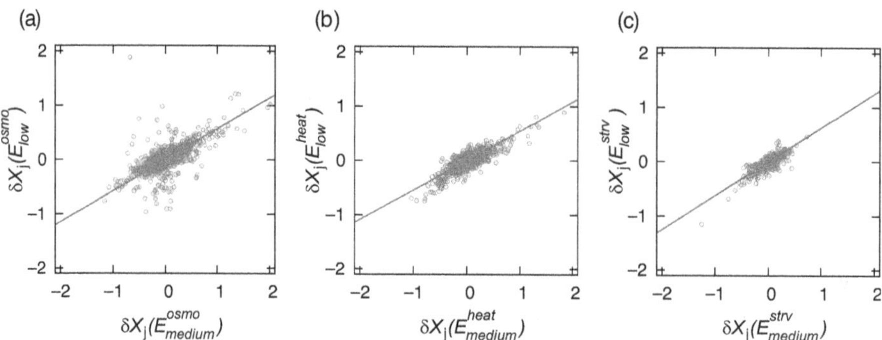

Figure 2.6 Examples of the relationship between changes in gene expression $\delta X_j(E_{s_1}^a)$ and $\delta X_j(E_{s_2}^a)$ for genes in *E. coli*. δX_j represents the difference in the logarithmic expression level of a gene j between non-stressed and stressed conditions, where s_1 and s_2 represent two different stress strengths, that is, low and medium. (a), (b), and (c) show the plot for a = osmotic pressure, heat, and starvation stress, respectively. The fitted line was obtained by the major axis method, which is a least-square fit method that treats horizontal and vertical axes equally. The slopes are 0.57, 0.54, and 0.62 for (a), (b), and (c), respectively. These slopes agree with those estimated from the growth rate and the theory. Reproduced from Kaneko et al. (2015).

reduces the growth rate to below 20 percent or as compared to the standard. Such expansion of the linear regime goes beyond the theoretical expectation.

This result means that the macroscopic quantity of the cell, the proliferation rate, can be linked to the amount of change extracted from the large number of degrees of freedom, that is gene expression levels (mRNA concentrations). In this way, establishing a theory to relate multidimensional state variables of cells to macroscopic quantities is one of the basic themes in this book. In Chapter 3, we consider not only the growth rate and gene expression but also nutrient consumption (activity) as state variables, and thereby consider the state theory of cells. In fact, by specifying specific chemical essential for the growth, such linear relationship is extended, as was first uncovered by Shaechter et al. (1958), and also developed recently by Scott and Hwa (2010), which will be discussed in Chapter 3.

The law discussed here is based on the exponential growth state. In unicellular organisms, there generally exist the transitions between "phases" of cells, such as exponential growth, dormancy, and death, as the supplied nutrients are decreased. Characterization of such transition will also be discussed in Chapter 3. In addition, from the macroscopic description of cellular state, we can extract the degree of plasticity and understand which direction the cellular states are feasible to change. This issue will be essential to discuss adaptation and cell differentiation, as will be discussed in Chapters 4, 5, and 7.

2.6.6 Homeostasis Principle in a Strong Sense

So far, we have discussed that strong constraints work for cells that are able to grow at a steady rate. The cell must be able to maintain a large number of components, which must be balanced with the growth of the entire cell. Given the complexity of the reaction process, such state with balanced growth would be rather restricted within all possible high-dimensional state space. (The balanced growth across cellular components was reported experimentally and its importance was discussed theoretically by the pioneering study by Campbell [1957]).

In the discussion in Section 2.3.2, the concentration of each component was treated as a continuous variable, but there are components with low concentrations in the cell, while the cell volume is not so large that the number of some molecules in it is sometimes only a few or a few dozen. For example, bacteria have a volume of only a few μm, while molecules with an intracellular concentration of about $1nM$ are often present, which are necessary for a cell. In this case, the number of molecules would be about 60. If the concentration drops by further two orders of magnitude, the (average) number of molecules will be less than one. In other words, for low-concentration components, their concentration (number/volume) varies greatly around the average, both in time and on a cell-by-cell basis. In some cases, a change in the external environment can reduce the number of molecules to zero. If such a component is needed for a cell to replicate, then the consistency between the growth of the cell as a whole and the growth of individual components, where all components can grow at the same rate, would be lost. In general, in the whole state space, the range in which consistency can be maintained without the loss of many components will be limited. In the presence of environmental change then, as many components as possible will try to maintain their relationship with each other, to keep the consistency.

Therefore, we might expect a long-term response that mitigates the short-term changes in the internal state caused by environmental changes. In fact, such an *adaptive process* is often seen in cellular responses. It can occur in changes in the abundance of a few components, but also across many components, that is, a process in which the gene expression of thousands of components returns toward its original level over time. We will see examples of this in Chapter 5 for adaptation, and Chapter 9 for evolution.

2.7 D: Consistency among Processes across Hierarchy of Time Scales

2.7.1 Multiple Time Scale in Life

At the end of Section 2.6.6, we discussed how rapid changes are subsequently compensated by slower changes. In general, many different time scales form a

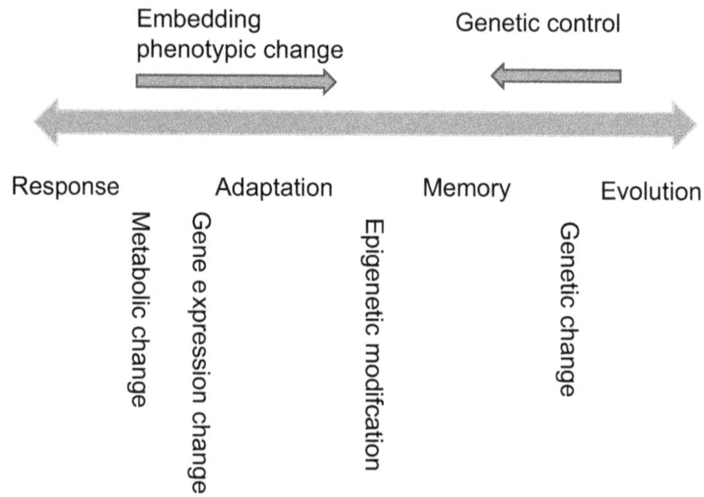

Figure 2.7 Hierarchy in time scales: phenotype, epigenetic modification, and genetic change.

hierarchy in life phenomena (Fig. 2.7). Of course, micro- and macro-scales differ in both space and time. Starting with the fast time scale of molecular changes, changes in the abundance of components range from the scale at which signals are transmitted from the outside world, to the slower scale of gene expression, which is then passed on to the much slower epigenetic modifications of DNA. Because living systems are dynamic, the hierarchy of time scales is of paramount importance. Therefore, the hierarchy of time scales is deliberately discussed in each chapter as an essential theme in the macro-micro relationship (see Fig. 2.7).

First, at the molecular level alone, there are reactions with time scales that vary by orders of magnitude. The turnover time that is, the time between its synthesis and degradation ranges from 10 milliseconds to 1 minute for metabolic components, from 1 to 10 minutes for mRNA, and several hours for proteins, and these vary by orders of magnitude depending on the molecular species, cell types, and cell states (Milo & Phillips, 2015). Since gene expression is a process of synthesizing mRNA and protein from genes, there are different time scales: faster synthesis and degradation of mRNA, and slower synthesis and degradation of protein as described above.

Despite these different time scales, the cell maintains a stable state. Furthermore, changes in its macroscopic state (phenotype) occur on a slower time scale than changes at the molecular level. The time scale of cell division is tens of minutes in bacteria, and often an order of magnitude slower depending on culture conditions and cell type, which is many orders of magnitude slower than the time scale of chemical reactions.

If a cellular state (protein composition) deviates from the average, this deviation could be passed on to the next cell when it divides into two. However, if the cell divides repeatedly, this deviation will be reduced, and the cell will gradually return to its original state. If we naively assume that each component is probabilistically split into two cells by cell division, then on average the deviation will be reduced, to 1/2 per division. If this is the case, then the half-life of such a deviation would be one division time, which is the timescale of this naive memory of phenotypic change. On the other hand, in microorganisms such as bacteria and yeast, epigenetic memory, that is, non-genetic memory by modification of DNA, often persists more than 10 generations (10 divisions), which is an order of magnitude longer than cell division.

Secondly, genetic mutations occur at a much slower rate, and the time scale of genetic change typically spans more than 100 generations. Evolution is a much slower phenomenon, as it is the process in which genetic variation occurs and is fixed in a population. In material terms, this means that the DNA is stable and mutations do not occur very often, or that some mechanisms have evolved to repair the errors (proofreading). In addition, between these slow genetic changes and much faster phenotypic changes, there is a group of phenomena called epigenetic processes, involving the modification of DNAs molecules, such as methylation or the binding of histone proteins in the case of eukaryotic cells. These change the feasibility of the expression of genes. In addition, it is also possible that the structure of the cell membrane maintains the cell state for a long time.

2.7.2 Interference and Regulation of the Phenomena with Different Time Scales

As mentioned, the theory in the present volume often relies on a dynamical-systems approach, the standard mathematical method for the phenomena with temporal changes. Recalling the argument of Section 2.7.1, the theory of biological systems then requires a dynamical system with distributed time scales. No matter how many processes with different time scales there are, it is still a dynamical system. Nevertheless, dynamical systems with distributed time scales have their salient characteristics and require special methodologies.

In general, phenomena in a faster time scale can quickly settle down to the state, so that one can first obtain the steady state of the fast variables under given slow variables fixed as parameters. Then by replacing the fast variables by the stationary solution, one can get the dynamical system to govern by the temporal change of the slow variables only. In this way, if the time scales are separated, the slow variables work as control parameters for faster ones.

This structure, that a slowly changing component controls a fast-changing component is often seen in biological systems. For example, DNA takes a long time to replicate and a long time to change; genetic information is encoded in DNA, and genetic changes are on the slow side. The gene controls the abundance of protein components, which are produced and degraded on a much faster time scale. Accordingly, genetic changes can be thought as controlling phenotypes. In biological terms, genes that are mutated on a slower time scale control changes in phenotypes, while the fitness is determined by phenotypes, which are slowly changed by genes.

Generally, microscopic changes are faster, whereas macroscopic ones are slower. If the time scales are separated in this way, the microscopic changes are controlled according to the conditions given by the macroscopic changes. When the two sides form a consistent and stable relationship, the separation of time scales is maintained. On the other hand, as we will see later, if this separation of time scales is not maintained, many variables interfere with each other, resulting in a non-stationary state in which the consistency between micro and macro is broken. The perspective of this book, on one hand, is exploration of the consistency between macro and micro scales, but on the other hand, the consequences of its collapse.

Let us briefly summarize the mathematical issues of the fast-slow time scales:

(1) In general, the slow variable acts as a "control parameter" for the fast variable, because on the time scale where the fast variable changes, the slow variable does not change.

(2) When the slow variable changes, the fast variable x has already settled down, so we can use its value to determine the change. For example, by solving the equation $dx/dt = 0$, (i.e., the fast variable x is settled to a steady state), and the fast variable are represented by the slow variable (adiabatic elimination method) (Haken, 1979; Kaneko, 1981).

(3) If the fast variable does not reach a constant state, but oscillates in time, the time average of the fast variable(s) over periods can be used to determine the dynamics of slow variables. (Averaging method).

As described above, if the time scales between variables are separated, the way they change is simplified (principle of time scale separation). On the other hand, if many variables interfere with the same time scale, unstable and complex changes may arise that are difficult to control. Therefore, for the formation of a stable system, it would be effective to have the time scales separated so that a few slow variables control many others. Furthermore, it would be easier to create a stable system if many elements are divided into groups that are controlled by (nearly) independent slow variables. This leads to the formation of functional modules that are often found in living systems.

Of course, control by slow variables is not unique to life phenomena. However, in the case of living systems, such time scales themselves are not always fixed, but vary depending on the situation. For example, the dilution of each component by cell growth directly affects the change in the concentration of components in the cell. Since the rate of cell growth depends on the nutritional status of the cell, the time scale of the change in the concentration of each component also changes depending on the state of the cell.

In addition, many reactions in cells are catalyzed by enzymes, whereas the reaction rate is regulated by the abundances of enzymes. Note that enzymes themselves accelerate both forward and reverse reactions in the same way and do not alter the equilibrium condition itself, but in nonequilibrium conditions where there is an external flow and the reaction occurs mainly in one direction, the resulting reaction rate is accelerated by the enzyme. In fact, in the cell, enzymes can accelerate the reaction rate by a factor of 10^9–10^{15}.

Since the enzymes themselves are produced inside the cell, the amount of enzymes is not constant, but can be changed autonomously depending on the state of the cell. As a result, the reaction time scale in the cell is regulated by the state of the cell and can vary greatly. This kind of time-scale control by the regulation of enzyme abundance will be discussed in Chapter 5 on adaptation and in Chapter 6 on memory.

In this case, since the time scale can be changed for each component, the relative relationship between slow and/or fast processes can also change. For example, if the separation of slow and fast variables is no longer valid, the relationship of the slow variable controlling the fast variable will also break down, and the two variables will interfere with each other, resulting in complex changes and new state transitions.

2.8 E: Experimental Approaches

2.8.1 Experimental Approach to Macro-Micro Consistency

Of course, theory alone is not sufficient to extract the universal properties of living organisms from the perspective of macro-micro consistency, and experiments are essential. However, conducting experiments from this perspective requires a different experimental standpoint from the conventional molecular biology methodology in which molecules (genes) that perform a certain function are searched for.

(A) Macro-side: First, experiments from the macroscopic viewpoint are, in a sense, a classical approach. Rather than following each molecule in the cell with advanced technology, it is a phenomenological analysis from a macro perspective that measures quantities such as cell growth rate, activity (e.g., the degree of consumption of nutrients and ATP), and cell population growth. However, it is also useful to distinguish between the population average and the state of a single cell by using techniques such as long-term single-cell measurement (see Chapter 3).

(B) Micro-side: On the microscopic side, the aforementioned intracellular transcriptome analysis has made it possible to examine the expression levels of thousands to 10,000 different genes. This is not limited to the measurement of average gene expression levels in a cell population; it is now possible to measure the expression levels of each gene at the single-cell level by using single-cell RNA-seq method. However, in spite of all the progress in acquiring such high-dimensional data, often only a portion of the data is used. From the standpoint of looking at all of the high dimensional quantities in this volume, it is necessary to look at the statistical distribution of the gene expression levels as a whole and their changes, rather than to examine the expression of specific individual genes.

Another important issue in the micro side is the **fluctuation**. For example, there exist relatively large cell-to-cell fluctuations in protein concentrations arising from molecular noise, whereas they are related to the macroscopic growth rate of the cell and the process of adaptation to the external world, as will be seen in later chapters. From the macro-micro consistency, this fluctuation is related to the degree of response on the macro side, as mentioned. Therefore, it is important to experimentally investigate the relationship between fluctuations and adaptation to the external environment or those and evolutionary change. Furthermore, if we consider a higher level of hierarchy, in which the micro side is cells and the macro side is individuals (tissues), we should explore the relationship between fluctuations in the number of cells and adaptation of an organism.

(C) Macro-micro consistency: Based on the above (A, B), it is necessary to connect micro and macro levels, that is, to conduct experimental analysis from the viewpoint of connecting the expression distribution on the micro side and the cellular state on the macro side. Here, the measurement on the micro side involves information with large degrees of freedom. It is necessary to compress the information and extract few macroscopic quantities from it. In this extraction, it is important to take the viewpoint that the macroscopic and microscopic states are changing to keep consistency.

(D) Time-scale hierarchy: With the hierarchy of time scales in mind, we seek the hierarchy of temporal changes, such as the dynamics of expression levels in cells and of the number of cell numbers. Furthermore, it is necessary to conduct experiments by focusing on whether the time scale of change in protein levels varies with the state, or whether the rate of change in internal states differs between the exponential growth phase and the non-exponential stationary phase of the cell, and whether memory occurs as a longer time scale phenomenon, and so forth.

2.8.2 Constructive Biology

Earlier we proposed *constructive biology* (Kaneko & Tsuda, 1994; Kaneko, 1998a), in which we aim at understanding life by constructing it. It is an important methodology for universal biology. This is because it is difficult to identify universal properties that an organism must satisfy only by examining the current organisms that have been a result of evolution that happened on this Earth. In contrast, it is necessary to set minimum conditions from our side and seek what properties emerge in the system. In this case, one should note that even if we construct a system at a certain hierarchy (e.g., molecular level), the construction of life system should not be confined to a single layer of hierarchy, but is related to the upper layer (e.g., cellular level). Constructive biology focuses on the linkage between the microscopic and macroscopic sides. The focus is on the collective behavior of a group of constitutive elements that exhibit properties that were not anticipated at the level of each individual element. For example, when we incorporate a gene to introduce certain protein expression, the interference between molecular-level dynamics and the growth of the cell as a whole may bring about a result that is not simply expected at a microscopic level. Exploration of such emergent behavior is an issue in the constructive biology.

Similar to constructive biology, there is a field called synthetic biology (Elowitz & Leibler, 2000; Benner & Sismour, 2005), which is currently being actively pursued. Let us discuss some differences between the two.

Synthetic biology has an engineering orientation to design certain properties with a goal in mind. The circuit of a reaction is embedded in a cell, as in the design of an electrical circuit or transistor. Here, the blueprint is given by the genes, which are used to design the reaction and give the system certain properties.

Let us compare synthetic and synthetic biology using the example of a study in which a new gene network is introduced into a cell. For example, in the groundbreaking study in synthetic biology, a gene network (circuit) is introduced into *E. coli* in which gene A represses expression of B, B represses C, and C represses A. When A is expressed, then B is repressed, so that the repression to C is weakened, and C is expressed. Then the expression of A is suppressed, which next leads to the expression of B. Hence, successive change of states from high A expression to high C expression, and then to high B expression occurs.[5] These changes continue, and as a result, an oscillating state that cycles through the expressions of A, C, B, A, ... is realized. In fact, if the proteins A, B, and C are made to have different colored fluorescence, one can detect such oscillation as color change in a cell in sequence (Elowitz & Leibler, 2000). In synthetic biology, gene circuit consisting of few genes is designed and incorporated to provide a certain function to a cell.

On the other hand, constructive biology has a heuristic orientation in which we give simple setups and see what logic of life can be drawn from them. In this sense, more weight is put on finding a behavior that is unexpected at the time of design. For instance, we will discuss later the experimental result in which a gene circuit is embedded to have a bistable state, but only one of the states that lead to a higher growth rate of a cell is selected (Kashiwagi et al., 2005), thereby extracting the logic of adaptation even without using a signaling system (see Chapter 4).

Why, then, is there a difference between the designed behavior and the behavior of constructed system in a cell? This is closely connected to the perspective of the macro-micro consistency. When a design is made at the molecular level of gene expression, and if this behavior is confined within a single level in the hierarchy, it would behave as designed. This is the position of synthetic biology, which designs at one level of the hierarchy. However, changes in gene expression at the molecular level and the upper level of cell growth influence each other in the aforementioned experiment. Therefore, to be consistent with the upper hierarchy (cell growth), gene expression dynamics can behave differently from that originally designed. It is the position of constructive biology to extract the nature of the consistency between levels in hierarchy, from the deviation from the designed behavior.

In fact, synthetic biology sometimes faces difficulty in making a cell to behave as designed. This is because the assumption that the embedded circuits work independently as other components does not necessarily work, and then we need to face the macro-micro hierarchy. To understand the essence of a living system, one level of hierarchy is not sufficient. It is necessary to construct a system with a large number of degrees of freedom and explore how the fundamental properties of life manifest themselves. This would be the position of constitutive biology. The fundamental difference between synthetic biology and constitutive biology lies in whether one takes seriously into account that many degrees of components

---

[5] A certain parameter condition for the reaction rates has to be satisfied.

are involved to form a hierarchical system, and this is probably one of the reasons for the difficulties that synthetic biology faces.

To begin with, to what extent is it possible to make a design within a system in which numerous components influence each other? To do so, it would be necessary to know in advance their influence on each other. However, in reality, it would be difficult to know all of the effects. And even if we did, it would be computationally difficult to find a good design among them. In fact, in the case of living organisms, it is possible that this solution is inherently dependent on evolution.[6] Therefore, evolution experiments will be indispensable for exploring the principles of living systems. Again, from the standpoint of constitutive biology, the main emphasis is on the search for some laws of evolution, such as the relationship between evolvability and plasticity through laboratory evolution, rather than in the engineering direction to create highly functional molecules or cells by evolution (see Chapters 8 and 9).

2.9 F: Collapse of Consistency and Recovery of Plasticity

So far, we discussed the logic of life, noting the conditions and consequences of macro-micro consistency, achieving a robust system. On the other hand, when a life system is forced into a harsh environmental condition, the stable macro-micro relationship will be sometimes collapsed. Such a break down of consistency needs to be discussed in order to understand living organisms. In each chapter, the significance of the collapse will be discussed, but a brief sketch will be given below.

(i) **Collapse of consistency for reproduction**: If cells can grow exponentially with sufficient nutrients, the theory of Section 2.3 will work. However, when nutrient conditions are poor, unicellular organisms generally enter a stationary phase or dormant state, in which they cannot grow, as mentioned. In other words, the consistency between growth and internal state is broken. How can we describe this growth-arrested state? This will be discussed in Chapter 3.

(ii) **Switch from low-growth to high-growth cellular states**: Upon some environmental change, cells may not be adapted to the novel condition, and the cellular growth may be suppressed, that is, the consistency between cell reproduction (macro) and molecular-replication reactions (micro) is broken. Such state with loss of consistency turns to be non-robust to perturbations. With an underlying nose, then, the transition to a different cellular state will occur that can achieve higher growth. We will see this generic adaptation by attractor selection in Chapter 4.

---

[6] On the contrary, looking back at what evolution has created, it might be possible that it has created a combination of modules that are as isolated as possible, limiting their effects on each other.

(iii) **Collapse of the consistency between multicellular populations and single-cellular states**: In development of multicellular organisms, each cellular state and cell population (organism) shapes a stable state keeping mutual consistency between the two. However, there are cases in which such robustness with mutual stability is lost. In metamorphosis, stability of cell types and cell-population that is sustained up to a certain stage in development breaks down, and switches to novel cellular states occur, by resetting the previous state. Although this dynamic mechanism is not yet well understood, it may be regarded as a loss of consistency at some stage of development.

On the other hand, cancer can be regarded as a "pathological state" due to the collapse of consistency. Cancer cells will be "selfish" cells that proliferate without keeping stable relationships with other cells present in the tissue. In this respect, cancers may be regarded as the loss of consistency between cellular and tissue (organism) levels. This will be discussed in Chapter 7.

(iv) **Evolution of novelty**: In Chapter 8, we will discuss that phenotypic changes due to environmental changes and those due to genetic changes will evolve in a stable manner while maintaining consistency. Under harsh environmental conditions, such consistency may no longer be sustained. In such a case, cells lose robustness, and the original cellular state is no longer sustained, giving rise to large fluctuations in the cellular state, from which evolution in a new direction will arise.

In these instances, when the system is put into a harsh condition, the separation between fast and slow scales is broken, and many degrees of freedom change rather independently, and, the stable state is disrupted, leading to the increase in fluctuations. This may lead to a pathological state on the one hand, but on the other hand, it can also lead to acquisition of novel properties needed for the survival in the harsh conditions.

2.10 Summary

In this chapter, we briefly look at the standpoint of the present volume, which consists of (A) macro-level phenomenology, (B) micro-level statistical laws of large-degree-of-freedom systems, (C) macro-micro-level consistency, (D) time-scale hierarchy, (E) experimentation, and (F) collapse of consistency. With these six points in mind, the following chapters will look at basic issues of reproduction, adaptation, homeostasis, memory, differentiation, development, and evolution. In order to make it clear which of the items (A)–(F) is the main concern in each section, each section is labeled A, B, C, D, E, and F. This will clarify the ideas and standpoint that run through the present volume – macro-micro consistency across levels.

3

Reproduction of Cells: Dynamic and Thermodynamic Characteristics, Dormancy, Fluctuation, and Requisites of Protocells

3.1 Introduction

The living state consists of diverse components. Here we focus on cells. They contain diverse molecules within a certain region bounded by membrane, and maintain themselves, and can even grow and reproduce themselves. The question of this chapter is how such a state is possible and if there are universal characteristics therein.

Nonequilibrium: First of all, it is often taken for granted that the state out of thermal equilibrium is essential for the generation of such a biological system. It is true that it will be impossible to maintain a living state in a perfect thermal equilibrium, and the characteristics of life, such as growth under external flow of resources, do not satisfy the requisite for equilibrium thermodynamics. In fact, the importance of maintenance of nonequilibrium condition by external flow is manifested by the famous phrase "life feeds on negative entropy" in the latter part of Schrödinger's book "What is Life" (Schrödinger, 1946). In the 70's, there was a trend to study "life as nonequilibrium," centering on Prigogine and others, which generate extensive studies for pattern formation and temporal rhythms in nonequilibrium open systems, termed as self-organization and dissipative structures (Nicolis & Prigogine, 1977).

The origin of such studies of pattern formation can be traced back to Turing (1954). Turing (1954) proposed that, in a system consisting of chemical reactions and diffusion of substances, the symmetry of spatially uniform states is broken, to generate spatial stripe patterns, which are proposed to explain morphogenesis in the developmental process (see also Rashevsky [1940; 1960]). However, although a series of studies that originated in Turing showed remarkable advances in the areas of pattern formation in physics and chemistry, not much attention was paid in biological sciences, which were originally the target of these studies. There, detailed analysis of genes that concerns with the reactions for morphogenesis were advanced. Specific sets of genes that governed morphogenesis were uncovered, known, for instance, as *homeobox*.[1]

Let us look back at Schrödinger's phrase that a life system feeds on "negative entropy." To eat negative entropy presupposes that the organism is divided into inside and outside, and that there is some kind of flow between the inside and outside. In the viewpoint of dissipative structure, this flow maintains a special nonequilibrium state within the organism. In the view, there are external systems A and B with different equilibrium states that are heat bath with different conditions (say, with different concentrations of some chemical components or temperature), and a system is placed between them to maintain a flow (say, of chemical components or heat). For instance, when a fluid is heated at a high temperature from the bottom and attached with a bath with a low temperature at the top, fluid convection is generated, which can generate a spatial pattern when the temperature difference is large. When one chemical component is flowed in with a certain rate and another component is continually removed with a certain rate, a nonequilibrium state that shows rhythmic change in some other component can be generated by a certain reaction, for example. Relevance of such dissipative structure to a life system has been put forward.

In the case of living systems, however, such nonequilibrium conditions with flow are not externally imposed. It is not the case that life cannot exist without external gradients of chemical components or temperature gradients. The conditions under which nonequilibrium conditions are maintained must be created by the organism itself and is propagated to the next generation through cell growth and division. How is the autonomous sustainment of "non-equilibrium state" possible?

If we consider the origin of life, of course some kind of external energy flow must have been necessary first, but then an internal structure must have been generated to maintain such flow, say by a membrane that distinguishes the internal and external system. Then a novel question how nonequilibrium conditions are maintained and reproduced has to be addressed, which is different from the problem studied in dissipative structure (Nicolis & Prigogine, 1977) in which structures are created in a nonequilibrium open system where the flow from outside is given.

Let us now consider the characteristics of cells. First, the cell membrane separates the internal environment from the outside. Inside, there are molecules that carry genetic information. However, this alone is not enough for life to exist. More important, as mentioned in Chapter 2, are the enzymes that control reactions in

---

[1] However, since the 1990s, some changes in the study of developmental process have occurred: we will return to this in Chapter 7.

the organism. These are proteins that catalyze many reactions that proceed in a cell. Usually, the catalyst C accelerates both the forward and backward rates of $R + C \leftrightarrow P + C$ of the raw material (Resource) and the product (Product). Then, one could often assume that the equilibrium conditions for the concentrations of R and P do not depend on the presence or absence or concentration of the catalyst, but that the catalyst only accelerates the velocity to fall into equilibrium. If this is the case, even if the catalyst exists only within a cell, the ratio of concentrations R to P inside of the cell would be equal to that outside of cells, after enough time. The question, however, is the degree of acceleration. The acceleration of the reaction rate by the enzyme often ranges from 10^9 to 10^{15} times. Then, in the normal timescale we are concerned with, the above reaction would not occur without the enzyme. Only when the enzyme is present, the reaction progresses and the approach to equilibrium can occur within the scale. If the chemical potentials are μ_R, μ_P, and $\beta = 1/(kT)$, the concentration ratio of R and P will be toward an equilibrium distribution where $[R]/[P] = \exp(-\beta(\mu_R - \mu_P))$ inside the cell with the enzymes, whereas outside the cell equilibration is not achieved as there are no enzymes. At the outside of cell, $[R]/[P]$ ratio can be arbitrary.

In other words, the cell membrane separates inside from the outside Paradoxically, one can say that the cell, with the aid of enzymes, unveils the "nonequilibrium" situation on the outside and then promotes equilibration within it, by taking into account of enzymes. Further, as a result of this enzymatic reaction, the production of the enzymes and of the membrane separating the inside and outside is promoted. The growth dilutes the concentration of each component and also brings in nutrients from the outside, thus moving the state away from equilibrium. In other words, cells work as an "equilibration promoter," by encapsulating enzymes within, and such structure is sustained with catalytic reactions, and lead to reproduction. How such self-consistent structure is achieved is the fundamental question to be addressed.

It is important to note that even though equilibration is accelerated, the cell is never in a state of equilibrium. In order for the equilibration accelerator itself to work, a nonequilibrium process of taking in nutrients and producing catalysts (enzymes) is necessary. If the cell is completely in equilibrium, the process will not work well. Furthermore, the degree of proximity to equilibrium itself differs depending on the components. For example, the ratio of Na^+ to K^+ is closer to equilibrium outside the cell, while it is farther from equilibrium within the cell. In any case, it is important to keep in mind that approach to equilibrium depends on which reactions are possible within an appropriate time scale. With this in mind, it would be meaningful to take the perspective of an equilibration accelerator working in nonequilibrium.

Secondly, these catalysts are produced in the cell as a result of catalytic reactions, and the time scale of the approach to equilibrium depends on their concentration. Then the dependence of thermodynamic efficiency on flow is not so simple, since enzyme concentrations change with the flow of nutrients (substrates) from the outside. Furthermore, the catalytic reaction results in an increase in the cell-volume growth, which leads to a dilution of each component. Then the timescale of equilibration process depends on catalyst concentration, replication by autocatalytic processes, and dilution by growth, which are essential properties for all cellular systems. Hence, it is important to understand the basic properties that such systems satisfy.

Here, the questions to be asked are:

(1) What is the relationship between growth rate and efficiency in such a thermodynamic system in which cell growth is accelerated to equilibrium by encapsulated catalysts? Does such system have a different characteristic from that of Carnot engine, in which quasi-static processes are most efficient?

(2) What is the significance of growth? Does a cell grow because of nonequilibrium nature, or does it grow so that nonequilibrium can be maintained?

(3) Cells have many components, and for them to continue to grow and multiply at a steady rate, all components must increase in the same way. In Chapter 2, we discussed the law of the average concentration of each component that such a state must satisfy, but is there a statistical law that includes the fluctuation and growth rate of each component?

(4) When nutrition is poor, cells stop growth. This does not mean that the cell dies. In fact, when the nutrition supply is restored, the cell resumes growth after some time. In other words, besides growth and death, cells generally have a "sleeping" state. Indeed, in considering the origin of cells, if cells die under depletion of nutrients (which can occur sooner or later), they cannot survive. Then, how can the transition to sleeping state is achieved? Can we understand the transition between these different phases and express it in terms of the equation of cellular state?

(5) How can we fill the gap between a (proto)cell and just an ensemble of chemicals that react with catalysts, which is essential to understand the origin of cells? The protocells can have diverse components, and can grow and divide, whereas they go to a sleep state under nutrient deficiency. Also, the chemical components within the cell eventually separates the roles between those that work as function for catalysts for growth (e.g., proteins) and those as templates for information (e.g., DNA). Is such separation of roles, also termed as the central dogma of molecular biology, a necessity of all possible cells? Can we reach a coherent view to fill the gap between ensemble of chemicals and cells?

3.2 Macroscopic Phenomenology of Growing Cells (A)

3.2.1 Equilibration and Growth by Autocatalysis

Attempts have long been made to consider the minimal model of the cell; Ganti (1975) proposed a proto-cell model consisting of metabolism, self-replicating systems, and membranes. Self-producing systems consisting of nutrients, catalysts, and membranes have also often been discussed. As mentioned in Chapter 2, Campbell (1957) noted the importance of balanced growth across many components, whereas a model with balanced exponential growth of RNA, DNA, and protein was discussed by Bremer (1981).

Here, based on the discussion in the introductory section, let us first consider a minimal setup that takes into account the following three points. (1) There are enzymes inside the cell that act as catalysts. With their presence, reactions that almost never happen outside (or take a very long time) can happen fast enough inside the cell. In the example in Section 3.1, R and P were arbitrary outside, but inside they approach equilibrium, that is, the cell rather acts as an equilibrium accelerator by encapsulating the enzymes in a membrane, and (2) this enzyme itself is synthesized as a result of a reaction inside the cell. Therefore, the presence of the catalyst accelerates its own synthesis, that is, the autocatalytic property exists. (3) As a result of the reaction, the membrane that encloses the product is synthesized, while the increase in volume due to the membrane synthesis dilutes the concentration of the chemicals inside (Fig. 3.1).

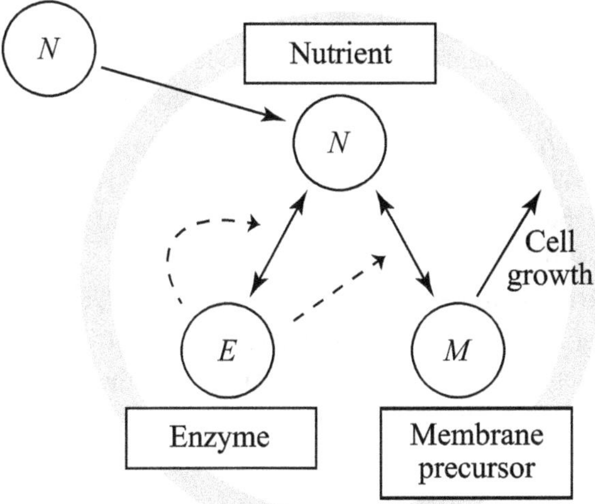

Figure 3.1 Schematic diagram of an autocatalytic cell model consisting of nutrients, enzymes, and membranes. Based on Himeoka & Kaneko (2014). (Courtesy of Yusuke Himeoka).

As a simple setup, we assume that there are nutrients N and N', an enzyme E that converts them to other components, and a membrane that separates the inside of the cell from the outside. Here E is synthesized from N, and the membrane is synthesized from membrane precursor M, which is synthesized from N' with the aid of E. Then the reaction is written as

$$N + E \leftrightarrow E + E$$
$$N' + E \leftrightarrow M + E$$
$$M \to Membrane,$$

The last reaction creates a membrane M from the membrane precursor M, which induces the growth in cell volume and leads to the dilution of the concentration of each component. For simplicity, let us not distinguish N and N', and let their concentration be just a parameter representing nutrient concentration.[2] The time evolution of the system is given by

$$\frac{dx}{dt} = \kappa_x x(ks - x) - x\lambda, \quad \frac{dy}{dt} = \kappa_y x(ls - y) - \phi y - y\lambda, \tag{3.1}$$

where the variables x and y denote the concentrations of the enzyme (E) and membrane precursor (M), respectively, whereas $\lambda \equiv \frac{1}{V}\frac{dV}{dt}$ denotes the cell volume growth rate to be determined. Our protocell model grows by consuming membrane precursor molecules with rate ϕ, and all chemical species are diluted by volume expansion of the cell in proportion to its concentration with rate λ.

Now, since volume growth is caused by an increase in membrane, it is proportional to ϕy and $\lambda = \gamma \phi y$.[3] Note that the reaction rate constants are given by $k = exp(-\beta(\mu_x - \mu_{\text{nut}}))$, $l = exp(-\beta(\mu_y - \mu_{\text{nut}}))$, by using $\mu_{\text{nut}}, \mu_x, \mu_y$ as their standard chemical potentials.

We then consider the entropy production rate σ. It is generally given by $\sigma = \sum_i J_i \frac{A_i}{T}$, where J_i is the chemical flow and A_i is the affinity for each reaction i and for temperature T. Here the affinity is given by $A_i = \partial G / \partial \xi$, by using the free energy G and the rate of reaction progression ξ in each reaction i (e.g., (Kondepudi & Prigogine, 2014)).

Now, since the volume growth rate of this cell is λ, the cell grows exponentially in the form of $exp(\lambda t)$. So we can write $\eta \equiv \sigma / \lambda$ to indicate how much entropy is produced per cell growth. Roughly speaking, the smaller this η is, the more efficient the cell growth is.

---

[2] Even if we treat them separately, essentially the same results are obtained.

[3] We assume here that both component volume and flow are proportional to V as the cell grows in a rod shape. If the cells were spherical, the flow of S would be 2/3 power of the volume, and so on, but the following conclusion remains essentially the same.

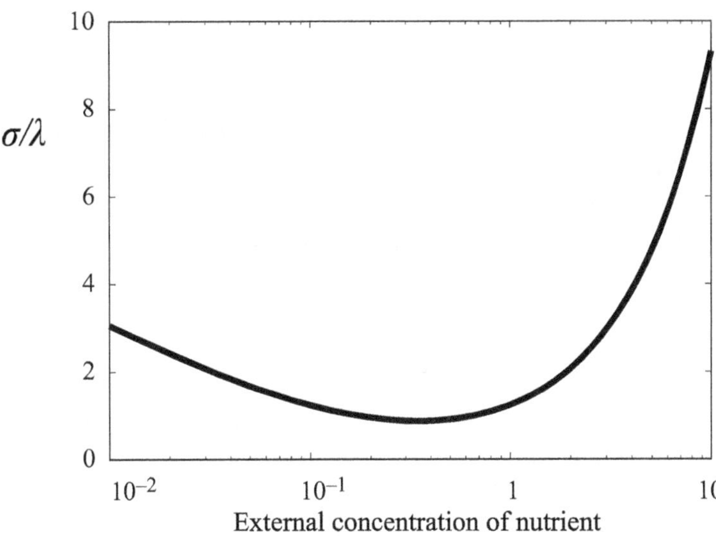

Figure 3.2 Entropy production per growth $\eta = \sigma/\lambda$ plotted against nutrient influx rate (degree of nonequilibrium). Based on Himeoka & Kaneko (2014). (Courtesy of Yusuke Himeoka).

Figure 3.2 plots this η as a function of nutrient flow rate (which is proportional to the growth rate λ). As can be seen, the thermodynamic loss for cell growth is minimized at a certain value of flow. In other words, the loss is not minimized when the flow is infinitesimal, as in the quasi-static process in thermodynamics, but the efficiency is maximized when there is a moderate flow and the growth is at a certain finite speed.

This increase in efficiency at nonzero flow values is due to the fact that the equilibration is accelerated by making the enzyme itself. First, the reaction from nutrients to other components takes an enormous amount of time because there is no catalyst outside the cell. Hence, the concentration ratios of these chemical components outside the cell can be far from equilibrium. On the other hand, if there are sufficient enzymes inside the cell, the reaction proceeds fast enough to approach equilibrium. Then, the faster the nutrient influx, closer to equilibrium. In contrast, if the amount of catalyst were fixed externally instead of being synthesized internally, the lower the flow, the better the efficiency. Equilibration by the enzyme and the requirement to produce the enzyme itself lead to an increase in the efficiency due to increased flow.

On the other hand, as the flow increases, the dilution due to the increase in volume is greater, and the reaction proceeds to compensate the concentration decrease by the dilution. The entropy production σ is the product of flow J and affinity A, both of which increase in proportion to nutrient flow, and therefore increase as the square of the product, whereas the growth rate λ is linearly proportional to the nutrient flow. Therefore the entropy production per growth rate σ/λ increases with the flow.

Up to a certain point in the flow rate, the decrease due to equilibration is greater, whereas when the flow rate is too high, the original increase wins. Thus, there is a certain finite flow rate that minimizes the loss. On the other hand, if there is flow but no membrane formation and no increase in volume, equilibration by enzymatic reaction is accelerated, and the cell moves toward equilibrium. In this sense, the dilution by cell growth maintains a nonequilibrium condition.

The above properties are valid for an autocatalytic system in which an external nonequilibrium state is equilibrated internally by a catalytic (enzymatic) reaction, while the enzyme and membrane are generated by the reaction. Therefore, the property of maximizing efficiency at a certain growth rate is generally considered to be true for cells.

3.2.2 *Laws with Regards to Cell Growth*

Experimental results have shown that the loss of nutrient is minimized at a certain cell growth rate. Here, we briefly explain the law of Pirt (1965). First, the mass of cells produced per unit mass of nutrient source is defined as the yield Y (which varies with the state of the cells). Next, the cell growth rate is defined as μ(cells/sec). Here, not all of the mass produced by the nutrient source is used for cell growth, but there is a portion m that is used for cell maintenance. (This is assumed to be independent of μ). Consider a hypothetical situation where cell growth is sufficiently large (infinite), and let Y_∞ be the value of the yield at that time (which is a constant). Then, in this situation, the maintenance part can be ignored, so we can consider maintenance and growth separately. If the two can be added independently, the expression μ/Y can be written as

$$\mu/Y = m + \mu/Y_\infty, \tag{3.2}$$

(or equivalently $1/Y = m/\mu + 1/Y_\infty$). Therefore, if we plot the experimental data with $1/\mu$ on the horizontal axis and $1/Y$ on the vertical axis, we would obtain a linear relationship, the slope of which gives m, and the Y-intercept should give $1/Y_\infty$. This is shown in Fig. 3.3. The assumption above that the maintenance term is constant is not so bad in this example. (Note that if Eq. (3.2) is difficult to understand, it may be easier to understand to write $q = m + \mu/Y_\infty$ with $q = \mu/Y$ as the nutrient consumption per unit time and q as a function of the growth rate μ and the Y-intercept giving the term m for maintenance).

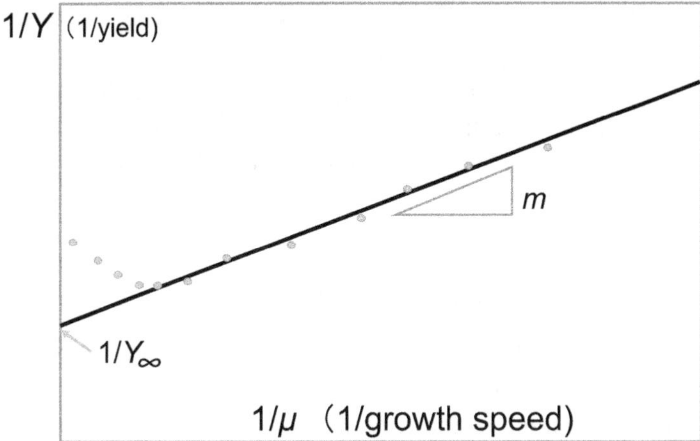

Figure 3.3 The relationship between nutrient consumption and growth: a law by Pirt. There is often a proportional relationship between the inverse of the bacterial cell yield and the inverse of the growth rate (Pirt, 1965), which gives the basic linear equation. However, as indicated by the dots in the figure, there are cases in which the relationship deviates from a straight line and exhibits non-monotonic behavior (Russell & Baldwin, 1979). Schematic diagram based on the above paper.

This may be a pioneering linear law in "steady-state growth systems." Another basic law is Monod's (1949) law, which states that the growth rate as a function of nutrient concentration increases linearly at low concentrations but saturates at higher concentrations. The linear law with the restriction of steady-state exponential growth has already been discussed in Section 2.3.2.

As for the balanced growth discussed in Chapter 2, Schaechter et al. (1958) were the first to uncover the linear relationship between the RNA abundances and the growth rate of bacterial cells, across different conditions. Recently, the linear relationship between ribosome abundance and the growth rate has been demonstrated by Scott and Hwa (Scott et al., 2010). Noting that ribosomal proteins work for autocatalytic growth (that corresponds to E in Section 3.2.1), they showed that the law is derived from the optimal growth strategy for the balanced growth (as in Chapter 2), irrespective of the details of the molecular implementation. (See Section 2.3 and Chapter 9 for a general linear law of the abundances across components as a result of steady and robust growth).

Let us return to Pirt's law. Note that the experimental results do not always yield a neat linear relationship. This may be because the amount needed for maintenance depends on the nutrient flow and the cell growth rate. For example, there may be an optimal flow rate at which $1/Y$ is minimal for a finite μ, that is, yield is maximal, as indicated by the dots in Fig. 3.3 (Russell & Baldwin, 1979). Since all of the above formulations are expressed in terms of mass, this example means that the material loss is minimal at a given value of flow. It is not the thermodynamic loss as entropy production. Since the flow of matter is caused by reactions, however, the two can be considered related, while no general rule linking them has yet been established.

3.2.3 Toward Thermodynamics of Growth, Nutrient Consumption and Loss

The cellular model of Section 3.2.1 is not sufficient to consider Pirt's law of cell growth losses from nutrient flows. This is because in this model, all nutrients are converted to membrane or intracellular components, and there is no loss for maintenance.

Of course, in the actual reaction, not all of the nutrients turn to be intracellular components, and some components go out of the cell, implying a loss of material. Furthermore, protein macromolecules often lose their catalytic activity due to misfolding caused by slight differences in their sequences. The synthesis of macromolecules is not always correct, and incorrect "waste" molecules are produced. These waste molecules stick to other molecules or aggregate and precipitate. The cell must then somehow dispose of these "garbage" macromolecules.

In addition, because cells need energy to make complex macromolecules, there is a process called catabolism, in which external nutrients are broken down and energy is extracted from them. This produces molecules that serve as energy currency, such as ATP, which is then used in a process called anabolism to make complex macromolecules. Since these are nonequilibrium processes, the flow creates waste in both entropy generation and material. This waste exists at a certain rate even when cells are not growing and can be considered a cost of maintenance.

If we can get macromolecules as food in the first place, we can use them as they are. In an extreme case, would it be the best nutrition if cells ate their neighboring cells of the same type? In reality, however, cell cannibalism is rarely observed. This is probably because even if macromolecules are taken in as they are, the proportion of waste macromolecules is so high that they have to be broken down before they can be used.

Cells are composed of enzymes with catalytic activity, which are quite rare components, encapsulated in the membrane. On the other hand, errors are inevitable in the synthesis of these enzymes, and as a result, the accumulation of waste macromolecules is considered inevitable. As a result, cells are forced to pay the cost of processing and disposing of this waste component. This results not only in the entropy production in Section 3.2.1, but also in material loss. Thus, not all external nutrients become intracellular components, but there is also a loss of mass.

In this way, for the description of the macroscopic stable state of the cell, the following formulation is required:

- Nutrient influx from outside = macromolecules with activity + membrane + garbage
- Growth rate = volume increase due to membrane synthesis
- Nutrient conversion efficiency $\propto$ activity

Here, this growth rate μ serves as a macroscopic state variable, since it dilutes each component in the cell. On the other hand, the influx of nutrients creates a catalyst that participates in the overall synthesis of the others. The amount of molecules performing this autocatalysis is the basic macroscopic state variable responsible for the activity of the cell. Therefore, activity (A) is also a macroscopic state variable, as in the following example. Activity determines the internal reaction rate, the degree of nonequilibrium, and cell growth through membrane synthesis. In other words, similar to the relationship between free energy and energy, Nutrient influx = growth rate – activity $\times$ waste production rate. We will return to this topic in Chapter 9. On the other hand, the "phase transition" caused by the accumulation of garbage components is discussed in Section 3.4 as a state in which the consistency with the growth and abundances of each component is broken.

3.3 Distribution Laws and Macroscopic Variables in Catalytic Reaction Networks (B and C)

In Section 3.2, we investigated a protocell system with few components. In contrast, let us examine a large-degree-of-freedom multicomponent model. Previously we introduced a cell model with a collection of stochastic chemical reaction dynamics, in which catalysts are synthesized with the aid of catalysts, by using nutrients that flow into the cell (Furusawa & Kaneko 2003, Kaneko, 2006). We investigate the characteristic properties that allow a cell to maintain reproduction.

Suppose there are many catalytic components x_i and a network of catalytic reactions is formed among them in the form of $x_i + x_j \rightarrow x_\ell + x_j$ (x_i is the ith type of molecule). The catalytic reactions form a complex network, but here the network is randomly assigned. Several nutrients flow in and are sequentially converted to catalysts through this reaction network. Once the catalytic reaction network is determined, how the amount of each molecule changes is given, and this change in amount is represented by a dynamical system with stochastic fluctuations (Furusawa & Kaneko, 2003). This cellular model shows optimal growth at moderate values of external nutrient influx, where cells with nearly identical composition are reproduced. The results of this model were briefly discussed in Chapter 2 and in detail in Chapter 6 of (Kaneko, 2006), so we will only briefly summarize the results here. (See also Chapters 5, 8, and 9 on adaptation and evolution for this type of cell models).

First, if growth of cells is optimized, and cells of approximately the same composition are reproduced, a special distribution rule holds for the amount of each component. In other words, if the components are arranged in order of increasing quantity, then

$$\text{abundances} \propto (\text{abundance-rank})^{-1}$$

which is also called the Zipf's law. In statistical physics, such a power-law relationship is often seen at the critical point of transition of states.

Of course, even if cells with almost the same composition are reproduced in this state, each reaction does not produce exactly the same amount, resulting in fluctuations in the abundances of each component, even among cells sharing the same network (i.e., the same genes). As discussed in Chapter 2 and discussed in (Kaneko, 2006), the fluctuations of each component do not follow a Gaussian distribution, but the logarithm of the component concentrations across cells exhibits roughly the Gaussian distribution.

The origin of this distribution is attributed to the multiplicative nature of the reactions in the cell. Since changes in the abundance of each component are due to other catalysts, the change in abundance is proportional to the product of the concentrations of catalyst and substrate. Thus, the fluctuations of each component are multiplicatively influenced by other components. Therefore, by taking the logarithm, the noise is added up, successively. Then the central limit theorem is applied to the logarithm of the concentration, and a Gaussian distribution is expected for the logarithm of the components. Thus, a lognormal distribution arises from the multiplicative nature of the component concentrations (Furusawa et al., 2006).

This lognormal distribution is an ideal form for the multiplicative stochastic process, whereas in real cells there may be several factors of deviation. In the ideal limit to derive the lognormal distribution, the cascading multiplicative stochastic process continues infinitely (or sufficiently many steps). In real cells, sometimes this successive process is not so large. Moreover, if the average concentration of some molecules is low, then the concentration easily reaches zero by fluctuation. (For a given cell volume V, the concentration can only take $0/V$, $1/V$, $2/V$, so if the average number of components is small, the distribution easily reaches the cutoff at $1/V$, and the distribution of the log concentration cannot be extended below it). Thus, the lognormal distribution is a first-step approximation form for the components with not-too-low concentration.

In contrast, if we consider multiplicative stochastic steps, it is well known that the gamma distribution: $P(x) = \frac{x^{a-1}exp(-x/b)}{\Gamma(a)b^a}$ often follows. In fact, such a gamma distribution is observed for the protein in a cell (Taniguchi et al., 2011).

In addition to the multiplicative aspect of the catalytic reaction, dilution of the concentration of each component by cell volume growth also introduces such a multiplicative nature of the fluctuation, whose experimental example is discussed in Section 3.5.

3.4 Transition to Dormant State: Emergence of Slow Dynamics with the Collapse of Consistency (D and F)

3.4.1 Phenomenological Model with Active Components, Wastes, and the Complex (A, D, F)

In Section 3.2, we considered a minimal cell model comprising nutrients, an auto-catalytic, and a membrane that provides compartments and growth conditions. This model has enabled us to realize a cell that grows at a constant rate (exponential growth). However, bacteria and other microorganisms have states other than exponential growth, under a starved condition. The following properties are known in general:

- Most unicellular microorganisms, undergo a phase transition to a dormant state (or stationary phase, as an ensemble of cells), in which the growth rate drops to nearly zero when nutrients become scarce.
- Once a microorganism falls into such a dormant state, it takes some time, called lag time, before the growth resumes, after sufficient nutrient is supplied again.
- This lag time increases with the length of time the organism has been starved.
- Apart from this dormant state, there is a death state in which growth cannot be resumed even if nourishment is provided.

Now, the models mentioned in Sections 3.2 and 3.3 do not exhibit such a transition to this dormant state. In an autocatalytic system, in which the components (population) take charge of their own replication as catalysts, the amount X grows in the form of $dX/dt = \gamma \times X$, which gives an exponential growth. On the other hand, if these components are decomposed or leaked, their decrease is also proportional to their own amount, so $dX/dt = -d \times X$. If $\gamma > d$, exponential growth leads to cell replication, and if $\gamma < d$, exponential decay leads to death. Hence, the dormant state with neither growth nor death cannot be derived from a model with an autocatalytic process that consists of an active component and a membrane component.

Now, as mentioned at the end of Section 3.2, cells are not able to convert all nutrient components into components that contribute to growth. Macromolecular replication is error-prone, and components are often created that cannot contribute to replication. They are often referred to as "parasitic molecules" in the question of the origin of life. Not only do they not contribute to the growth, but as they increase in number, the active molecule may be attached to the component to form a complex. Then the catalytic activity of the active molecule no longer works, thus the existence of parasitic components results in inhibition of the autocatalytic process. Even if they do not directly inhibit the process, the presence of such "waste components" can be considered to inhibit growth indirectly as a result.

Therefore, it is worth considering whether the transition to a dormant state can be understood by considering such "waste" components. For it, let us consider a simple model in which waste is introduced into the autocatalytic system for replication. Specifically, a catalytically active protein P (e.g., ribosomal protein) is produced from a nutrient (substrate) S, but waste W (misfold or e.g., inactive protein) is produced in the process, and the waste and active protein P form a complex C. The model is that garbage and active protein P form a complex C. In other words, the reaction is $S + P \rightarrow P + P$, $S + P \rightarrow W + P$, and $W + P \leftrightarrow C$ (Fig. 3.4(a)). The rate equations giving the respective concentration changes (with S, P, W, and C also used to denote the respective concentrations) are given by

$$\frac{dS}{dt} = -F_P(S)P - F_W P + P(S_{\text{ext}} - S) - \mu S$$

$$\frac{dP}{dt} = F_P(S)P - G(P, W, C) - \mu P$$

$$\frac{dW}{dt} = F_W(S)P - G(P, W, C) - \mu W, \tag{3.3}$$

$$\frac{dC}{dt} = G(P, W, C) - \mu C,$$

where $F_P(S)$ and $F_W(S)$ are the rates at which P and W are synthesized from the substrate S, and $G(P, W, C)$ is the rate at which the complex C is formed, given by $k_p PW - k_m C$, because C is synthesized by combining P and W, while they are separated into P and W at a certain rate. S_{ext} is the concentration of external nutrients. μ is the growth rate, and $\mu = F_P(R)P$ by assuming that the membrane is synthesized from the raw material based on the catalyst of the active protein P (the above reaction rule can be slightly modified, but the result will not change by adding more details). Furthermore, in order to study phenomena on a longer time scale, we can take into account the fact that each component of P, C, W decomposes at a certain rate, where C is harder to decompose because it forms a complex.

The ratio $F_P(S)/F_W(S)$ is expected to decrease as the nutrient concentration S decreases, since the selective synthesis of active molecules instead of the waste component requires more energy. Recall that the waste component is produced at

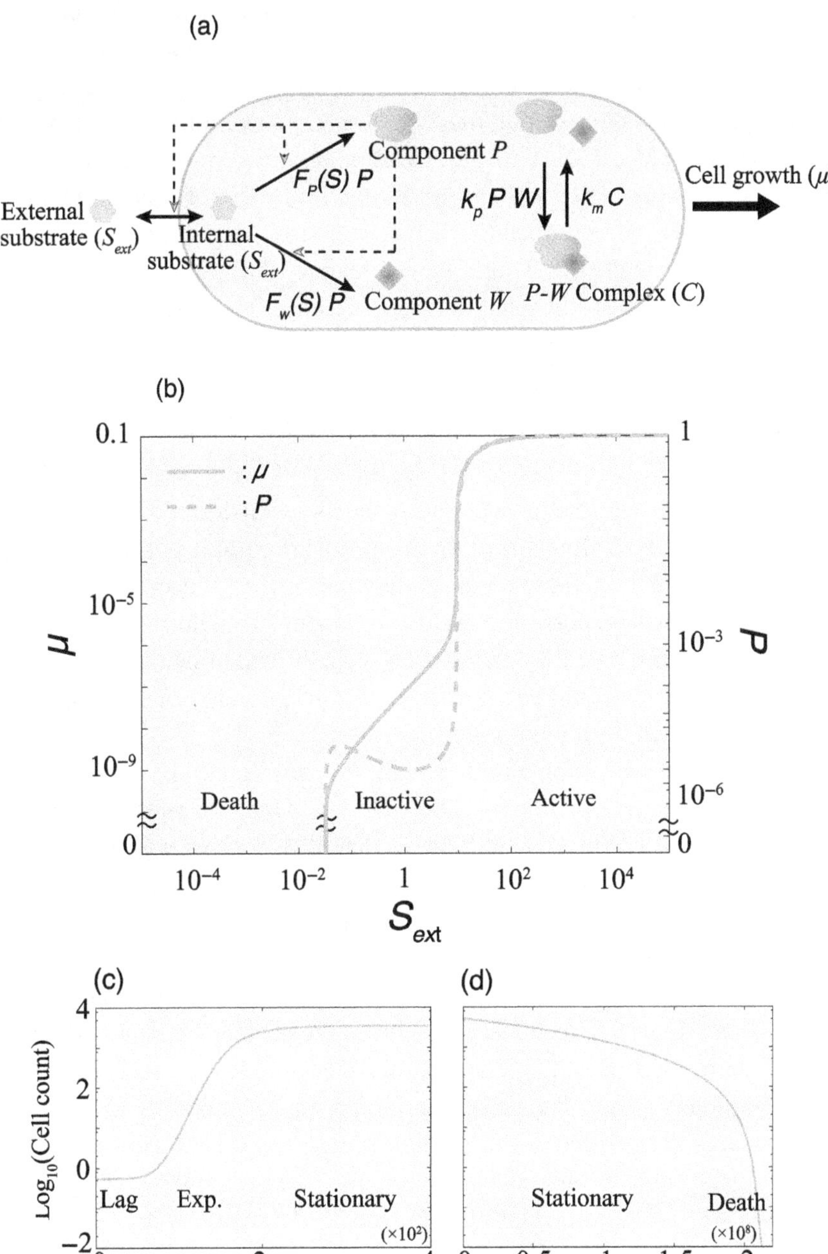

Figure 3.4 (a) Schematic of model with inactive (garbage) component (b) Nutrient S dependence of the active component P obtained by the model (c, d) Growth curves obtained by the model. Modified from figure 1 in Himeoka & Kaneko (2017), see it for details. (Courtesy of Yusuke Himeoka).

a certain rate due to an error that may occur in the synthesis of P. Therefore, there are generally processes in the cell that selectively reduce the error. For example, a process called Kinetic Proof Reading (KPR), a kinetic calibration process, reduces the proportion of error polymers by performing a multi-step catalytic reaction process (Hopfield, 1974). This proofreading requires a nonequilibrium flow of nutrients to selectively increase the proportion of the "correct" component P compared to that in thermal equilibrium. In fact, the error correction process of KPR requires such a nutrient cost. Therefore, we expect the rate of production of W to increase as the flow of nutrients decreases. Hence, the value of $F_P(S)/F_W(S)$ is low when the nutrient abundance S is small, and rises with increasing S, and eventually approaches an value.[4]

Numerical and theoretical analysis of the model has been performed with this setup and the following results have been obtained (Himeoka & Kaneko, 2017).

- As nutrition decreases, the growth rate μ shows a transition to a state where the growth rate μ drops to almost zero. In this state, the complex of waste molecules and active protein P increases and the amount of free P decreases drastically. There, the synthesis process of internal components from nutrients almost stops. In this case, even if there is degradation, cell shrinkage does not occur easily because the complex is not easily degraded. Thus, a state of neither growth nor death is maintained for a long period of time.

- In the above state with suppressed growth, P is not exactly zero, so the number of garbage molecules increases, albeit slowly. Under a suitable approximation with $PW \propto C \sim constant$. It follows that W increases with time t by $\sqrt{t}$: If $P + W \leftrightarrow C$ is stationary, then $dW/dt \propto P \propto 1/W$, or $dW^2/dt \propto constant$, so $W \propto \sqrt{t}$.[5] As a result, a cell can say that the memory of how long a cell was starved is embedded in this slow change of W.

- If the nutritional condition is regained after starvation, the growth will not be restored immediately. This is because before the accumulated W is diluted and reduced, some of P is still hidden within the complex C, so it cannot work fully. This leads to the lag time. According to the model analysis, this lag time increases at the root of the starvation time T_{stv}. This can be understood from the results above. First, during starvation, the accumulated W increases in proportion to $\sqrt{T_{stv}}$, while during lag time W decreases at an approximately constant rate $dW/dt \sim -\mu W \propto -PW \propto -constant$. Hence the time it takes to decrease this accumulated amount of W and return to exponential growth

---

[4] The following transition is observed even if this ratio does not depend on S. However, the quantitative properties are different.

[5] However, this time dependence may depend on the form of $G(P.W.C)$. In general, this $\sqrt{t}$ can be replaced by t^{δ} with some exponent δ.

increases by the square root of the starvation time. (However, since the time dependence of W can depend on the form of $G(P, W, C)$, it will be better to weaken the statement, so that the lag time increases with starvation time as T_{stv}^{β} with some exponent β, which may depend on the specific choice of $G(P, W, C)$).

The above results are in good agreement with the experimental results mentioned at the beginning of this section, namely, the transition to the stationary phase with almost null growth, as well as the existence of lag time. In fact, the lag time increases with increasing starvation time in the experiment. According to a recent experiment (Gefen et al., 2014), it seems to increase approximately in proportion to the square root of the starvation time, whereas further quantitative experiments are needed.

The transition to a state with a growth rate of $\sim$0, described here, can be understood as a collapse of macro-micro consistency. When there is sufficient nutrition, a state of consistency between the reproduction of each component and cell growth is achieved. On the other hand, as nutrition decreases, such consistency is no longer realized due to the increase of garbage components, and the growth rate asymptotically decreases to zero.

In the models up to Section 3.3, all components are active and responsible for growth, and are regarded as "ideal gas" type models. In contrast, the model in this section incorporates the effects of other components that interfere with growth, which always occur in reality. In this sense, the model in this section can be regarded as a "van der Waals" type model, in the sense that it incorporates the effects of other components. As in the van der Waals model, the present model exhibits the phase transition.

In this chapter, this transition is expressed only in terms of nutrients, active proteins, garbage, and complexes. In order to make the growth clearer, it would be desirable to model this with a membrane component, as in Section 2. Furthermore, since there are many kinds of catalytic molecules and garbage molecules, it is desirable to consider a large-degree-of-freedom model that introduces garbage into the catalytic reaction network model. This will be discussed in the next subsection.

3.4.2 Reaction-network Model for the Transition to a Dormant State (C, D, F)

The phenomenological model in Section 3.4.1 assumes the existence of a component that acts as autocatalysis and the waste, as well as the complex formation between the two, and that the rate to produce the waste molecule increases as the nutrient supply is decreased. Can such structure emerge from a microscopic level, with different components that make catalytic reactions with each other? Recently, a simple cell model consisting of a catalytic reaction network with many

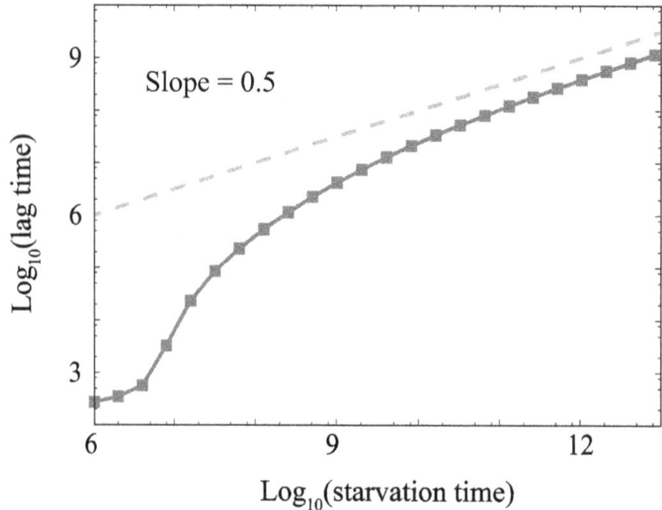

Figure 3.5 Starvation time dependence of lag time. Based on figure 2(a) in Himeoka & Kaneko (2017), and based on the experimental data by Gefen et al. (2014). (Courtesy of Yusuke Himeoka).

components as well as intermediate complex formation has been numerically studied (Yamagishi & Kaneko, 2024), which demonstrates the transition from the exponential growth phase to the growth-arrested dormant state with depletion of nutrients.

In the study, a randomly chosen catalytic reaction network model with intermediate complex formation in each reaction is adopted, where two-step elementary reaction processes with the formation of an intermediate complex C are introduced as follows:

$$X_i + X_j \leftrightarrow C_{ij}$$

$$C_{ij} \rightarrow X_k + X_j.$$

Here, each elementary process proceeds according to the law of mass action, where X_i, X_k, and X_j denote the substrate, product, and catalyst. The rate of the second reaction is assumed to be v times faster than the reversible complex formation reaction. In the limit with $v \rightarrow \infty$, the product is formed as soon as the complex is formed, so that the above reaction processes are reduced to the single mass action kinetics as $X_i + X_j \rightarrow X_k + X_j$, without intermediate complex formation, and thus are reduced to the model mentioned already (Furusawa & Kaneko, 2003; Chapter 6 of [Kaneko, 2006]), as well as described in Chapter 5. In contrast, when v is small, the intermediate complex C remains formed for some time, and the complexes may accumulate. Then the abundances of free reactants that are not bound into complexes will decrease, which can hinder the reaction processes.

By numerically solving the above model in the case that v is not so large, it is found that the growth rate drops by orders of magnitude at a certain value of nutrient concentration as it is decreased, thus demonstrating the transition from exponential growth to dormant phase (Yamagishi & Kaneko, 2024). In addition, when the nutrient concentration is increased starting from the dormant phase, the transition to an exponential growth phase occurs at a larger value of nutrient concentration, thus demonstrating hysteresis. Within this intermediate value of concentration, there is bistability between the growth and dormant phases. (This bistability is also observed in experiments on microbes (Kotte et al., 2014)).

Now, the suppression of growth at the transition can be understood as a competition between autocatalytic sub-network and subsidiary parasitic networks: In the complex reaction network with many components, there exists a set of components that catalyze each other for the cell growth, that is, autocatalytic subnetwork, whereas some other subsidiary components attached to them can grow with the aid of the autocatalytic subnetwork (see Fig. 3.6 for schematic representation). This autocatalytic subnetwork corresponds to the component P and the subsidiary one to W in Section 3.4.1. In the exponential growth phase, the abundances of the components are concentrated on the autocatalytic subnetwork. This concentration is sustained when the nutrient supply, and thus the reaction flux, is high enough to allow the reaction to proceed before the complexes accumulate. However, when the nutrient supply is reduced, the flow is reduced and the complexes begin to accumulate. The reaction flow then stagnates in the pathways within the autocatalytic network, so that the fraction flowing out to the non-autocatalytic subnetwork increases. Therefore, the assumption of an increase of the rate to W in Section 3.4.1 naturally appears in this reaction network model. This accumulation of complexes occurs in other pathways, so that the jamming in the reaction flow occurs in a cascade.

The result here, then, suggests that the growth-dormant transition will be inevitable for cells that grow with catalytic reactions via complex formation, and appear without tuning by evolution or adaptation.

In fact, the following characteristics are common to many of the complex networks that are randomly generated: (i) growth-dormant transition against changes in the external nutrient (ii) hysteresis against the changes in the nutrient concentration (iii) increases in diversity of chemical components (for instance, as measured by Shannon entropy over fraction distribution of each component abundance) at around the transition to the dormant state. In addition, for most networks, the lag-time to recover the growth increases with the starvation time with some power as $t_{lag} \propto T_{strav}^{\beta}$ with $\beta \sim (0.3 \sim 0.7)$.

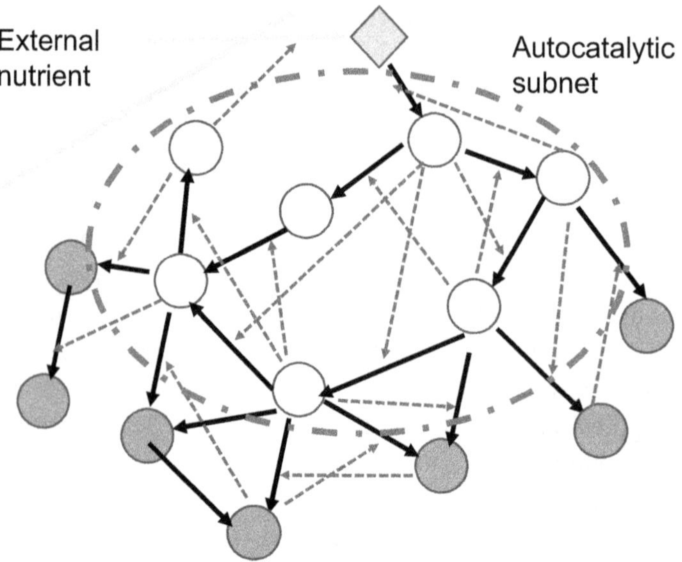

Figure 3.6 Schematic representation of the autocatalytic subnetwork and other components. Each component is represented by a circle, which can act as a substrate, catalyst, and enzyme. A thick arrow indicates the reaction process from substrate to product, which is catalyzed by the component shown by the corresponding dotted arrow. This model cell takes up a nutrient (shown by the diamond at the top) from the environment, which is catalyzed by a component shown by the dotted arrow. In this schematic example, seven components within the dotted ellipse form an autocatalytic subnetwork. Within the subnetwork, the components catalyze each other and produce all the components from the nutrient. The model includes several other components and reactions. Such components with gray circles are subordinate to the autocatalytic subnetwork and are produced parasitically to the autocatalytic subnetwork.

Thus, the present model at the microscopic level not only supports the result of Section 3.4.1, but also explains the hysteresis observed in experiments and predicts the increase in chemical diversity. Furthermore, the present results suggest the breakdown of the consistency between autocatalytic components and global cell growth, resulting in the increase of diversity at the transition to dormancy. This increase in diversity in the dormant state suggests a violation of dimensional reduction in an exponential growth state (see Chapter 9 for dimensional reduction).

3.5 Experimental Approaches: Growth Rate, Division Time and Fluctuation Law of Component Concentrations (E)

3.5.1 (Macro)growth Rate and Fluctuation Law

Now, let us return to the steady exponential growth state of the cell and discuss the properties measured experimentally: as mentioned in Section 3.3, in a state where

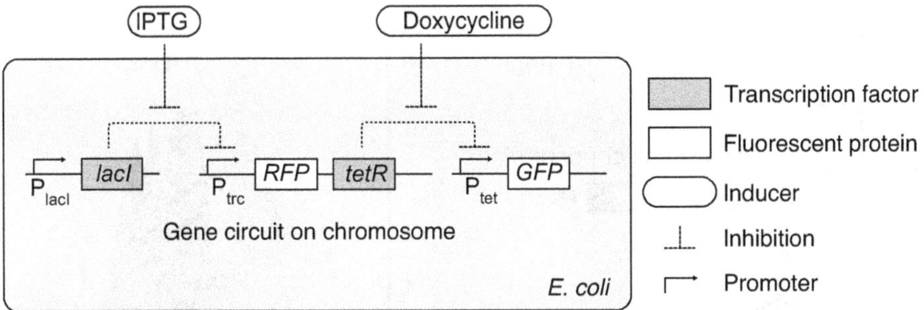

Figure 3.7 Gene circuit used in an experiment to investigate the relationship between growth rate and protein level fluctuations. This gene circuit is embedded into *E. coli*. Based on Tsuru et al. (2009). (Courtesy of Saburo Tsuru).

the abundances of each component grow consistently, there is a power law over all components with respect to the average abundances,[6] whereas the distribution of each abundance across cells is approximately lognormal. Such a statistical law for individual components has been experimentally verified (Kaneko, 2006). The existence of such fluctuations is natural, since the reaction itself is caused by stochastic molecular collisions, and the increase is due to the amplification process.

On the other hand, we are interested in macroscopic cellular states. So, let us investigate how the fluctuation of the growth rate, which is a macro-state quantity rather than an individual component, behaves and how it is related to the protein concentration.

To see the relationship between the cell growth and protein concentration, Tsuru et al. (2009) embedded the gene network shown in Fig. 3.7 in *E. coli* and examined how the protein concentration is distributed in each cell (Tsuru et al., 2009). In the setting of this experiment, *tetR* expression represses GFP transcription. In addition, an inhibitor of expression, doxycycline, is introduced, which suppresses this repression. Raising the doxycycline concentration increases the concentration of GFP protein. Then, they measured how the average concentration of GFP, as well as the distribution of the concentration across cells, changed as the doxycycline concentration was increased. On the other hand, the relationship between cell growth and fluctuations in protein concentration can be examined by measuring the cell-volume change (length change, since *E. coli* is almost rod-shaped and its length increases without growing in width). The experimental results show that

(1) The fluctuation of protein concentration is far from a Gaussian distribution, with a tail on the large side. Rather, the distribution of logarithm of the concentration is closer to be Gaussian (Fig. 3.8).[6]

---

[6] See also (Elowitz et al., 2002; Kaern et al., 2005; Bar-Even et al., 2006; Taniguchi et al., 2010) for studies on fluctuations in protein levels.

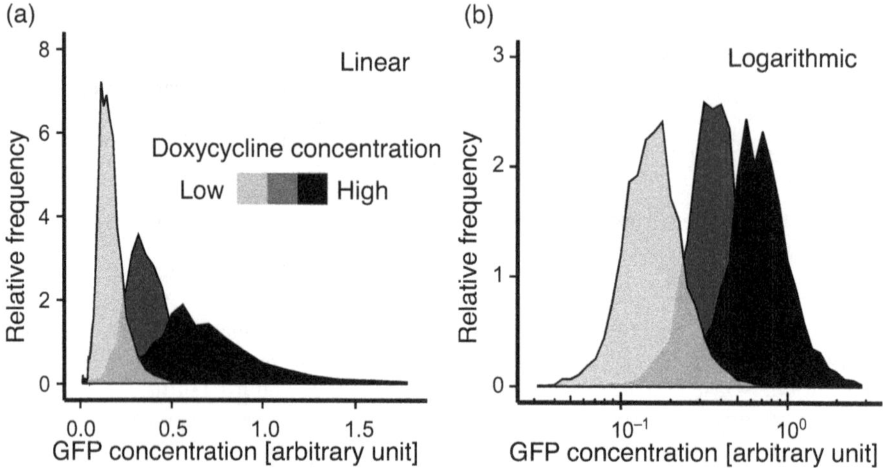

Figure 3.8 The distribution of GFP concentrations obtained in the experiment with the circuit shown in Fig. 3.7. Different lines are distributions obtained with different amounts of doxycycline. Results when the amount of doxycycline was reduced from left to right. (a) Linear GFP concentration scale. (b) Logarithmic scale. Based on Tsuru et al. (2009). (Courtesy of Saburo Tsuru).

(2) By changing the concentration of the expression inhibitor doxycycline, the mean and variance of protein expression level changed. The measurement showed that the variance of protein concentration is approximately proportional to the square of the mean.

(3) According to the single-cell measurements, the distribution of the cell growth rate μ was approximately Gaussian (Fig. 3.8). The variance was relatively large, for example, CV (=standard deviation/mean) of more than 10 percent.

(4) The correlation time of the growth-rate was about one generation. That is, cells with a high growth rate generally keep the high growth until they divide, but this trend is not maintained after division. In contrast, fluctuations in protein concentration are correlated over several generations.

Here, the protein measured here is synthesized with a certain rate k within the gene circuit, while it is diluted by cell growth. Thus, the concentration of the protein x approximately follows

$$dx/dt = k - \mu x + \eta(t), \tag{3.4}$$

where μ is the growth rate of the cell-volume (Fig. 3.9). If μ is constant, then mean of x satisfies $x = k/\mu$ as a stationary solution of the above equation. If the gene expression fluctuates with noise ($\eta(t)$ term in the above), the concentration x shows a Gaussian distribution around it. On the other hand, since the growth rate itself is distributed, $\mu = \mu_0 + \xi(t)$ around the mean μ_0. By noting that the fluctuations of the growth rate have a correlation time of about one generation and can be expressed

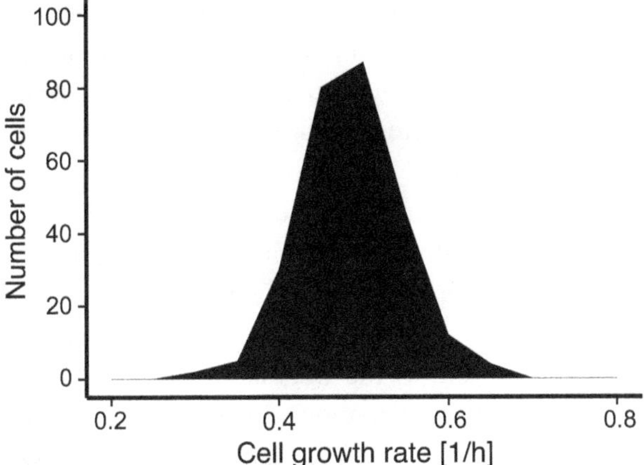

Figure 3.9 Frequency distribution of cell growth rates obtained from the experiments in Figs. 3.7 and 3.8: Based on Tsuru et al. (2009). (Courtesy of Saburo Tsuru).

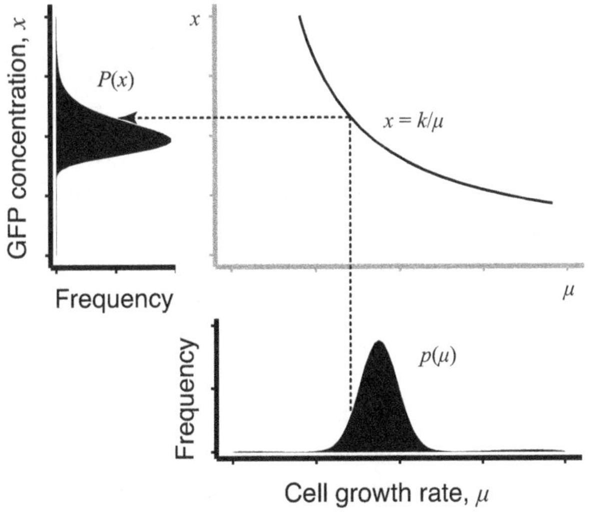

Figure 3.10 Relationship between concentration fluctuation and growth-rate distribution: Due to the dilution by the cell-volume growth, fluctuation in the growth rate is transformed into the protein-concentration fluctuation. (Courtesy of Saburo Tsuru (Tsuru et al., 2009)).

as a Gaussian distribution, $P(\mu) \propto exp(-(\mu - \mu_0)^2/\sigma^2)$ with σ^2 as the variance. Then, the distribution $P(x)$ of x can be obtained from this Gaussian distribution $p(\mu)$ by transforming $x = k/\mu$ as (Fig. 3.10).

$$P(x) = p(\mu)\frac{d\mu}{dx} \propto (1/x^2)exp(-(k/x) - \mu_0)^2/\sigma^2), \qquad (3.5)$$

This distribution differs from the Gaussian distribution in that there is a long tail in the larger side of x and is close to a lognormal distribution, and its variance is approximately proportional to k^2, or the square of the mean. The experimental distribution of GFP concentration agrees well with the distribution of x in the above Eq. (3.5).

Here the fluctuations in concentration are originated both in the growth rate fluctuations mentioned above and those due to the noise in gene expression reaction. Now, the former variance by the growth-rate fluctuation increases approximately with the square of the mean, while the latter increases in proporation to the mean, because it is simply due to the fluctuation in synthetic reaction, that is, the fluctuation in k in the above equation. Since the experimental result shows the square dependence, the main part of the concentration variation by cells is expected to come from the growth-rate fluctuation. In fact, detailed analysis shows that the contribution from growth rate fluctuations is more than 10 times larger than the fluctuation due to stochastic gene expression (see also Tsuru et al., [2009]).

Note that the growth rate fluctuation is due to dilution according to the cell volume increase, and thus provides variation common to all components, that is, provides common noise to all components. In contrast, since the fluctuations in each gene expression are due to the stochastic nature of the chemical reaction, such global correlation across all components does not exist.

The measurement of gene expression fluctuations as extrinsic or intrinsic noise was originated by Elowitz (Elowitz et al., 2002). The former is noise common to all components, such as environmental fluctuations that affect the entire cell, while the latter is noise in the individual expression process. According to the above experimental results, dilution by growth will be the most significant source of extrinsic noise.

The common effect of this growth fluctuation is clearly demonstrated by the correlation across components. In the gene circuit of the above experiment, the upstream protein (measured by red fluorescence by embedding Red Fluorescent Protein (RFP)) suppresses the downstream protein (measured by green fluorescence by embedding Green Fluorescent Protein (GFP)). If the expression of the upstream protein goes higher by fluctuation, the expression of downstream protein should be repressed. Therefore, a negative correlation would be observed between the expression levels of the two proteins. On the other hand, since the growth rate-derived fluctuation works in the same way for both proteins, there should be a positive correlation between the two, that is, both concentrations should decrease (increase) if the growth dilution is higher (lower). The experimental results show a clear positive correlation, as shown in Fig. 3.11. This indicates that the growth rate fluctuation has a much larger effect.

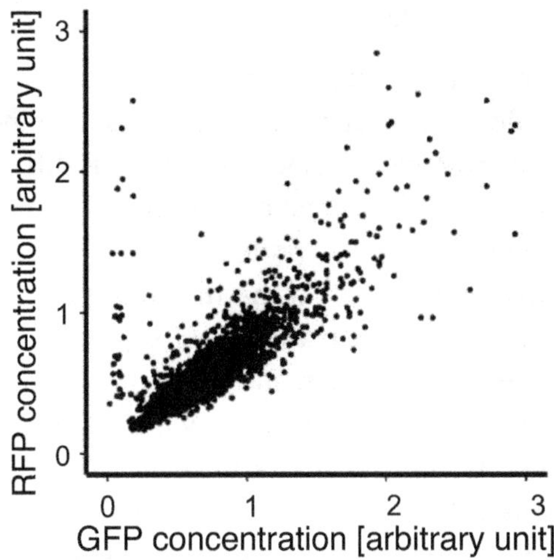

Figure 3.11 Correlation between GFP and RFP concentrations in *E. coli* from experiments in Figures 3.8 and 3.9: Based on Tsuru et al. (2009). (Courtesy of Saburo Tsuru).

As described above, the fluctuations indicate that the growth rate is a macroscopic quantity that affects all the components. On the other hand, growth itself is a result of the overall reactions of a variety of intracellular components. In other words, the growth rate, which is the overall result of each local reaction process, is a macroscopic variable that has a common dilution effect for all components. This macro-micro consistency shapes the steady-state growth of the cell and determines the nature of the fluctuations.

So far, we have only looked at the effect of growth rate on each component, that is, the effect of decreasing concentration by the cell-growth dilution. However, if the growth is a result of the synthesis of the components, then it is conceivable that there could be a positive correlation between the rate of synthesis of the components and the rate of growth, that is, a proportional increase in the rate of growth. If the latter process, the increase due to the growth of synthesis does not occur at all, and then the concentration would be inversely proportional to the growth-rate. In contrast, if the latter compensation effect fully works and each component is synthesized in proportion to the growth rate, then each concentration is independent of the growth rate because the synthesis and dilution are balanced. The experimental data show that many of the components lie between the two cases. In other words, synthesis is increasing to compensate for dilution to some extent, but it is not 100 percent (say 50 percent). Moreover, the degree of this increase of synthesis can vary from component to component. This will be discussed again in Chapter 4 in relationship with the issue of adaptation.

3.5.2 Single-cell Measurement of Cell-division Process

In the above experiments, in tracking cell-by-cell fluctuations during the growth-replication process, a flow cytometer was used to determine cell size and the amount of protein by fluorescent protein. This makes it difficult to keep track of time development in individual cells one by one. It is then necessary to track the cell growth, the amount of (fluorescent) protein, and size under a microscope while single cells grow and divide. However, if this division is continued, the number of cells will soon increase and they form a large colony. Then other factors, such as the lack of nutrition and cell–cell interactions, start to come into play. (Finally, the cells reach a dormant (stationary) phase in which they are unable to increase much.) The growth of cells in the center of the colony is also dependent on their spatial arrangement, with cells at the periphery of the colony growing less.

Therefore, it is necessary to construct an experimental device that can measure cells at the level of single or a few cells over a long period of time by removing the increased number of cells. To meet this need, microfabrication techniques have been developed for long-term measurement of single cells, such as the Mother Machine by Jun and colleagues (Wang et al., 2010) and the Dynamic Cytometer by Wakamoto and colleagues (Hashimoto et al., 2016; Wakamoto et al., 2001). For example, in the latter, as shown in Fig. 3.12, a flow channel is generated within which (bacterial) cells are arranged in a row, and the cells are enclosed in the channel. The top is covered by a semipermeable membrane that allows nutrients to flow in. There, the cells grow and divide along the channel, and when they reach both ends, they are drained out. Microscopic observation of the process of cell growth and division at the center over several hundred generations allows one to measure fluctuations in the growth and division time of a single cell. If a fluorescent protein is introduced into the cell, the distribution of the amount of the protein in each cell and its correlation with the growth rate can also be measured.

Figure 3.12 shows the results obtained by using this device to measure one-cell growth divisions over a long period of time. From these measurements, the division time and its distribution at the one-cell level can be determined. For example, from these long-term measurements, a linear relationship between the mean and variance of single-cell division times has been found, as shown in Fig. 3.13 (Hashimoto et al., 2016). This is a linear relationship between the mean division time and the variance of the distribution of cell division times under various environmental conditions.

The proportional relationship between variance and mean follows naturally from the central limit theorem when the quantity can be expressed by the addition of independent random processes, whose distribution is Gaussian. However, in the present case, this relationship is not obvious because the distribution of division

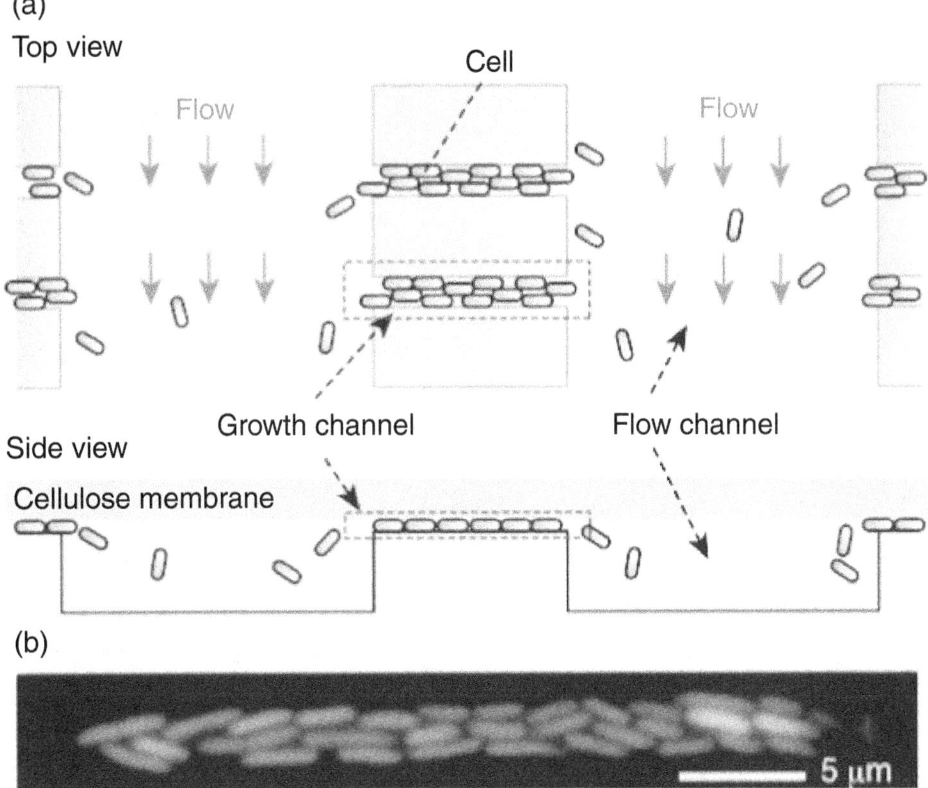

Figure 3.12 The schematic diagram of Dynamic Cytometer (a) and an example of measurement of *E. coli* cells (b). (Courtesy of Yuichi Wakamoto).

time is non-Gaussian, with a tail on the longer side. For example, it can be derived by assuming that there are several steps leading up to cell division, and that each step involves a checkpoint, and that at each step there are extra paths that are taken off and back on due to noise. Then the variance in the time through each step will be proportional to the step length. However, no conclusive evidence of this argument has been obtained at this time, and the basis of this law is not yet completely clear (Fig. 3.14).

Another interesting point here is that by extrapolating this relationship, the variance goes to 0 at a certain division time $T_d^{lm} > 0$. Since the variance is greater than zero, this value could be considered to provide the minimum possible division time. This minimum division time (maximum growth state) is probably the limit at which the cell grows without any fluctuation, where the reaction proceeds like a perfect machine. In fact, no condition or no cellular state with a faster division time than this is known so far for *E. coli*. In this sense, this "machine limit" with 0 fluctuation may correspond to the maximum possible growth rate.

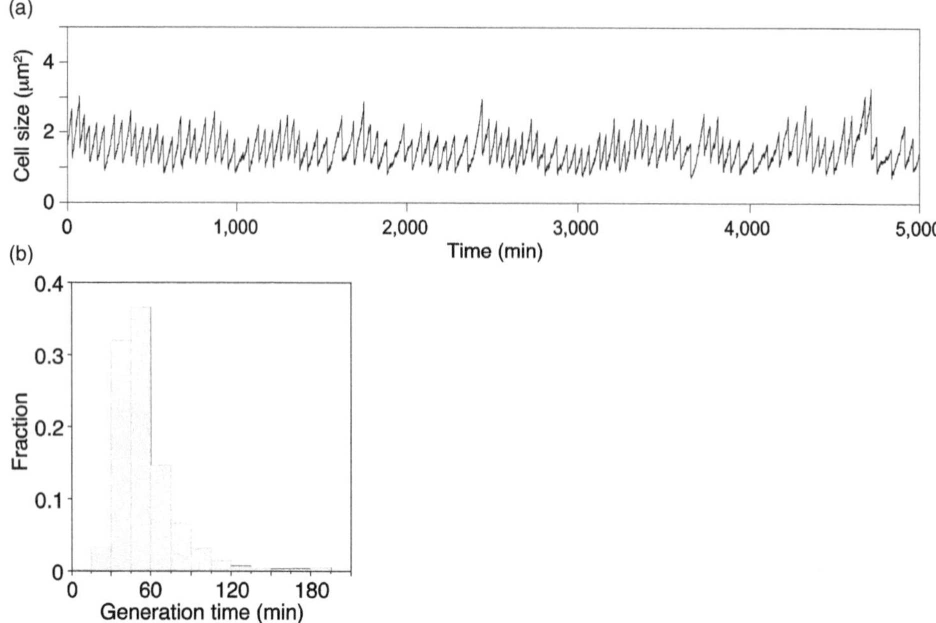

Figure 3.13 (a) Long-term measurement of cell length by Dynamic Cytometer in Fig. 3.12. When the cell division occurs, the size decreases almost by half. (b) Displays the distribution of division time thus measured (Hashimoto et al., 2016). (Courtesy of Yuichi Wakamoto).

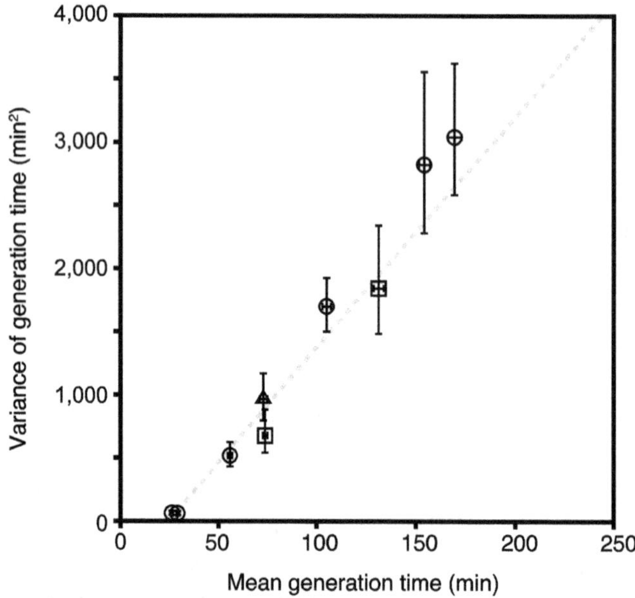

Figure 3.14 Relationship between the average and variance of cell-division time of *E. coli* obtained by the experiments in Figs. 3.12 and 3.13. Results from different culture (nutrient) conditions (Hashimoto et al., 2016). (Courtesy of Yuichi Wakamoto).

If we take the above interpretation of surplus paths, then shortening the division time by improving nutritional conditions corresponds to reducing these surplus reaction paths and reducing the number of wasted paths as much as possible. If so, it is understandable that the variance of the division time asymptotically approaches zero, as the surplus paths asymptotically approach zero at the minimum division time limit.

3.6 Basic Questions on the Origin and Construction of Protocells: Parasites, Central Dogma between Functional and Information Molecules, and Origin of Heredity

3.6.1 Difficulty in Construction of Reproducing Protocells

The current cells are a result of evolution and have acquired various properties through this. Some of these properties may be fixed by chance through evolution, and it is sometimes difficult to tell whether they are inevitable for a living system or not. Universal biology, on the other hand, seeks to understand the properties that a life system must inevitably satisfy, including those found in the universe and those that are artificially constructed. It is then an important direction for universal biology to construct minimum artificial cells that can replicate, and to discuss the properties they must satisfy. This is why we take a "constructive approach" to cells.

At the present time, the study to construct artificial cells (protocells) is focused on solving many experimental difficulties, rather than deriving a general concept for universal biology. However, there are basic theoretical problems that should be considered as universal biology within the difficulties themselves. To achieve stable replication, to what extent a system has to be tuned up? Is it a rare phenomenon, such as threading a needle, or does it occur in a rather broad set of conditions? Is it possible to reach that state by starting with a suitable reaction system and by selecting robust replicating system? How large is the region of possible "survival" of cells within all the possible space of reaction systems?

In addition to these issues on the degree of "difficulty," there are several conceptual issues regarding construction of protocells. Let us briefly discuss them in the following sections.

3.6.2 Conflicting Multilevel Evolution between Molecules and Cells: Problem of Parasitic Molecules

For a cell composed of molecules to continue to proliferate and divide, the molecules within the cell must replicate. As discussed in this chapter, such replication is the result of catalytic reactions, and the catalytic molecules must also replicate. These catalytic molecules are polymers consisting of monomers. Note that the replication of such polymer is not always perfect, and the sequence of monomers can be slightly altered due to thermodynamic fluctuation. Then the catalytic activity or proliferation rate can be changed, and some polymers may lose the catalytic function. Now consider the situation that the resource (monomers) needed for the replication is limited and the polymers will be degraded and the polymers compete to increase their number. Accordingly, the molecules that can be replicated faster can have a higher chance for survival and growth, so that their fraction is increased. Faster-replicating molecules are selectively favorable. In other words, the molecules themselves are subject to Darwinian evolution because they are under mutational changes and selection by the fitness (i.e., the rate of replication).

Since the replication reaction of polymers needs catalysts, we need to consider a system consisting of molecules that mutually catalyze their replication. However, while the molecule is working as a catalyst, it does not replicate. Therefore, it would be advantageous for the molecule to reduce its own catalytic activity rather than to maintain or increase it, in order to increase the degree to be replicated by the aid of other catalysts. However, if mutational changes to decrease the catalytic activity are selected for all molecules, eventually every molecule would lose the catalytic activity. In other words, everyone would be selfish and no one would help others to replicate. Then the molecules will no longer be replicated and the protocell consisting of them will stop the growth. This is an example of the "parasite" problem.

This problem can be solved by considering a multilevel selection, that is, to consider selection at a cell level. As molecules replicate, they aggregate, and will form a structure of concentrated molecules, which will constitute a protocell. If somehow, these protocells divide after the growth,[7] such protocells can continue the reproduction to survive. On the other hand, if protocells are filled with parasitic molecules that do not help the synthesis of others, they stop the growth and eventually will die out. Hence, such protocells will be extinguished. Only if cells have not yet been dominated by parasites, the replication of molecules to catalyze each other continues, which leads to the growth of the protocells, leading eventually the reproduction of them. Hence, by considering the selection at the level of cells, the parasite problem could be resolved (Eigen, 1992; Szathmary & Maynard-Smith, 1997; Altmeyer & McCakskill, 2001; Boerlijst & Hogeweg, 1991; Hogeweg, 1994).

However, if the total number of molecules in a protocell before division is large, the error (mutation) to lose the catalytic activity would occur with high

---

[7] We discuss in Section 3.2.6.6, how this cell-level reproduction is achieved later.

frequency, and such protocells would be dominated by parasite molecules, and the reproduction of a cell would not be achieved. Here, molecular-level selection works in the direction to decrease the catalytic activity, whereas cellular-level selection supports to increase (or maintain) the average catalytic activity over molecules within. If the molecule number in a cell for division is too large, the molecular fitness dominates, whereas if it is small, the cellular fitness dominates, and the selection by cellular fitness wins over the molecular one, so that the catalytic activity is sustained at least in some cells. Can we formulate the conflict between the two levels and find an appropriate cell size (molecule number) below which protocells survive (and evolve)?

To answer this question, Takeuchi et al. (2016, 2022) considered a model in which a protocell consists of molecules that are replicated by using the nutrients S with the aid of other molecules that catalyze. Molecules catalyze each other, and if this process continues, the total molecule number in a protocell is increased. For simplicity we assume that when the total number of molecules reaches N, the protocell is divided into two with equal numbers of molecules. Now each molecule X has a parameter specifying the catalytic activity that assists the replication of others, and this parameter value may vary slightly during replication. This is the evolutionary process at a molecular level (see Fig. 3.15(a)). On the other hand, cells compete for resources. Here it is simply assumed that the total cell number is fixed, so that after one cell is divided, another cell randomly chosen is removed. This leads to selection process at a cellular level, in favor of a higher growth (lower division time) (Takeuchi et al., 2016, 2022).

Here the replication occurs as

$$X_i + X_j \rightarrow C_{ij}; \quad C_{ij} + S \rightarrow 2X_i + X_j,$$

where the molecule j works as a catalyst for the replication of i. The replication occurs after the formation of complex. Accordingly, the net replication process requires some time.

The rate of the complex formation is given by the catalytic activity of j, k_j. Within this process of complex formation and replication, the catalytic molecule is occupied in this process, so that it cannot be replicated. The molecule is disadvantage in terms of molecular replication because while it is working as an enzyme, it cannot be replicated with the help of other enzymes. If the catalytic activity is larger, the probability that the molecule is involved in the reaction is higher. Hence, for the sake of increasing the chance for replication, lowering k_j will be preferable. As a result, evolution at a molecular level occurs so that the molecule loses catalytic activity as a molecule.

As mentioned, however, if all the molecules in the cell evolved in that direction, replication would not proceed. In fact, when the total number of molecules in a cell

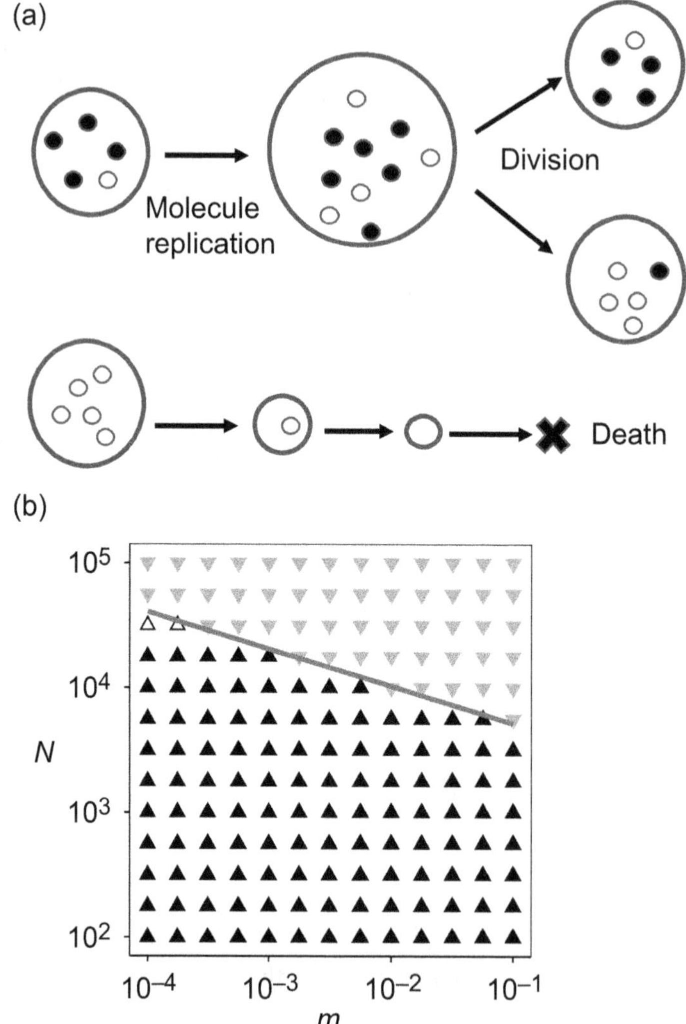

Figure 3.15 (a) Schematic of the multilevel model of cell reproduction. Molecules (small circles) have different levels of catalytic activities (in the picture, darker ones have higher activities and white ones lose activity). In the catalytic reaction, molecules replicate, and accordingly, cell growth and division follow if the molecules retain sufficient catalytic activities. (b) In the region with ▲ in N and mutation rate m, the molecules keep catalytic activities and cell division continues, whereas in the region with ▽, molecules lose the catalytic activities and cell division stops. Based on Takeuchi et al. (2022).

N is sufficiently large, this loss of catalytic activity occurs for all cells, until no replication progresses any more. In contrast, if the number of molecules N at the time of cell division is small, this loss of activity is avoided, and cells that retain their catalytic activity remain and continue to divide.

Hence, there is a critical system size N_c below which the cell can keep catalytic activity, sustained by the cell-level selection. This N_c depends on the error (mutation) rate m in the replication of molecules. If m goes to zero, even for the case with $N \to \infty$ the catalytic activity can be sustained. As the mutation rate m is larger, the critical size N_c decreases (see Fig. 3.15(b)). By using theoretical analysis of stochastic process with multilevel evolution, it is shown $N_c \propto m^{-1/3}$ (Takeuchi et al., 2022).

Furthermore, there is an interesting region in the intermediate size at around N_c. There, most molecules in a cell lose their activity, and such cells stop reproduction, whereas for some cells, a small percentage of molecules regain their activity by mutation, so that cell division is restarted and maintained. This recovery occurs with a certain frequency in time, so that the activity of the molecules oscillates in time, repeating the decline and recovery.

3.6.3 Origin of Central Dogma as Symmetry Breaking

So far, we have shown that catalytic ability is lost when the number of molecules in a protocell is large. For an intermediate level of number, temporal changes between gain and loss of catalysts are observed, whereas the state to keep catalytic activity is generally unstable if the number of molecules in a protocell is large. Then how can the catalytic activity be sustained for a protocell with a larger number of molecules?

In the present cell, there exists division of labor in molecules, between those carrying out catalytic function and genetic information for replication. Proteins correspond to the former, and nucleic acids (DNA or RNA) to the latter in the present cells. Except for few special cases, the genetic information flows only from DNA to proteins, and no backflow exists. This is termed as *central dogma in molecular biology* by Crick. Now, is such differentiation of roles a general consequence of reproducing cells?

Recall that if the number of molecules N is greater than N_c, the catalytic activity is lost and the cell stops reproducing. Then, if there are two types of molecules, is it possible that one type loses the catalytic activity while the other retains it, so that cell reproduction continues?

For this purpose, we consider a protocell with two types of molecules (polymers) (let us say P and Q) that are synthesized using the template information

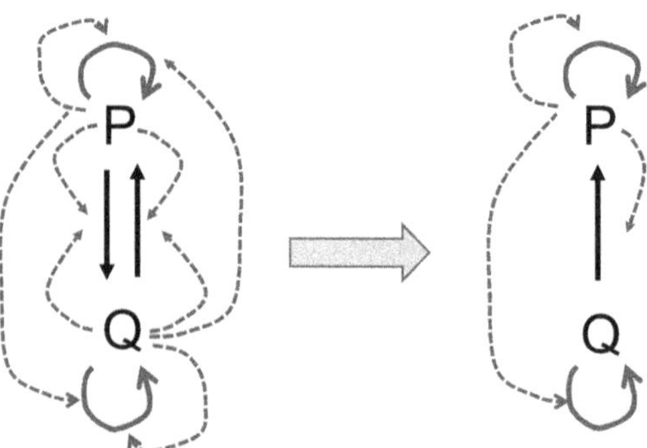

Figure 3.16 Schematic of the origin of central dogma as symmetry breaking. Initially both the molecules, *P* and *Q* work as catalysts and templates. Through the multilelvel evolution of replicating molecules and reproducing cells, one type of molecule species (say *Q*) loses catalytic activity and works as template for information, whereas the other (say *P*) works as a catalyst and loses the capacity of templated. Based on Takeuchi & Kaneko (2019).

of one molecule and the catalytic activity of the other, encapsulated in a protocell. Initially, we assumed that there is no role separation, so that both act as templates and catalysts. Here we consider a hierarchical molecule-cell replication system as mentioned in Section 3.6.2. Each molecule has a set of parameters as a template capacity (i.e., whose information is used as a template for the synthesis of molecules) and the catalytic activity for each possible template-mediated replication reactions.

When the number of molecules is small, the two types of molecules keep the same degree of parameters, that is, both of them keep catalytic activities. When the total number of molecules N is large, the molecules tend to lose their catalytic activity in order to increase the chance of replication. Interestingly, it is found that one type of molecule (say Q) loses its catalytic activity while keeping the capacity for information as template. In contrast, the other type of molecule keeps the catalytic activity and loses the capacity for information carrier (see Fig. 3.16).

This can be intuitively explained as follows. When the number of molecules is larger, the evolution to decrease in catalytic activity progresses. Since the mutation process to lose it is probabilistic, the rate of decrease in catalytic activity is different between the two species. Assume that the catalytic activity of Q decreases more. Then, Q can be used more as template information. When it is used as template information, the selection pressure works more strongly, so it can evolve faster to increase the template capacity and decrease the catalytic activity.

In contrast, the other (P) has a larger chance to work as catalyst, during which the template information is not transferred from P, so that evolution to decrease catalytic activity does not proceed much. Thus, the evolutionary pressure at the molecular level works less for P, and the pressure at the cell level dominates for P, suppressing the decrease in catalytic activity and accelerating the loss of capacity to work as template information. Eventually P loses the template information capacity. Then, it is no longer a replicator to have the information of itself to transfer. Its synthesis is determined by Q. Now, P turns out to be a *worker* determined by Q.

Thus, Q becomes the molecule that transmits information to the next generation, while P maintains its catalytic activity but does not transmit information to the next generation. This division of roles between molecules is the result of spontaneous symmetry breaking. This is consistent with the separation of roles between information molecules (DNA in the present cells) and catalytic molecules (proteins in the present cells). In other words, the "central dogma" of molecular biology, the unidirectional flow of information from DNA to protein, is considered to be an inevitable property, which is universally satisfied by replication systems with a molecule-cell hierarchy with a sufficiently large number of molecules in the cell. The above discussions are confirmed in multilevel evolutionary simulations as well as in the analysis of the evolutionary equation (Takeuchi et al., 2017; Takeuchi & Kaneko, 2019).

It should also be noted that as a result of multilevel evolution, the number of P molecules that act as catalysts will increase because they can increase the speed of cell reproduction. Then the number of Q molecules that act as carriers of genetic information will decrease.

3.6.4 Minority Control: The Origin of Heredity

As discussed in Section 3.6.3, the molecules responsible for inheritance and the components responsible for catalytic activity are separated. Then the number fraction of the latter tends to decrease. In fact, the relevance of molecules with minority in number to carriers of heredity was discussed earlier (Kaneko & Yomo, 2002).

We considered a mutually catalyzing reaction system between two molecular species X_i and Y_i, where $i = 1, \cdots, k$ and only those with $i = 1$ can help the synthesis of others, so that the species $i = 2, \cdots k$ are parasites, which are synthesized with the help of Y_1 or X_1, but cannot help the synthesis of others. These molecules are put into a protocell, and the cell is split into two if the number of molecules it contains is greater than a given threshold N. Here, the number of parasitic molecules $k-1$ is much larger than 1, so molecules in a cell may lose catalytic activity and growth may stop, as discussed in Section 3.6.2 when N is large.

Here we assume that the replication of Y molecules is slow, so that only a few (e.g., 1~3) Y molecules exist within the protocell. In this case, it was found that such protocells have one or two active Y molecules (Y_1) without Y_i at all for $i > 1$, while the majority of molecules are X_j, which include several active X_1 molecules and many inactive molecules X_i ($i > 1$). In this case, the Y_1 molecules regulate the synthesis of all the other X molecules. If the Y_1 molecule turns to be more active by mutation, all the X molecules in it are affected, and the cell can grow much faster, whereas if it loses such activity, the cell cannot grow. Therefore, some machinery will evolve to maintain the Y molecules well. In other words, evolution is controlled by this Y molecule that is preserved over generations. In this sense, the molecule acts as a carrier of genetic information. In contrast, due to the large number of X molecules, they follow the statistical law, so that possible mutational changes in X molecules are blurred out, and the directed evolution is not generated. The activities of X molecules just follow the statistical distribution, and many inactive X molecules are maintained.

In summary, the molecule with slower replication speed and minority in number controls the properties of the cell and acts as a carrier of genetic information in the sense that it is transmitted to the next generation and thus has evolutionary potential (Kaneko & Yomo 2002a; Matsuura et al., 2002). This is consistent with the fact that DNA molecules, as carriers of genetic information, are generally in the minority compared to protein molecules.

In fact, an experiment with a protocell showed that reproduction can only continue when the number of DNA molecules in it is small (Matsuura et al., 2002). More recently, a 144-component reaction-based replication system called the Pure System (see also Section 3.6.7) was introduced into an oil droplet, and it was shown that the reproduction of such protocells is possible when the number of RNA molecules is small (Matsuura et al., 2011). Furthermore, experiments on the development of artificial cells, in which the RNA replication system is enclosed in the oil droplet, are progressing by using minority RNA molecules (Mizuuchi & Ichihashi, 2018).

3.6.5 Discreteness in Molecule Number

Reaction systems with such a small number of molecules pose new theoretical problems. As mentioned above, there is a large number of molecule species in a cell, whereas the number of total molecules contained in a cell is limited, so that the number of some molecule species is not large. On the other hand, the change in concentration of chemicals is often estimated theoretically by using differential equations with the use of reaction rates. Here, the concentration is a continuous variable. Then, the change in concentration is estimated by assuming that the rate of reaction A+B is proportional to the concentration of chemicals A and B. When the number of molecules is large, the estimate of reaction rate by such, continuous-variable concentrations works well. When the number is smaller, the concentration fluctuates around its average value. To account for the fluctuation, then, a mathematical method called a stochastic differential equation can be used, where the continuous concentration variable is still assumed (Takagi et al., 2025). However, as the number of molecules further decreases, a qualitatively different behavior may occur due to the discreteness in the number.

In particular, when the number of molecules reaches the level of 0,1,2, the behavior may be significantly or qualitatively different from that of the continuous-concentration-variable picture above. When the number of molecules reaches 0, reactions using it will stop, and reactions using n molecules will stop if the number of molecules is less than n. (e.g., to make a dimer of molecules, at least two monomers are needed). The qualitative change in reaction dynamics due to the smallness (discreteness) of the molecule number has been studied extensively in physics in recent years (Togashi & Kaneko, 2001, 2005; Awazu & Kaneko, 2007; Biancalani et al., 2012; Saito et al., 2016). Such discreteness may be a defect for smooth intracellular processes, because it stops the concerned reactions from time to time. On the other hand, such discreteness in number can be used to generate different discrete states or to sequentially execute one process after another, as in a "if-then" process in computer program.

3.6.6 Crowding and Compartmentalization

In a cell, large macromolecules, whether proteins or nucleic acids, are synthesized and accumulated. They are generally packed together in a cell, like people in a crowded train (Goodsell, 1998). However, unlike humans in a crowded train, macromolecules can move around by diffusion, possibly because such macromolecules can deform under energy input. In this crowded situation with energy input, molecules interact strongly with each other, which can facilitate or suppress their movement. How reactions leading to cell replication proceed in such a situation is an important question.

Let us consider a system in which molecules catalyze each other to replicate, and these molecules are crowded together. Following the discussion in Section 3.6.4, let us assume that some molecules replicate more slowly than others, whose numbers will be small. As mentioned in Section 3.6.4, a minority molecule tends to play the role of heredity. Now combine this with molecular crowding and discuss how a cluster of molecules will divide at a certain size, where the replication of minority molecules can trigger the replication and division process of the cluster (cell).

Now consider the simple catalytic reaction mentioned in Section 3.6.4: $X + Y \rightarrow 2X + Y$ (rate γ_X), $Y + X \rightarrow 2Y + X$ (rate γ_Y), decomposition $X \rightarrow \phi$ (rate d_X), $Y \rightarrow \phi$, rate d_Y) (Kamimura & Kaneko, 2010), where $\gamma_X \gg \gamma_Y$ and $d_X \gg d_Y$. In other

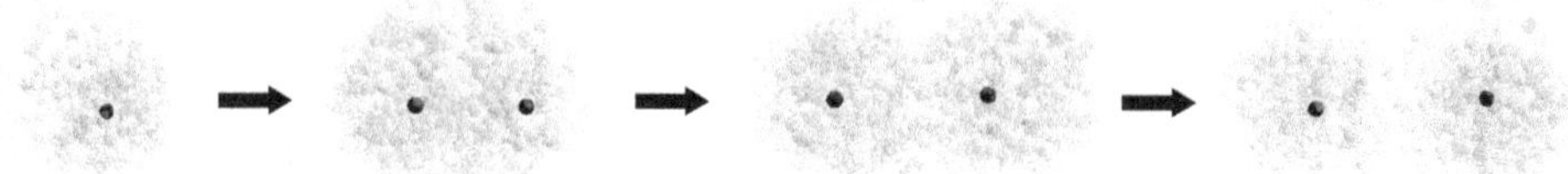

Figure 3.17 Time evolution of the mutual replication model consisting of X and Y described in the text. The gray sphere is the majority component X, and the black sphere in the center is the component Y that takes time to replicate. The replication of Y triggers the breakup of the cluster. Based on Kamimura & Kaneko (2010). (Courtesy of Atsushi Kamimura).

words, both synthesis and decomposition of molecular species Y are much slower than those of X. Then Y is not easily replicated, and its number should be small, as discussed in minority molecule in Section 3.6.4. Here X and Y are macromolecules diffusing slowly in a crowded situation with macromolecules around them.

First, consider the situation where there is only one molecule of Y. Here, X is synthesized much faster than Y. Once synthesized by the Y molecule, X molecules will diffuse around it. Then X molecules, whose number increases much faster than the replication of Y molecule, cluster around Y molecule. As X is decomposed, the cluster reaches a certain size where synthesis at the center (where Y exists) and decomposition are balanced. As a result, a spherical cluster is formed around the Y molecule. Now, after a longer time, Y is also replicated, so that two Y molecules emerge from the center. They slowly diffuse by a Brownian motion while X molecules are synthesized around each one, forming a dumbbell-like structure. Eventually, the two Y molecules move far enough apart, and the dumbbell structure finally splits into two clusters of X around each Y. Thus, the division of the protocell occurs with this simple setup.

As mentioned earlier, in a mutually catalyzing system, minority Y molecules with slow synthesis are responsible for heredity, that is, they control the cellular process and are well preserved, leading to evolutionary potential (Kaneko & Yomo, 2002). The above simple spatially structured model shows that the replication of such minority Y molecule and cell replication are synchronized, in a similar manner that DNA molecule replication and cell division occur synchronously, even though no special mechanism is imposed. In terms of the theme of this volume, the consistency between molecular and cellular replication has been achieved (Fig. 3.17).

3.6.7 Diversity

In every cell we know, there is a huge diversity of components. In contrast, today's artificial protocells contain far fewer chemical species. Perhaps this is why such protocells cannot sustain themselves and replicate unless they are supplied with highly controlled amounts of given resources. So far, there is a large gap to be bridged between the artificial protocol and robust multicomponent reproduction, which will be the basis of a living system.

As an attempt to achieve this goal, there is the already-mentioned Pure system, which consists of 144 components. This is a complex system with several thousand reactions, constructed from extracts of *E. coli* cells. Note that all chemical species and reactions are fully specified. This reaction system is responsible for metabolism and replication. As such, it is an important system for the synthesis of an autonomously replicating protocell (Shimizu et al., 2005). If this system is enclosed in a membrane, and if the contents of the membrane are replicated with autonomous reproduction of the membrane, and if this reaction process continues, we come closer to an artificial protocell. Unfortunately, the construction of such a protocell that continues to reproduce has not yet been achieved, but experimentalists are moving in that direction (Ichihashi et al., 2010).

On the other hand, if we consider a simpler replication system with fewer components, the speed of replication could be increased. In other words, there would be a pressure for simplification, and how to maintain diversity in this situation is a major issue. In Chapter 1, we discussed the possibility that diversity is maintained, for example, under a limited supply of diverse nutrient components (Kamimura & Kaneko, 2015, 2016). Experimental confirmation of this condition and the search for a "minimum diversity" to maintain a stable replication system will be necessary in the future. Theoretically, fundamental questions remain to be answered: How is the diversity of molecular species further increased? Is there an optimal level of chemical diversification? Is there a transition point to diversification beyond which a positive feedback process for diversification occurs?

3.6.8 Garbage Disposal

When the system is not simple, it will be harder to optimize all the components for growth. Functional biopolymers are rather rare among all polymer sequences, and it is difficult to synthesize only functional polymers that are a product of many steps of polymerization: as these are thermodynamic processes, it is not possible to add only "correct" monomers successively without errors, as already discussed as error catastrophe (Eigen & Schuster, 1979) and mentioned in Section 3.6.4. In the synthesis of macromolecules, the appearance of polymers without catalytic activity is inevitable, at least to some extent. Then, within a cell, some "unnecessary" *garbage* components would coexist, hindering the reproduction process. However, if cells are occupied with such garbage molecules, cellular activity and reproduction cannot be maintained. In fact, in the construction of artificial cells, such inactive ("garbage") molecules would accumulate. To solve the problem, garbage disposal is essential.

One way to deal with the garbage problem is to prepare a space that allows cells to grow. In this sense, the achievement of the growth-division process of protocells will be necessary to suppress the accumulation of garbage. Under what conditions can cells continue the dilution of garbage through growth? Will there be an evolution of the processes that perform waste decomposition? Naturally, this question is related to the "growth, sleep (dormancy), death" transition discussed in the present chapter Section 3.4 (Himeoka & Kaneko, 2016). There, the transition is discussed as a model for the present cells. Probably such a transition would exist at the primitive stage of life. At present, however, experiments on artificially replicating cells are aimed at synthesizing a system that grows and divides. It is desirable to construct artificial cells that can maintain a "dormant" state with some waste disposal process, instead of dying immediately when the supply of nutrients is limited.

Note that in considering the primitive stage of life, protocells would soon encounter with the nutrient limitation, once they increase the number and would be crowded. It is then important for the cells not to die but to enter the dormant state under such poor-resource condition, so that they will survive. In this sense, the result of Section 3.4.2 will be promising, since a randomly chosen catalytic reaction network can exhibit the transition to the dormant state without any evolutionary optimization.

3.6.9 Maintaining and Reproducing a Nonequilibrium State

To continue garbage disposal, some molecules must be selectively diminished. Also, to continue cell growth, a flow of nutrients from the outside is required. Of course, these are not possible in a state of thermodynamic equilibrium, and a nonequilibrium flow is required. How can this nonequilibrium state be maintained and reproduced? This requires a membrane separating the inside and the outside, but the nonequilibrium flow is necessary to create the membrane. As a result, the compartmentalized structure by the membrane and the nonequilibrium state with spatial inhomogeneity must be mutually stabilized. How is such a "self-consistent" system possible, that is sustained at least for a time period corresponding to the life span? These are also questions that should be answered by creating artificially reproducing cells.

3.6.10 Jamming and Sequence of Stepwise Reactions

Intracellular reactions depend on catalysts and nutrients. However, nutrients are not inexhaustibly supplied from the outside, and since catalysts are made from them by using catalysts, they are often in short supply. As a result, each of the reaction components becomes rate-limiting. Then, the reaction often gets stuck at some reaction path, and jamming of the reaction often occurs somewhere in the reaction network, as discussed in Section 3.6.2.

For continuous growth and reproduction of cells, the reaction must proceed without getting stuck. On the other hand, as a consequence of such traffic jams, the reactions in the cell may not proceed smoothly at a constant rate, but instead stall for a while before proceeding again via some bottlenecks. In fact, such a step-by-step process is sometimes observed in cells (or in biological systems in general). In protocells so far, the system often collapses when the right reactions stop. It will be important to find a condition under which a system stalls for a while, but then resumes the process. Such sequential dynamics may have something to do with "programmed behavior" in cells, such as after state A is reached, then the input X leads to state B, and so forth.

3.6.11 Endo-Embedding of Exo-Physical Processes

Life phenomena are now physicochemical phenomena realized under given initial and boundary conditions that are provided by evolution. In considering the origin of life, however, it may be possible that the proto-form of such conditions was provided by certain external environmental conditions, and that such physicochemical ("generic") phenomena that appear by external conditions was later embedded in reactions and genetic information within a cell (see also [Newman & Comper, 1990; Newman, 1994] for such a discussion.) Of course, in the early stage of the process, such behavior is not so stable and fluctuates greatly. Later, through evolution, such loose behavior with large fluctuations was stabilized. How such consolidation of generic physical phenomena by genetic evolution proceeds, which could be a generalization of genetic assimilation by Waddington (1957), needs to be elucidated.

For example, (i) the catalytic function itself is generally realized by the molecular structure of the catalyst, but such structure could originate from confinement within a physical space, for example, porous media with small holes in it, thereby increasing the reaction rate. (ii) nonequilibrium conditions could originally be supplied by thermal convection at the hydrothermal vent, which may have later been replaced by the cell membrane structure and endogenously maintained. (iii) The role of minority molecules, stepwise behavior by small numbers, or jamming, mentioned above, may have been realized to some extent in the exogenous physico-chemical state, supplied by the environment and later consolidated. Theorizing such consolidation will be important to fill the missing link between physics and biology.

3.7 Summary: Universal Properties and Laws Discussed in This Chapter

- Cells encapsulate enzymes that accelerate reactions inaccessible outside. Therefore, the cell can be regarded as a device that takes components from a nonequilibrium environment into its interior and directs it toward equilibrium through catalytic reactions (A).

- As a consequence of the above nature of enzymatic reaction and cell-growth dilution, the entropy production rate per cell growth is minimized at a finite growth speed, not quasi-statically (A, B).

- Nutrient uptake = cell growth + cell maintenance, each term of which varies approximately linearly with the environmental conditions (E).

- When the environmental conditions are changed, there is a common linear change of each concentration across all components (see also Chapter 2, Section 2.1) (A, C, E).

- The average amount of each component follows a power law distribution across all components, whereas the cell-by-cell fluctuation of each component shows an approximately lognormal distribution (A, C, E).

- With the decrease in nutrients, a transition occurs from exponential growth to a state in which growth has almost stopped, termed as dormant state. This transition under the nutrient depletion occurs through the accumulation of waste (non-autoctalytic) components, as a result of jamming in reaction by the accumulation of complex.

  When the nutrition is regained for the dormant cells, growth resumes, for which a lag time is needed. This lag time increases with the starvation time during which the cells stay at the dormancy. These are explained by considering the interaction between the autocatalytic growth component and the waste accumulation (A, B, C, D, E, F).

- Since the cell growth rate fluctuates, the corresponding dilution of concentration fluctuates. Thus, the concentration of each component is made to fluctuate in proportion to its amount. This gives a dominant part of the concentration fluctuations. Accordingly, each intracellular component fluctuates in correlation, whose distribution is close to lognormal (E).

- How cell replication and molecular replication can proceed in a consistent manner is a fundamental issue both in experiments to construct replicating cells and in the origin of life. Relatedly, relationship between minority molecules and genetic information, synchronization of minority molecular replication and cell division, separation of genetic information and catalytic function, and acquisition of evolutionary potential are discussed as universal properties that must be satisfied by cell reproduction systems (A, C, E).

4

Generic Adaptation of Cells

Attractor Selection, Stochasticity, and Consistency

4.1 Introduction

This and the following chapters consider adaptation. In biology, when we speak of adaptation in organisms, we are interested in two aspects. One is the aspect of "plasticity," in which the state of an organism changes in response to changes in the external environment so that it can survive and grow under such changes, whereas the other is homeostasis (Cannon, 1932) or robustness (de Visser et al., 2003; Barkai & Leibler, 1997; Alon et al., 1999; Kaneko, 2007; Wagner, 2007), in which the internal state of an organism that may have changed in response to external changes returns to the original state after a while. The former describes what changes, while the latter emphasizes the aspects that do not change. At first glance, these two aspects may appear to be antithetical. In fact, however, life systems somehow succeed in reconciling these two opposing characteristics.

Since it is difficult for a system with a single component (or a single state variable, in terms of dynamical systems) alone to reconcile these different properties, organisms achieve this by using several chemical species (state variables) and different time scales. For instance, when the external environment is changed, concentrations of some components change so that the cell can survive in the changed environment, while those of other (many) components remain in their original state without much change in concentration, or return to their original state after a change.

However, the roles of each component are not always so clearly divided, and one of the keys to understanding life is how this complementarity between "change and non-change" is realized. Still, it may be difficult to understand both sides at the same time, so in this chapter, we will first deal with the aspect how the state of an organism changes in response to environmental changes so that it can survive and grow.

When external conditions change, organisms actively change their internal conditions to survive. In general, when the environmental condition is changed, the internal composition of the cell (the amount of each protein) changes. This increases the abundances of components necessary for survival in that environment, so that the cell survives. Usually, such changes are thought to be achieved by the interplay between signal transduction and protein synthesis: information about the environment (e.g., nutrient concentrations in the outside world) is transmitted through the signal transduction network to the gene expression system, resulting in the synthesis of the necessary proteins.

Specifically, environmental information is transmitted from receptors on the cell membrane to the gene expression system through a chain of chemical reactions constituting the signal transduction system. As a result, the amount of synthesis proteins changes. With the evolution of signal transduction system and gene-expression regulation system, cells would survive and grow faster in the new environment. Such appropriate signaling system to adapt to a given environment is thought to be acquired through evolution, and in fact, how different signaling systems function in response to different environmental conditions has been studied extensively.

However, are these signaling systems prepared against a huge variety of possible environmental changes? Signal transduction network for certain environmental conditions should be acquired through evolution over generations, but environmental conditions are diverse, and novel environmental conditions may be encountered that rarely occur or have never been encountered before. Still, organisms have survived and thrived under such diverse conditions. For example, some bacteria manage to survive from new environmental conditions or novel antibiotics. Is it possible, then, for a signaling system to be prepared in advance to produce the necessary proteins for all of these conditions? Can such huge variety of signal networks be acquired through evolution? Would it be possible that a signaling system be prepared for conditions that have never been encountered before?

Once some individuals survived the new environment, genetic mutation could accelerate their growth, but if they could not survive in the beginning, there would be no room for evolution to work. If the organism cannot adapt to new conditions at all without a signaling system, then it is mysterious how organisms have survived so far in the face of various environmental variation. Then, it is expected that organisms can adapt to the external world to some extent without a preprepared network.

In bacteria, a phenomenon called persistence is known. When bacteria are given antibiotics, many of their cells die, but some (about the fraction 10^{-4}) survive, or die only very slowly (Balaban et al., 2004). It is observed that the surviving bacteria are not genetically altered, but are genetically identical (clones). Some cells survive as a result of phenotypic variations (Wakamoto et al., 2013), that is, the variation in the amount of a certain protein that is relevant to the survival in the condition.

Now, there are two possibilities for this variation by cells, either passive or active. In the passive case, because of the variation due to the noise that always exists between cells, some of them happen to survive. As already discussed, the amount of a protein varies from cell to cell that share identical genes. Because of the variation in protein expression, some cells may be resistant to antibiotics and survive. In this case, the cells do not actively change their state, but some of them accidentally have an adaptive state to the stressed environmental condition. In this case, the question is how much the fluctuation facilitates survival as a *group* of cells. This will be discussed in Section 4.5 with reference to the experiment of Wakamoto's group (Hashimoto et al., 2016).

Before getting to this topic, we will discuss the latter possibility in the following three sections, where each cell can actively adapt to the environment even without a specific network prepared in advance. Again, noise will be important. In this case, however, triggered by the noise, the cellular state actively shifts to an adaptive state. We present a theory that the consistency between cellular growth and protein expression level leads to an adaptive growth state, even before cell reproduction, in the absence of an evolutionarily designed network. We then test this theory both by simulation and experiment.

4.2 Phenomenology of General Adaptation by Activity and Fluctuation (A)

A major factor determining the state of a cell would be the composition of each component within it, as already stated. Consider the change in the concentration of protein x_i, which is represented by $f_i(\{x_j\})$ as a function of any other components j which gives the synthesis and degradation of the component i. Including the dilution by an increase in cell volume (whose rate is $\mu_g(\{x\})$), the change in x_i is represented by

$$dx_i/dt = f_i(\{x\}) - \mu_g x, \qquad (4.1)$$

as already mentioned in Chapter 2 Now, the growth rate μ_g generally depends on the protein concentration x as well as the environmental conditions. With the environmental switch, which x value gives a higher growth rate μ_g alters. Then, if there are several stationary cellular states (attractors) with different x-values, is it possible that the state (x-value) with a higher growth rate μ_g is selected for a given environmental condition? This is the question we will address here.

Now, to keep the cell growth in a balanced manner, the dilution by the cell growth need to be somehow compensated by the synthesis. (If the concentration is diluted more by a higher growth, there will be more synthesis). For simplicity, let us assume that as the cell grows, synthesis also progresses in proportion to its dilution; f_i increase along with μ_g. In fact, it is natural that each component is synthesized faster in a fast-growing state. Although we do not know whether this is strictly proportional to μ_g or not, let us first consider the case of proportionality as a first approximation, which implies the perfect compensation of the dilution by the synthesis. Then one can assume $f_i(\{x\}) = \mu_g \hat{f}_i(\{x\})$, and Eq. (4.1) turns to be $dx_i/dt = \mu_g(\hat{f}_i(\{x\}) - x_i)$. Then the steady state is given by $\hat{f}_i(\{x\}) - x_i = 0$ which does not depend on μ_g. Put another way, if the synthesis is proportional to μ_g, then another property of adaptation (homeostasis) is realized: the concentration of each component "does not change" depending on growth conditions.[1]

Now consider flow in dynamical systems. A simple illustration of the flow for the case with only a single variable x is shown in Fig. 4.1, where the direction of the arrow is determined by $\hat{f} - x$, and the steady state (fixed point) is determined where it becomes 0, which does not depend on μ_g. As mentioned above, the steady state of the cell does not depend on the growth rate. In other words, under this simplification, the growth rate μ_g only gives the speed of relaxation to each attractor, but does not change the nature of the attractor. Thus, even if there are several steady states (attractors) in this dynamical system, for each of which the growth rate μ_g is different, its stability does not depend on μ_g. Thus, the exit from a slow-growing state and the entry into a fast-growing state is not expected. Selection of an attractor with higher growth is not possible.

Let us now recall that fluctuations are inherent in the chemical reactions of cells. Considering this, the noise term of $\eta_i(t)$ is added to the end of Eq. (4.1)

$$dx_i/dt = \mu_g(\hat{f}_i(\{x\}) - x_i) + \eta_i(t), \tag{4.2}$$

where $\eta_i(t)$ is the noise term that represents the fluctuation around the average concentration change (to formulate it theoretically, we can use the stochastic differential equation (chemical Langevin equation).[2] This noise perturbs the concentration x away from the steady state (attractor), whereas the first deterministic term drives the state back to the original attractor.

Now let us consider how the noise term perturbs the attractor state. When μ_g at the state is large, the first term is much larger than the noise term, so that the dynamics are not much affected by noise, and the state returns to the original attractor immediately. On the other hand, when μ_g is small, the magnitude of the first term is often smaller than the noise. Then the return to the fixed point is perturbed, and the state can be easily kicked out of the attractor by the noise. As a result, it is

---

[1] If the synthesis is not perfectly proportional but correlated to μ_g, the state is not exactly the same, but it will be close to the original state. This corresponds to the partial adaptation in Chapter 5.

[2] In this book we do not go into details of the theoretical formulation of the Langevin equation. See for example (van Kampen, 1992) for the Langevin equation and (Gillespie, 2000; Takagi et al., 2025) for the chemical Langevin equation, if one is interested.

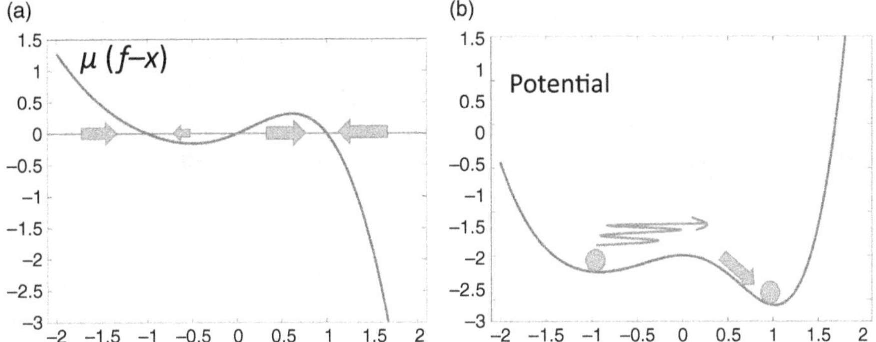

Figure 4.1 Schematic of attractor selection: (a) $\mu(f - x)$ is drawn as a function of x. (b) On the other hand, this can be expressed as $-\int \mu(f - x)dx$ potential and the ball rolling low on that landscape.

expected that the state is jumped out by noise from the low-growth-rate state and moves to the high-growth-state. (This mechanism is called attractor selection (by noise) in the following).

If this argument holds for real cells, then we can expect the existence of a general adaptation process in which the cell spontaneously chooses a state of higher growth rate due to fluctuations. For this to be the case, we must have

(i) Multiple attractors (steady states) exist.
(ii) Faster growth leads to more synthesis, to compensate the dilution by the growth (to some degree).
(iii) Fluctuations of moderate magnitude are inherent in the dynamics of protein expression changes; if the noise level is too small, it cannot kick the state out of even a nonadaptive (low-growth) attractor, whereas if it is too large, it can kick the adaptive state out so that the state cannot stay at the adaptive state (Fig. 4.1).

These three points are necessary. Let us first discuss the latter two points a little more, while the first point will be discussed later. Consider a system with steady states with different growth rates. In order for a cell to grow, reactions are needed to synthesize components within the cell. It is natural that the synthesis reaction will be faster in the fast-growing state. Even if the synthesis is not necessarily a simple proportional relationship $\mu_g \hat{f}$, as long as the synthesis term is written with the "activity" A as $A(\mu_g)\hat{f}$ and this activity increases with the growth rate, the above argument holds (of course, in this case the state (concentrations of chemicals) may differ from the original $f - x$ term, the steady state will be shifted. However, as long as the attractors continue to exist, the above argument still holds).

For example, in the discussion in Chapter 3, the cell reaches a growth state in which the flow of nutrients, the synthesis of enzymes for cellular activity, and the membrane components that determine cell volume must be balanced. Since most of the reactions in the cell are controlled by catalysis, it makes sense that in a state of high flux and growth rate, every process would be fast, and in a state of lower growth rate, every process would be slowed down.

The next question (iii) concerns the noise term. As already mentioned, the rate of a reaction is given by the average value, and there are fluctuations around it (as long as the number of molecules is finite), so the existence of fluctuations is quite natural. The question is its magnitude. If it is too small, there will be no transition from the slowly growing steady state (attractor). On the other hand, if it were too large, the transitions would continue incessantly, and the cell would not remain in the adaptive state above. Roughly speaking, selection by noise for the more active (higher growth-rate) state will work if the (average) change in concentration around the steady state (i.e., the first term) is larger than the noise term for the high growth state, but not so for the low growth state (see Section 4.3 for specific examples). Conversely, if the difference in activity between the two states (ΔA, or $\Delta \mu_g$ if it is determined by the growth rate) is small compared to the noise, there is no precision in this attractor selection mechanism to distinguish between them.

Of course, because the adaptation process we have shown here is stochastic, it sometimes takes time to reach the adapted state. If a certain environmental condition is encountered frequently, it would be advantageous to evolve a signaling network system for it. In contrast, the mechanism of adaptation here is valid even for environmental conditions that the organism encounters rarely or never encountered. The necessary conditions for the present mechanism are just that "cells grow and gene expression processes are subject to fluctuations," which is a natural condition for cells. In this sense, the mechanism of adaptation discussed here is quite general.

4.3 Example of Attractor Selection with a Two-Component System (B and E)

As an example from the section 4.2, consider a gene regulatory system in which two genes repress each other's expression (Hasty et al., 2000). This is called a toggle switch. Protein 1 synthesized by gene 1 suppresses the expression of gene 2, whereas protein 2 synthesized by gene 2 suppresses the expression of gene 1. In this gene-regulation circuit, mRNA1 is synthesized from gene 1, and protein 1 is produced at a certain rate from mRNA1, and similarly for gene 2, protein 2 is produced from its mRNA2. Let x_1 and x_2 denote the concentrations of both proteins.

The synthesis rate of each mRNA and thus of the corresponding protein decreases toward zero as the concentration of the other protein increases. Suppose that the synthesis rates can be expressed as $f_1 = S/(1 + x_2^2)$ and $f_2 = S/(1 + x_1^2)$.[3] Now consider that the amount of each protein is diluted by the cell-volume growth, and assume that the rate of protein synthesis increases to compensate it, so that S is proportional to the growth rate μ_g. Then we obtain the following dynamical system;[4]

$$dx_1/dt = f_1(x_1, x_2) = \mu_g(\alpha/(1 + x_2^2) - x_1) + \eta_1(t)$$
$$dx_2/dt = f_2(x_1, x_2) = \mu_g(\alpha/(1 + x_1^2) - x_2) + \eta_2(t), \tag{4.3}$$

where η_1 and η_2 are the noise term with the strength of ϵ.[5]

First, we discuss the dynamics without noise term in Eq. (4.3). If $\alpha > 2$, the system has two stable fixed points as shown in the figure. They are represented by the solutions (x_1^*, x_2^*) of $f_1(x_1^*, x_2^*) - x_1^* = 0$ and $f_2(x_1^*, x_2^*) - x_2^* = 0$, and they are located symmetrically with each other, (a, b) and (b, a). Here $a \gg b$, that is, one concentration is much higher than the other.

Now consider the case where different attractors have different growth rates (Kashiwagi et al., 2006). Here, each of the two stable fixed points in the figure remains an attractor, since the direction of the flow does not change with the change of the growth rate, since $\mu_g(x_1, x_2) > 0$. The question then is whether an adapted attractor with a high growth rate can be chosen, as we discussed in Section 4.2. So let us assume that protein 1 is necessary for the cell to grow in a certain environment. Then μ_g will be small for an attractor where x_1 is small (even though it remains an attractor). However, based on the discussion in Section 4.2, under moderate noise intensity, a transition from an attractor with $x_1^* < x_2^*$ to an attractor with $x_2^* < x_1^*$, is expected.

In the above model, we set the magnitude of the noise in η_1, η_2 be ϵ, and numerically calculated the model by using the Langevin equation (see the previous footnote). If the noise amplitude ϵ is moderate, indeed, $x_1^* > x_2^*$ attractor is selected for the condition in which the growth is mediated by the synthesis of protein x_1. In the example of Fig. 4.3, the environmental condition is switched between those requiring protein 1 and 2. Then, following the environmental switch, the corresponding adaptive attractor is selected (Fig. 4.2). Furthermore, when the percentage of attractors with high growth rate is plotted against the noise intensity, the

---

[3] We assumed that the amount decreases as a power of 2, considering, for instance, the following situation: two proteins attach to form a dimer, which attaches to the promoter of the other gene and inhibits the synthesis of mRNA. (It can be said that the Hill coefficient is two, and in fact, there are many such cases where the Hill coefficient is 2 or higher). However, even if we do not use strictly this form, the following argument is valid for most (sufficiently fast) decreasing functions.

[4] For simplicity, we have assumed that the parameters of the two gene expression reactions are the same, and the parameters are normalized accordingly so that the synthesis rate is given by a single parameter α.

[5] Mathematically, η is Gaussian white noise with $<\eta_i(t_1)\eta_j(t_2)> = \epsilon\delta_{i,j}\delta(t_1 - t_2)$.

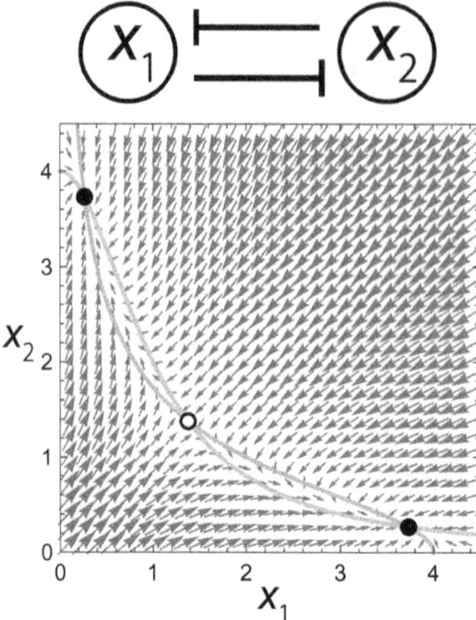

Figure 4.2 Toggle switch Eq. (4.2) flow $(dx_1/dt, dx_2/dt)$ in state space for the change in (x_1, x_2) in the gene regulatory network in the upper column, represented as an arrow. ($\alpha = 4$). If the initial condition is $x_1 > x_2$, the state is attracted to a fixed point on the side satisfying $x_1^* \gg x_2^*$, and if the initial condition is $x_2 > x_1$, it is attracted to a fixed point (attractor) on the side satisfying $x_2^* \gg x_1^*$, which is the mirror image of the first fixed point attractor. The solid line is a line with $dx_1/dt = 0$ and $dx_2/dt = 0$, called the nullcline, and its intersection is the fixed point, where the black circle is the attractor (stable fixed point) and the white circle is the unstable fixed point. (Courtesy of Jumpei Yamagishi).

adaptive attractor is selected almost 100 percent, over a certain range of noise, as shown in Fig. 4.4. This result is also valid even if the composite is not simply proportional to μ_g, but rather involves a nonlinear function that saturates and approaches a constant value as μ_g increases (Kashiwagi et al., 2006).

In fact, Kashiwagi et al. performed an experiment corresponding to the above model by incorporating an artificial gene network with the above-mentioned toggle switch into *E. coli* and experimentally examined whether this artificial gene network adapts to environmental changes by changing its expression state (Kashiwagi et al., 2005) (see Fig. 4.5). In this system, as described above, there are attractors with a high concentration of protein 1 and attractors with a high concentration of protein 2, and both states are realized with roughly equal probability.

Here, two environmental conditions are set up: condition 1 requires synthesis of protein 1 for the cell growth, while condition 2 requires protein 2. Specifically, protein 1 is glutamine synthetase, and the environmental condition 1 medium lacks glucose, because the gene for glutamine synthesis has been knocked out in this *E. coli* strain. Therefore, gene 1 must be expressed in this artificial network in order

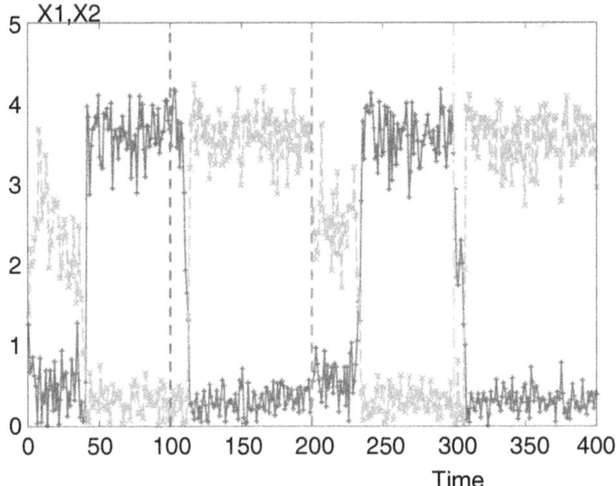

Figure 4.3 The time series of attractor selection. Plotted time series of x_1(solid line with +), x_2(dotted line with x). The toggle switch parameter $\alpha = 4$ is used and the growth rate is given by $\mu_g = 0.1A^4/(1 + A^4)$ with $A = e_1 x_1 + e_2 x_2$ where $e_1 = 1, e_2 = 0$ for environment 1, $e_1 = 0$ for environment 2, with $e_1 = 0, e_2 = 1$ in environment 2. In other words, the $x_1 \gg x_2$ attractor (attractor $X1$) grows faster in environment 1, and the $x_2 \gg x_1$ attractor (attractor $X2$) grows faster in environment 2 (but there is no signal to convey this information). In this simulation, the environmental condition is 1 up to the first 100 steps. Then it is switched to environment 2, and then to environment 1 at time = 200, and then to environment 2 at 300. Attractors with higher growth rates are chosen in response to environmental changes, as $X1 \rightarrow X2 \rightarrow X1 \rightarrow X2$). The standard deviation σ of the noise is set to 0.08.

for a cell to grow, and similarly for condition 2, protein 2 must be synthesized in medium 2 in order to grow (see the original paper for details).

The current two-gene system is an artificially embedded, novel network, so the existing signal transduction system in *E. coli* is not connected to it. Nevertheless, the experimental results show that many cells, achieve the state that synthesizes protein 1 in environment 1, and the state that synthesizes protein 2 in environment 2. In the current experimental system, this was confirmed by measuring the fluorescence of the cells, since we had introduced a green fluorescent protein on the side of gene 1 and a red fluorescent protein on the side of gene 2 (see Fig. 4.5).

Of course, an obvious, trivial possibility would be that both the $x_1^* < x_2^*$ state and the $x_1^* > x_2^*$ state exist, but only the latter can repeat growth and division, so the proportion of cells of the former type would be reduced. In fact, this possibility was rejected by the experimental observation that the proportion of green fluorescent cells increased before the time scale of cell division.

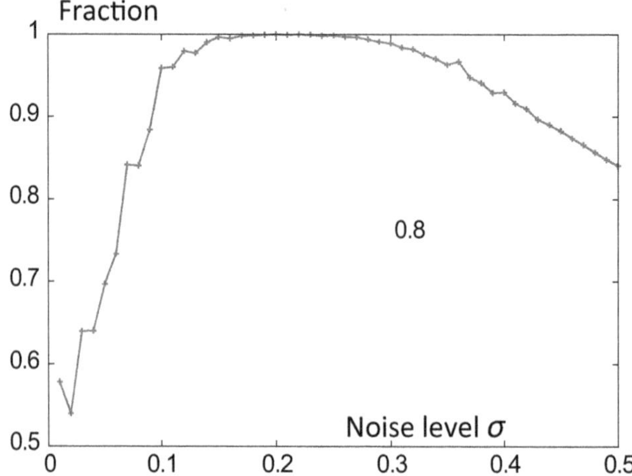

Figure 4.4 The percentage of adaptive attractors chosen is plotted against the magnitude of noise intensity. Using the example in Fig. 4.3, we plotted the fraction of attractors selected for the higher growth rate, calculated 100 times with different initial conditions and different random numbers to obtain that fraction. The fraction is plotted as a function of the noise intensity σ.

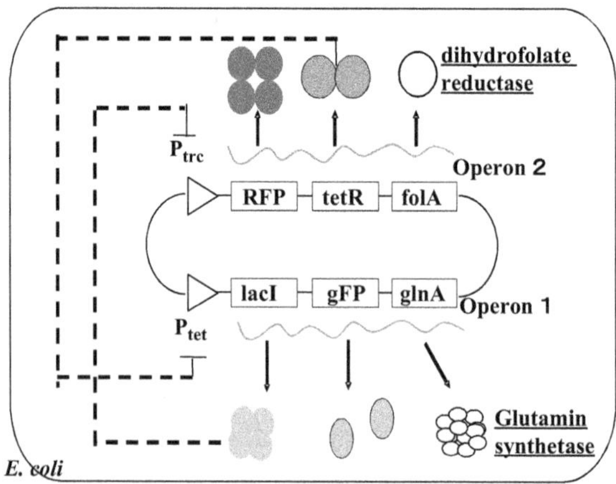

Figure 4.5 Experiment to embed toggle switch network: schematic of the gene network introduced (in the plasmid) in the experiment. Based on Kashiwagi et al. (2005).

The second possibility would be that a hidden signaling network, which happens to be present, transmits information about environmental conditions to the promoters (the parts that initiate the transcription of each gene) of this embedded gene expression network. This is an unlikely possibility, but it was clearly ruled out by swapping the promoter of gene 1 with the promoter of gene 2 in the current artificial gene network: if such a hidden signaling network were present and selection had occurred because of the existence of it, the information to express gene 1 would have been transferred to gene 2 when the promoters were swapped. Then the state $x_1^* > x_2^*$ would not have been selected for the environment 1. However, this experiment confirmed that cells with $x_1^* > x_2^*$ are also selected even in this swapped network. Thus, this assumption of hidden signaling system is ruled out.

In summary, the experimental results suggest that the mechanism of adaptive attractor selection by noise, as described in the theory, is at work.

4.4 Confirmation by a Many-Component Model (B and C)

In Section 4.3, the selection of the adaptive attractor by noise was demonstrated in a simple example consisting of two genes. Can this mechanism be applied to an intracellular process with a complex reaction network consisting of many components? To confirm this, the following gene expression-metabolism model was investigated (Furusawa & Kaneko, 2008).

(1) A gene regulatory network and a metabolic reaction network exist in the cell.
(2) The gene regulatory network is a generalization of the mutual regulation of protein expression in Section 4.2, to involve many genes. The expression of each gene is regulated by the binding of other proteins to its promoter, and the amount of each protein is given by the expression level of its corresponding gene, resulting in gene expression promoting or repressing each other in its amount. The relationship on promotion or suppression between gene i and j is given by the network shown in Fig. 4.6 (see also the Appendix at the end of the chapter for a model of gene expression network). The time evolution of the expression level (protein concentration) of each gene is given by the matrix representing mutual activation-repression. For example, the previous toggle switch of two genes given by Eq. (4.2) can be regarded as the simplest model with two genes, and the present model is its extension to many genes.

Here we use a network of many randomly connected genes, including positive feedbacks that promote their own expression and mutual repression. With time evolution, the expression of genes reaches a certain fixed pattern, an attractor of the dynamical system. We choose such a gene regulatory network that contains many attractors.
(3) Metabolic reactions sequentially transform nutrients into other metabolic components. This reaction forms a metabolic network, and each protein synthesized by the above gene regulatory network catalyzes each reaction in the metabolic pathway (Fig. 4.6).

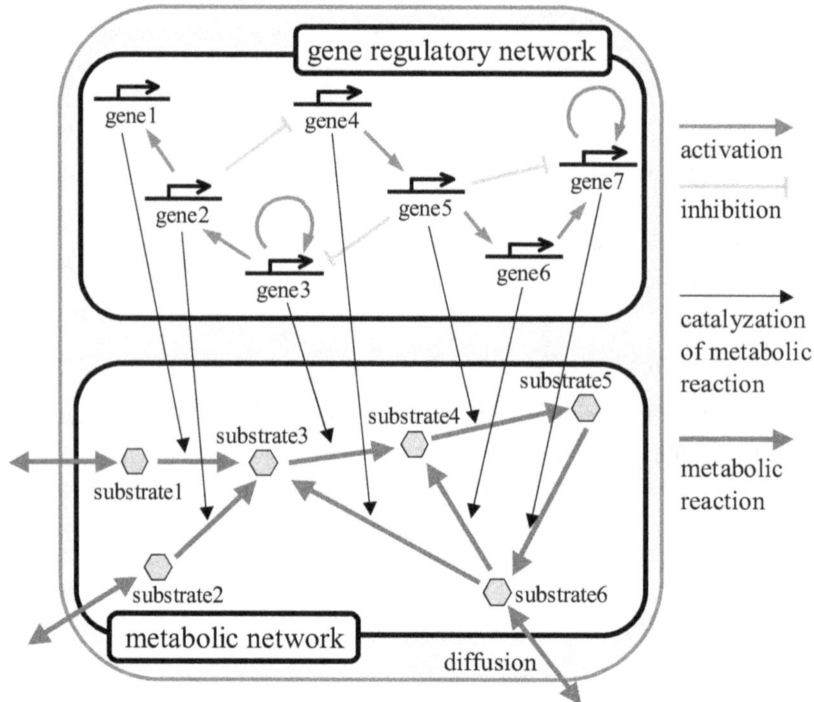

Figure 4.6 A model in which gene (protein) expression is regulated by a gene network, which in turn regulate metabolic system and results in cell growth. Based on Furusawa & Kaneko (2008). (Courtesy of Chikara Furusawa).

(4) Nutrients enter the cell by diffusion through the cell membrane, while other metabolites flow out of the cell.

(5) The growth rate of a cell is determined by the composition of metabolites. For example, the growth rate is set to be proportional to the minimum concentration of metabolic components necessary for growth.

(6) Protein synthesis by gene expression is proportional to the cell growth rate, while the concentration of each protein is diluted by the cell volume growth.

(7) There are fluctuations in each reaction process, represented by noise.

We have simulated the above dynamics (with noise) starting from randomly chosen initial conditions. In this case, it is rather rare for the gene expression pattern to reach an attractor with a high growth rate from such initial conditions, because to achieve high growth, all the necessary metabolic components must be present in abundance. Therefore, in most cases, the states fall into a low-growth-rate attractor when the noise level is low. However, since such attractor is easily kicked out by noise, transitions from low-growth attractors occur frequently. After this transition, the state may fall into a high growth state where the effect of noise

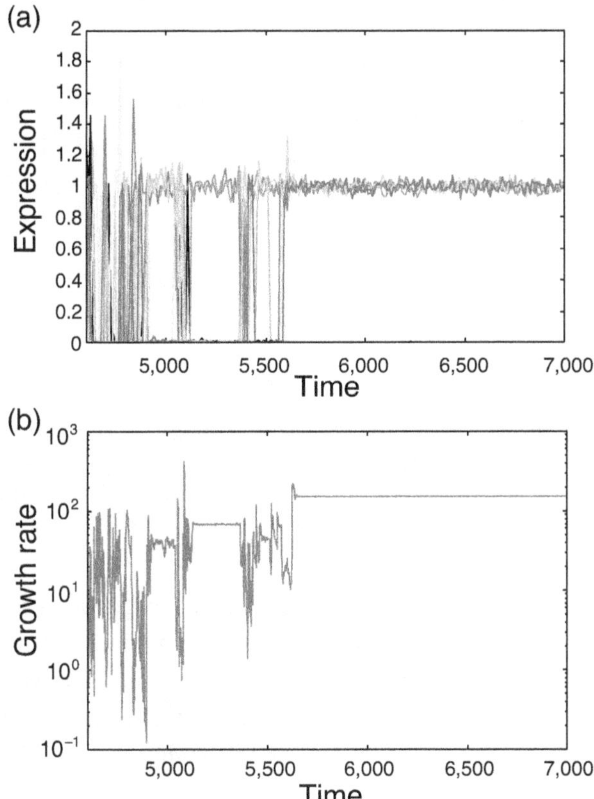

Figure 4.7 Attractor selection time series. Upper: time series of expression levels of some proteins. Lower: time series of the growth rate. As the expression pattern changes, it reaches a state with high growth rate and is stabilized therein. Based on Furusawa & Kaneko (2008). (Courtesy of Chikara Furusawa).

is less significant. This transition is repeated until the state of the cell falls into an attractor with a sufficiently high-growth rate, from which the state is not kicked out by noise. An example of this time course is shown in Fig. 4.7.

This transition to high-growth states will occur whenever the noise level is in the appropriate range. To see the dependence on the noise level, Fig. 4.8 is plotted, which shows how the growth rate of the final state depends on the noise, starting from a variety of initial conditions. When the noise level is moderate, a cell state with a high growth rate is selected.

Let us review the above results from the point of view of macro-micro consistency. There are many reactions in the cell, and the concentration of each component is changed accordingly. Although there is no optimization mechanism for the growth of the cell as a whole, the growth of cell volume has a global effect as it dilutes all components while compensating for the dilution, the synthesis

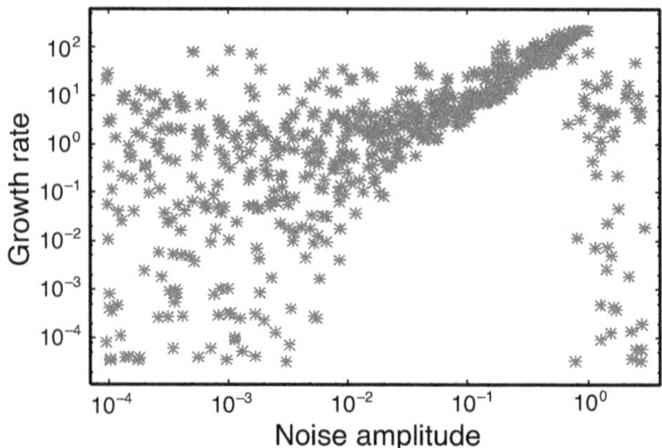

Figure 4.8   The growth rate at the final stationary state plotted against the noise strength. Ten simulations were run for each noise intensity, and the growth rate of the final state is shown after the steady state is reached. Based on Furusawa & Kaneko (2008). (Courtesy of Chikara Furusawa).

proceeds. As a result, a state is chosen in which the growth of cell as a whole is consistent with the reactions in the individual components. Here the noise itself acts randomly and has no directionality, but in the consistent state between cell growth and molecular reactions, the perturbation from the noise is relatively weak when growth is higher, and as a result, a high growth state is selected.

Significance, Limitations, and Extensions of Attractor Selection

To sum up, with this attractor selection mechanism, an attractor with a higher attraction speed (flow) is selected by the dynamical system and the noise. Since the growth rate (activity) and the speed in the dynamic process are correlated, an attractor with a higher growth rate (adapted state) is selected. However, if the difference in speed between adaptive and nonadaptive attractors is small compared to the magnitude of the noise, the present mechanism does not have the precision to distinguish the states with such a small difference in speed. Also, if the noise intensity is small, as shown in Fig. 4.8, the attractor will be trapped without being able to escape from the low growth rate.

So far we have assumed that the noise intensity is independent of the growth rate. Therefore, the dynamics falling on the attractor with a higher growth rate is less influenced by the noise in relative terms. If the variance of the noise were proportional to the growth rate,[6] the current mechanism would not work, because the

---

[6] For those (somewhat) familiar with statistical physics, it may seem natural that the variance of the noise would be proportional to the relaxation rate, according to the fluctuation-dissipation theorem. However, such a relationship between fluctuation and relaxation is only true in the vicinity of thermodynamic equilibrium. It would be too naive to assume such a relationship in a growing cell system.

relative rate of deterministic (average) relaxation to the attractor and the intensity of the noise to be kicked out of the attractor is independent of the growth rate. For an attractor with a low growth rate, the attraction rate would be weaker, but the noise intensity would be lower in proportion, so the state would not be kicked out.

In fact, this is not the case for a cellular system. Even if the growth rate (activity) is zero, there is still noise. This is because even in the nongrowing state, reaction processes are still going on, and temporal fluctuations in the concentrations of components persist. Therefore, even if the noise intensity could decrease with the growth rate, it would not go to zero: the noise intensity will not decrease proportionally to the growth rate. Then the attractor with the lower growth rate is relatively more sensitive to noise.

In the current model, for simplicity, the noise intensity is assumed to be constant regardless of the growth rate, but even if it decreases with the growth rate, as long as there is still a portion that is independent of the growth rate, the current mechanism works because it allows the transition from the low-growth to the high-growth state.

This attractor selection may remind of the simulated annealing method if one knows it (Kirkpatrick et al. 1983). In that case, the noise intensity (temperature) is initially set high (i.e., higher) so that the states are allowed to make transitions over metastable states, and then the noise is gradually reduced over time to select a lower-energy (or optimal or adapted) state. This method is widely used as an optimization technique. In attractor selection, on the other hand, the magnitude of the noise does not change over time, but instead the average flow rate to each stable state increases with the growth rate. Then the noise effect is relatively reduced. Comparing the two, the difference lies in which of the deterministic or stochastic terms changes, but the ratio of deterministic to stochastic terms increases in both cases. From this point of view, this attractor selection can be applied to optimization problems, and indeed it is used in some engineering problems. In the case of cell biology, it will not be easy to change the noise level depending on the cellular state. As we have seen, attractor selection that changes the average flow rate rather than the noise magnitude is more natural.

4.5 Significance of Noise to Cellular Adaptation (E)

4.5.1 Increase in Population Growth Rate Due to Cellular Fluctuations

Does noise actually work in real cellular adaptation? Before discussing relevance of "active" attractor selection studied in Section 4.4, here we first explain the "passive" role of fluctuations, which is confirmed experimentally. In this aspect, the mere presence of fluctuations leads to the increase in the growth rate of a cell "population" (i.e., at an ensemble level) from that of a single-cell level. After discussing this issue, then, we will discuss possible relevance of attractor selection, the "active" role of noise to adaptation, to the actual cells.

As described in Section 3.5, cells fluctuate not only in the concentration of each component, but also in their growth rate and division time. Here we show that these fluctuations the growth rate of a cell population will be greater than the average proliferation rate of a single cell. We consider the exponential cell growth, where the cell size increases by $exp(at)$. If a cell divides when its size doubles, the number of cells doubles by $log(2)/a$ from $exp(at) = 2$. Therefore, if there is no fluctuation in the growth rate a, the increase in the number of cells as a cell population also follows $exp(at)$.

However, if the growth rate of each cell fluctuates around the average growth rate a, the time for the population to double will be faster. This is because the faster the growth rate of a cell, the more offspring it will produce, and thus the faster the growth rate of the population will be. As an example, consider the case where the growth rate has two values, a_1 and a_2, each occupying half a fraction. Then the average growth rate is $a = (a_1 + a_2)/2$. Now, the increase in the number of cells in this case is given by $(exp(a_1 t) + exp(a_2 t))/2$, which is always greater than $exp((a_1 + a_2)t/2)$, as shown in Fig. 4.9. The point is that $f(x) = exp(x)$ is a convex function; in general, the average of $f(x_i)$ over $i = 1,..,N$, that is, $\sum_i f(x_i)/N$, is greater than $f(\overline{x_i})$, where the average is $\overline{x_i} = \sum_i x_i/N$ (in the figure, $f(x) = exp(x)$ is a convex function (Fig. 4.9 illustrates the case with $N = 2$, but this is true for any $N > 1$, no matter how $x(i)'s$ are distributed)). Thus, in general, cells with fluctuations in

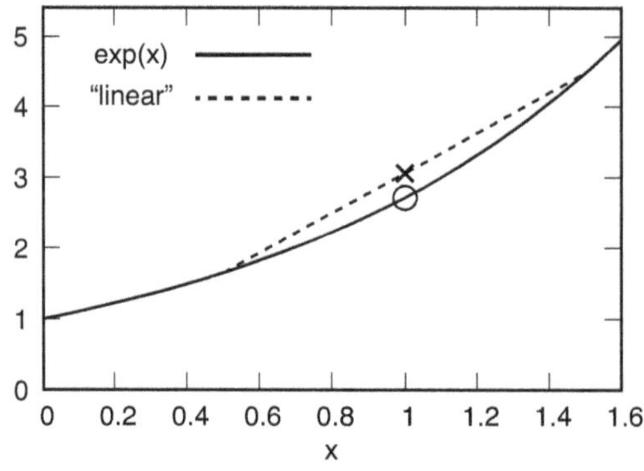

Figure 4.9 Because $exp(at)$ is a convex function, $(exp(a_1 t) + exp(a_2 t))/2$ is greater than $exp((a_1 t + a_2 t)/2)$. If there is no fluctuation in the growth rate at $a = 1$, the population growth rate is of course, given by $exp(1.0t)$. In contrast, if half of the cell has the growth rate of $a_1 = 0.5$ and the other half $a_2 = 1.5$, the average growth rate in a single cell is the same as 1.0, whereas the population increase as an ensemble of cells is given by $(exp(a_1 t) + exp(a_2 t))/2$, which is greater than $exp(1.0t)$.

their growth rate will show a higher growth in the number of cells as a population.[7] A more precise formulation of this increment in terms of the distribution function of the division time is given by Wakamoto et al. (Hashimoto et al., 2016), using the formulation of information theory.

It used to be experimentally difficult to measure a growth-division process at a single-cell level over a long period of time, but now that the single-cell measurement methods described in Chapter 3 have been established, the effects of the above growth fluctuations can be quantitatively measured. For example, the distribution of division times of *E. coli* is measured by using the single-cell long-term measurement system described in Chapter 3 (Fig. 3.11), and from this the difference between the average division time at a single-cell level and that of the population level can be obtained. Specifically, from the distribution of division times, the population division time $T_d^{ensemble}$ is calculated, and the ratio of this to the single-cell division time, $T_d^{single}/T_d^{ensemble}$ is obtained. From it, one can measure the gain $Gain = T_d^{single}/T_d^{ensemble} - 1$, which gives the increase in the growth rate in the population.

This gain was measured under various environmental conditions and plotted against the magnitude of division-time fluctuations in Fig. 4.10. Here, the horizontal axis is the CV, that is, standard deviation of the division time divided by the mean, and the vertical axis is the growth gain (*Gain*). The figure first shows that the fluctuations are relatively large, ranging from 20 to 40 percent CV, and that the growth rate gain in the population ranges from 5 to 10 percent CV. The fluctuations are significant enough to have the fluctuation-induced gain.

Hence, the fluctuations in cell growth (division) enhance the population survivability ensemble. Then, if the fluctuations in growth rate depend on environmental conditions, will it allow for the adaptation of cells to survive under different environmental conditions? Since the source of this fluctuation is due to the randomness of molecular collisions in chemical reactions, one might think that its magnitude would not vary so much by cellular states. However, how much the state fluctuates due to noise depends on the characteristics of the reaction dynamics in the cell. Indeed, negative feedback in the reaction network can reduce the fluctuations, while positive feedback can amplify them. As a result, the fluctuations depend on the cell type or state, as well as on the environment. In fact, as shown in Fig. 4.10, the fluctuation (CV) of division time is highly dependent on the environment.

---

[7] For example, this conclusion does not change even if the growth is t^2, and so on. Conversely, if the growth is concave, for example, $\sqrt{t}$, then fluctuations are detrimental to population growth.

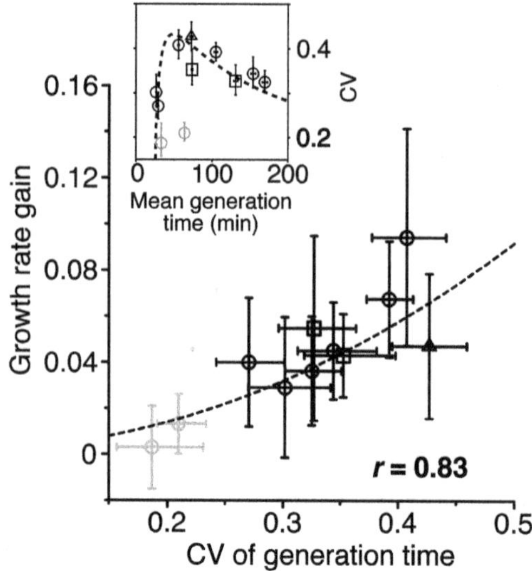

Figure 4.10 Fluctuations of cell division times and the incremental growth rate *Gain* in the population. Long-term single-cell measurements of *E. coli* were made using the dynamic cytometer described in Chapter 3 to obtain the division time distribution, and then *Gain* was obtained under various environmental conditions. Based on Hashimoto et al. (2016). (Courtesy of Yuichi Wakamoto).

4.5.2 Linear Relation between Increase in Population Growth Rate and Fluctuations

Let us then discuss the relationship between the fluctuation of the cellular state variable x and the growth rate. First, we set the mean value of x in the original steady state to 0 and consider fluctuations around it. If the steady state is stable, the state perturbed by noise from it will return to the original state. If the deviation is not too large, we can use a linear approximation: the strength of the return to the steady state is proportional to the deviation (like a spring returning to its natural length after being expanded or contracted, following Hooke's law). We set the rate to k, while adding the noise term $\eta(t)$ in the relaxation dynamics of x, as in

$$dx/dt = -kx + \eta(t). \tag{4.4}$$

(For example, if we consider the production and degradation of a protein at a constant rate, as in the example in Section 3.5, the time evolution of the concentration y is given by $dy/dt = a - dy$. By setting $x = y - a/d$, the above form is obtained). In this case, the stationary distribution of the state can be expressed by using the noise intensity D as $\exp(-kx^2/(2D))$. In other words, the variance of x, $< (\delta x)^2 >= D/k$, increases in proportion to the noise intensity and decreases in inverse proportion to the strength to return to the steady state.

Next, let us assume that this state variable x influences on the growth. (This can be seen in many cases. For example, suppose that this state variable is the concentration of a protein that facilitates the cell volume growth.) In other words, the growth rate μ depends on the state x. If the change of x from 0 is small, the change of the growth rate μ is approximated to be linear with x, so that it is written as $\mu = ax + \mu_0$. In this case, one can examine how the distribution of x is affected by the growth effect in which cells with larger x will produce more offspring. Formulating the change in probability distribution mathematically (by using the so-called Fokker–Planck equation derived from Eq. (4.4)), and combining the effect of proliferation depending on x (see (Sato & Kaneko, 2006) for the theoretical derivation), the stationary distribution of x is given by

$$P(x) = exp(-k(x - Da/k^2))^2/(2D)). \tag{4.5}$$

Then, the mean value of x is shifted to $\Delta x = Da/k^2$ from 0. Recalling the variance $< (\delta x)^2 >$ of x is given by D/k from the stationary distribution of x, and eliminating D, we obtain

$$\Delta x = \frac{a < (\delta x)^2 >}{k}. \tag{4.6}$$

Hence the displacement is proportional to the fluctuation (variance), and then, the according change in the growth rate is given by

$$\Delta \mu = a\Delta x = \frac{Da^2}{k^2} = \frac{< (\delta x)^2 > a^2}{k}. \tag{4.7}$$

This growth rate change is always positive (regardless of the sign of a), and the increase rate of growth as a population is proportional to the variance of x.

Now, if the environmental conditions get worse, the reaction process in the cell will slow down. Then the time scale for the state change will be larger (e.g., the relaxation process will be slower), so that k will be smaller. Accordingly, the variance of the state variable x will be larger.

In the example in Section 3.5, if the dilution by cell growth reduces x, then $k = d$ decreases with the decrease in $< \mu >$, the average growth rate. Hence, in a bad condition that lowering the growth rate, the variance $< (\delta x)^2 >$ will be larger. Then, the difference between the growth rate as a group and the growth rate of a single cell, $\Delta \mu$, will be larger. This result implies that when the environmental condition is worse, the fluctuations increase, which provides the survival strategy as a population even without any special mechanism.

In fact, in Fig. 4.10, when the environmental conditions is worse and the growth rate decreases, fluctuations increase, resulting in the increment in the growth rate from a single-cell level case. Although this increase in population growth due to fluctuations is "passive," it is effective as the environmental (nutrient) condition is worse.

4.6 Possible Extension of Attractor Selection for the Verification in Cells in Nature (E)

In Section 4.5, we discussed the possibility that fluctuations increase the level of adaptation as an ensemble of cells. In contrast, in Sections 4.2–4.4, we showed active role of fluctuations, in which fluctuations change the state of the cell, resulting in adaptation to increase the growth rate even at the single-cell level. Still, the experimental verification that this mechanism works in the natural cells is yet elusive. The experiment we demonstrated in Section 4.4 adopted bacteria with artificially embedded genes, and it is not yet sure whether this occurs in natural cells under physiological conditions.

The first question to be addressed here is whether the assumption in the attractor selection that a higher growth rate will also increase synthesis is reasonable or not (condition (ii) in Section 4.2). If the change in growth rate had no effect on synthesis at all, then the current mechanism would not work. In this case, only the dilution term due to growth changes, so that the protein concentration simply decreases inversely proportional to the growth rate. This is exactly what we discussed in Section 3.5. In fact, according to the experimental results presented therein, the concentration is diluted to some degree as the growth rate increases. In other words, protein synthesis does not compensate for the dilution caused by growth by 100 percent.

However, the experimental observation does not support that there is no compensation at all, either. The concentration does not decrease inversely proportional to the growth rate, and the experimental data show that the rate of decrease is about half of that rate. In other words, about 50 percent of the growth rate was compensated (see Section 3.5.1). In this sense, synthesis increased by about half as the growth increased. The attractor selection mechanism does not require that the synthetic term is fully compensated for the growth rate. It is sufficient if the synthetic term increases to some extent with the cell-volume growth. In this sense, the condition for the increase of synthesis for a higher-growth cell is satisfied.

Next, it seems to be a strong requirement that the attractors that can adapt to environmental conditions exist in advance ((iii) in Section 4.2). However, this is also related to the above question of whether perfect compensation occurs or not. In the studies of Sections 4.2 and 4.3, for simplicity, we assumed that synthesis is perfectly balanced with perfect growth compensation, so that the synthesis rate is proportional to the growth rate μ_g, and accordingly it is bracketed in front in the time evolution equation. Thus, μ_g only changed the time scale, but did not change the stability of the state, nor did it create a new stable state (attractor).

In fact, if this condition is relaxed, new attractors can be generated by changes in environmental conditions, and preparing attractors in advance is not necessarily required. Let us briefly discuss the generation of attractors and the selection of such attractors.

Consider the simplest case, where a cell has only one attractor (stable state) under the original condition, and assume that the growth rate is reduced under unfavorable environmental conditions. If the synthesis rate of proteins does not increase with the same rate as the growth rate, the dynamics equation for the protein concentration will be changed because the synthesis does not cancel out the change in dilution. For example, suppose the rate of synthesis (plus degradation) for a protein concentration x is represented by $f(x)$. This $f(x)$ has a constant value of x_0 for $x \sim 0$ and increases with x, approaching a constant value as x is increased (see Fig. 4.11). For example, $f(x) = x_0 + \frac{x^n}{x^n + K^n}$. Next, suppose that the growth rate μ also increases with x and approaches a constant as x increases. For example, $\mu = \frac{\mu_0 x}{1+x}$. The steady state is then reached at the point where $f(x)$ and the dilution $\mu(x)x$ are balanced. As shown in the figure, if the growth rate μ_0 is large, there is an attractor where x is small. However, when μ_0 is small, as shown in Fig. 4.11, there are three fixed points, two of which are stable and provide attractors.

Here, when the growth rate μ is changed from high to low due to environmental conditions, two attractors are generated. Initially, the cellular state that started from the branch with the large μ remains in the steady state with the smaller x. Since this state is an attractor, it will not move from the attractor, without noise. Now consider the effect of noise: Between the two attractors, the one with the larger x has more dilution, and has a larger absolute value of $f'(x)$. Hence, it has a stronger attraction.[8] Then, following the attractor selection mechanism as discussed already, the state with the larger x, that is, the higher growth rate $\mu(x)$ under this environmental condition, is selected.

In summary, even though there is only one attractor under the original conditions, due to the environmental condition change, a second attractor with a different growth rate can be created. Then, it is selected due to noise. In this way, the idea of attractor selection can be extended.[9]

In fact, Tan et al. (2009) carried out the experiment in which the above monostable gene regulatory network is embedded. When the environmental conditions

---

[8] Specifically, the larger the absolute value of $f'(x) - \mu < 0$, the stronger the attraction to the state. This is achieved by taking the appropriate $f(x)$.

[9] In fact, even without imposing adaptive attractors, one can introduce models in which such attractors are not provided, but they are generated by epigenetic mechanisms (Furusawa & Kaneko, 2013; Himeoka & Kaneko, 2018; Matsushita & Kaneko, 2023).

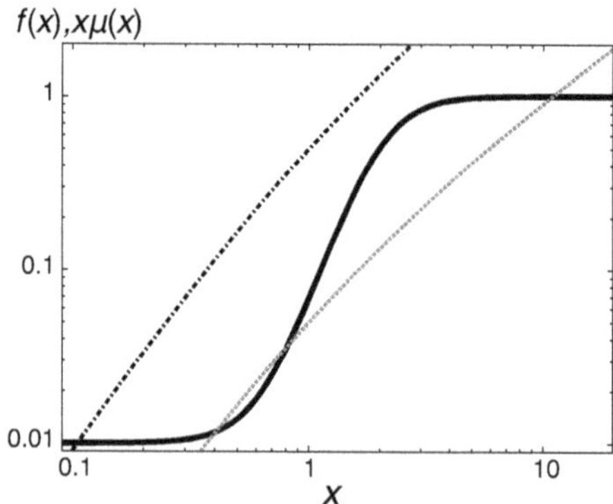

Figure 4.11 With the change in growth conditions, the dilution term is changed, and a new stable state is created, to which the state transition occurs by the attractor selection mechanism. In the figure, $f(x) = 0.01 + x^4/(K^4 + x^4), (K = 4)$ and $\mu(x) = \mu_0 x/(1 + x)$ are plotted against $f(x)$ (solid line) and $x\mu(x)$ (dotted line, right)1.0 (broken line, left). As can be seen in the figure, for $\mu_0 = 1$, there is only one cross-point between the two, and in fact only one stable state exists. On the other hand, for $\mu_0 = 0.1$, there are two cross-points other than the original fixed point with small x. Of these three fixed points, the largest (with x near 10) and smallest are stable fixed points. Here, when μ_0 is lowered from 1 to 0.1, the state initially falls into the fixed point with smallest x, but due to attractor selection by noise, it transitions to a state with larger x, leading to a higher growth rate. This is because x, that is required for growth, has been expressed more without a signaling system, and the noise selects a state with a higher growth rate.

were changed, bistable states of expression arise, and the highly expressed state was selected (Tsuru et al., 2011). Although it is not confirmed that the attractor selection mechanism works here, it at least provides a plausible explanation for the experimental result.

The experiment adopted embedded genes. At this time, no experiments have yet been conducted to directly confirm the attractor selection mechanism. Still, it will be possible that the attractor selection works in normal cells as well, because the gene expression dynamics described above exist in the normal cells. Interestingly, in *E. coli* metabolism Heinemann and colleagues (Kotte et al., 2014) found in two phenotypes, of which the state that can grow survives under fluctuations.

The existence of adaptation without signaling systems has also been suggested by Braun et al. (Braun, 2015). In their experiments, yeast cells were placed in a variety of environmental conditions that the signaling system could not cover. Nevertheless, the yeast could survive. The origin of the adaptation in these experiments remains elusive. However, recalling that significant levels of gene-expression fluctuations is confirmed in bacteria and yeast cells, the attractor selection by noise in this chapter is a promising candidate to explain the observed spontaneous adaptation without a signaling system.

Still, these are indirect evidence for the attractor selection. To verify the theory here, it is desirable to construct an experimental system that clearly demonstrates the correlation between the cellular growth rate and gene expression, rather than between signal and gene expression.

In immune system research, the hypothesis that antibodies to diverse antigens are prepared a priori was initially proposed, which was overturned. Now, it is known that they are generated a posteriori: this is a major paradigm shift in the immune studies. In the case of adaptation, if the hypothesis that all signaling pathways are prepared in advance is rejected and the existence of spontaneous adaptation is proven, this will also provide a paradigm shift.

Of course, as mentioned earlier, it would be more advantageous if signaling systems are acquired through evolution for frequently encountered environments, than to use attraction selection mechanism that is stochastic and takes more time. Therefore, it is conceivable that there had first existed spontaneous adaptation, and that the signaling system evolved later to tune-up the spontaneous adaptation, for the environment that cells frequently meet.

It is important to consider that phenomena that occurred by fluctuations are later fixed, evolutionarily so that they occur for sure. As will also be discussed in Chapters 8 and 9, this is a theme that often appears in this book: Evolution proceeds by fixing the properties that first happened to occur due to fluctuations. Darwinian evolution is good at making probable phenomena to occur deterministically and to tune them up, but it is not good at creating something out of nothing. On the other hand, if the attractor selection by noise was the first step in adaptation, which is later followed by the evolution of signaling systems for frequently encountered environments, then some traces of such evolution may remain in the current network or the properties of mutants.

4.7 State Selection by Time-Scale Interference (D)

Since adapted attractors are generally faster in both synthesis and decomposition, we can think of attractor selection as a process of choosing a state with a faster time scale for relaxation to the attractor.[10]

In general, cellular (life) systems involve processes with many time scales. Metabolic processes are relatively fast, mRNA synthesis and degradation processes are a bit slower, and changes in protein concentrations, which control these processes, are much slower. In general, these different time-scale processes interact with each other, but when one time scale is slower than the other, the slower

---

[10] This section is theoretical and one can skip it.

component acts as a parameter to control the faster component, and the faster side is driven by it.

In general, when the environmental condition is not good, processes that previously progressed faster than others will slow down due to lack of nutrients, catalysts, and other components responsible for activity. As a result, the time scale itself can change depending on the environment. If the system is adapted to an environment, the time scale of the controlled variables x, which represents metabolism, and so forth, is sufficiently faster than the time scale of controller variables y, which represents protein expression, etc. On the other hand, if the environmental condition is not good and the system is not yet adapted to it, the change of x slows down and will be at the same level as the time scale of y. It is expected that the timescale of fast variables x is slowed down when the system is not adapted. (Let us assume for simplicity that the time scale of y remains unchanged.)

For instance, the gene expression network regulates the metabolic processes through the control of enzyme levels, where the metabolic component constitutes faster variables x and the enzyme side gives y. In fact, the interaction between the gene expression and metabolism is not unidirectional; the metabolism affects the enzyme side as well. In the adapted state, however, the change in x is fast enough compared to y, so that the former is regarded to be controlled by the latter: the time evolution of x is given by y as given parameters. Then the time evolution in y is obtained by representing x as its time average value, which is a function of y at the moment. As a result, y is determined by dynamics with only y. (This is adiabatic elimination of fast variables.)

However, if the state is not adapted to the novel environment (with limited resources), the time scale of x slows down and becomes comparable to that of y. Then the time evolution of these variables is represented by a large degree of freedom dynamical system in which x and y are intertwined. Thus, the original attractor given by the y-only dynamics (with x as adiabatically eliminated) is destabilized, and the state will vary in time. Through such dynamic change, the state may reach an adapted high-activity state where x changes faster and is controlled by the slower y. Then the slower variables y (under the averaged x) will fall into an attractor determined only by the time evolution of y and stay there. Thus, we can expect directional transition dynamics to a state adapted to the environment, and once reached, the state stays in the adapted state (Hoshino & Kaneko, 2007; Kaneko, 2014).

Another possibility of separation between slow and fast timescales is seen in the epigenetic modification and gene-expression changes. As mentioned in Section 2.7, the transcription from genes is modified by epigenetic changes with a slower time scale. This change leads to generation of novel attractors in gene expression dynamics. By coupling this slow epigenetic change with gene expression dynamics, and assuming that the change in the timescale depends on the cellular activity, adaptive attractors are selected (Furusawa & Kaneko, 2013; Matsushita & Kaneko, 2023). This will be discussed in Section 6.4.

In these ways, adaptive attractor selection is expected to be a property of large-degree-of-freedom dynamical systems with different time scales. In this picture, the attractor selection by noise is extended to a generic property with fast-slow time-scale multidimensional dynamical systems, without relying on noise. In other words, we can think of adaptive attractor selection as the selection of consistent states on different time scales.

4.8 Collapse of Consistency: Adaptation Processes to a Novel Environment (F)

As shown in this chapter, gene expression dynamics determine the state of protein composition (phenotype) and cell growth, while the cell growth affects protein concentration through dilution. This results in a state of consistency between growth and composition of each component. If there is more than one consistent state, the state that is more stable against fluctuations is chosen. On the other hand, if the consistency is broken by large environmental changes, the state loses stability and exhibits larger fluctuations in gene expression levels over time. Under such fluctuations, a new state is generated that is consistent in its environment and has larger stability. In this respect, the temporal collapse of consistency is a preliminary stage of adaptation to the new environment. It will be important to follow such a transition process experimentally.

As we saw in Section 4.7, a consistent state can be described by fewer degrees of freedom. On the other hand, when consistency is broken, the transient dynamic state involves more degrees of freedom. Then, what are the characteristic behaviors of the adaptive process caused by the dynamics with these many degrees of freedom? This is one of the main topics of Chapter 5.

4.9 Summary: Universal Properties Discussed in This Chapter

- In adaptation, there are both plasticity and robustness. The former concerns the change to a state that is suitable (can grow) in a novel environment, and the latter concerns the maintenance (less change) of the state against the environmental change.
- Adaptation, that is, the selection of a state with a high growth rate, can occur even without specific signaling circuits (A, B, C, E), simply by considering the dilution of each component by cell growth, the homeostatic

process in which components are synthesized to compensate for this dilution, together with the presence of noise in the reaction process (stochastic gene expression).

- The above attractor selection can occur even if the adaptive attractor is not prepared a priori. For example, incomplete compensation of synthesis and changes in metabolic time scale can result in the generation of an adaptive attractor, and then the adaptive attractor will be selected by the noise (C, D).

- Adaptation by conveying information about the external environment to the gene expression system through a signal transduction system is thought to have later evolved in response to environmental changes that are frequently met, while cells can make spontaneous adaptation against a variety of environmental changes that are frequently met that are rarely encountered, for survival. Although direct experimental verification has not yet been made, attractor selection by noise is a strong candidate to explain such "spontaneous adaptation" (E).

- Besides the above adaptation with significant state change, there is "passive" adaptation mechanism by fluctuation: The average division time of a cell population will be reduced from that of a single cell simply as a result of cell-to-cell fluctuations (E).

4.10 Appendix: Gene Expression Network Model

First, mRNA is transcribed from gene i, and the expression of the protein synthesized from it is denoted by x_i. Here, x_i is 0 when the gene is not expressed and is scaled to be 1 when it is fully expressed: Each protein concentration p_i divided by the maximum possible value p_{max} may be considered as $x_i = p_i/p_{max}$. In general, a protein can mask the promoter of another gene or mask some other protein that masks the promoter. In the former case, the protein represses and in the latter case, activates the expression of the corresponding gene. This interaction is generally complex, and several proteins can affect others in concert. Let us simplify this effect to be represented by $J_{ij}x_j$, where j and i are the indices of genes. If $J_{ij} > 0$, j activates i, and if $J_{ij} < 0$ it represses i, and $J_{ij} = 0$ if it has no effect. (Usually there are many zero elements, and J_{ij} is a sparse matrix). Let us further simplify that the influences from different proteins are additive, and the input to gene i can be written as $\sum_j J_{ij}x_j$. This J_{ij} matrix, especially its positives and negatives, represents a mutual activation-repression network. This allows gene x_i to be expressed in response to input from other genes. Let us assume that gene i is expressed if this input is greater than some threshold θ_i. If the expression shows on/off response depending on if the input is larger than the threshold or not, the expression of i is represented by $S(\sum_j J_{ij}x_j - \theta_i))$ by using the step function $S(x)$ ($S(x) = 1$ for $x > 0$,

$S(x) = 0$ for $x < 0$), or instead of the step function $S(x)$, one can use such function $F(x)$ that takes ~ 0 at $x \sim 0$, and increases with x and asymptotically approaches 1 for larger x, with the form of $F(\sum_j J_{ij}x_j - \theta_i)$. One typical form often used is $F(x) = 1/(exp(-\beta x) + 1)$. It approaches a step function $S(x)$ as β is increased.[11]

Generally, each protein is decomposed at a certain rate. Let γ_i be the time constant for synthesis and decomposition; the time evolution of x_i is given by

$$\frac{dx_i(t)}{dt} = \gamma_i(F(\sum_j J_{ij}x_j - \theta_i) - x_i). \tag{4.8}$$

This equation (or its similar form) is often used as a model equation for simplified gene regulatory networks (Mjolsness et al., 1991; Salazar-Ciudad et al., 2000, 2001; Ishihara & Kaneko, 2005; Kaneko, 2007). (This model is also used in part in Chapters 5–8 in this book.). Note also that by taking x_i as the activity level of a neuron firing, and positive (negative) J_{ij} as activation (inhibitory) coupling by the synapse from neuron j to i, the model is used for neural network studies.

---

[11] β roughly corresponds to the Hill coefficient in molecular biology.

5

Adaptation and Cellular Homeostasis

5.1 Introduction

In Chapter 4, we focused on the ability of biological systems to change to fit new environmental conditions, within the two aspects of "adaptation." Biological systems, on the other hand, tend to preserve their internal states against environmental changes, and return to their original state, after they have changed for adaptation to new conditions. In other words, many variables (e.g., concentrations of components) that represent the internal state tend to return to their original values. As for such maintenance of the internal state of body, Cannon coined the term homeostasis (Cannon, 1932).

As for cellular behaviors, this tendency to return to the original state after a response to environmental change, has been classified as perfect and partial adaptation by Koshland (Koshland et al., 1982). In terms of dynamical systems, the two types of adaptation can be expressed as follows: The state of the system is represented by several variables, while the environment is represented as external parameters acting on the system, such as the concentration of external components. When the environment changes and these parameters change, the state of the system changes in response. In other words, the values of internal variables change. If the external parameters are left as they are, values of some variables are changed from their original values, but the values of rest (many) variables will return toward their original values. This is adaptation. That is, the external change is absorbed by some variables, and the rest variables tend to return to their original values. When a variable completely returns to its original value, it is called "perfect adaptation," and when it returns partially, it is called "partial adaptation."

As the simplest mathematical example of such perfect adaptation, consider a simple system consisting of two variables. In this case, we can consider an example in which one variable changes to absorb the external change and the other variable returns to its original state. For example, let S be a parameter that represents the external environment, and consider a system with

$$du/dt = f(u, v; S); \, dv/dt = g(u, v : S), \tag{5.1}$$

Now, consider (u^*, v^*) that satisfies this stationary state (fixed point state), that is, $du/dt = dv/dt = 0$, for $u = u^*, v = v^*$. Then, $f(u^*, v^*; S) = 0, g(u^*, v^* : S) = 0$. Suppose this steady state is stable and u is perfectly adaptive. It implies that u^* is independent of S: When S is changed, this u transiently changes in response to the change in S, but then following the change in v, u returns to the original value. As an example, consider

$$f = S - h(u, v); \, g = (uv - v)/\tau, \tag{5.2}$$

By solving $g(u, v) = 0$ for $v \neq 0$, the stationary solution satisfies $u^* = 1$, which is independent of S. On the other hand, from the condition $f(u, v) = 0$, v^* is given by $S = h(1, v^*)$, which depends on S in general. Let us now vary S from S_0, after (u, v) has reached the steady state for $S = S_0$. Recalling that $f = g = 0$ for $S = S_0$, after the change of S from S_0, $du/dt = f$ deviates from 0 to $S - S_0 \neq 0$. As a result, both u and v begin to change, and then eventually u returns to $u^* = 1$ again and v settles to a value satisfying $S = h(1, v^*)$. Thus, changes in the environment S are eventually absorbed by v, whereas u makes a transient response and then returns to the original.[1] For instance, we can take $h = uv + u$, then $v^* = S - 1$. This specific example will be discussed later.

Another example of this are given by

$$f = S - (u + v); \, g = (S - v)/\tau, \tag{5.3}$$

In this case $S = u^* + v^*, v^* = S$, so $u^* = 0$, which is independent of S. Furthermore, another example is provided by

$$f = S(1 - u) - uv; \, g = (S - v)/\tau, \tag{5.4}$$

In this case $S(1 - u^*) = u^* v^*, v^* = S$, so $u^* = 1/2$, which does not depend on S.

These might look like arbitrarily chosen unrealistic examples, but we can immediately think of corresponding examples for each of these in a living system. The form of h in the first example is given, for instance, by a combination of simple reactions such as synthesis $S \to U$ from S, autocatalytic reaction $U + V \to 2V$ catalyzed by V, decomposition $U \to \phi$, $V \to \phi$. In these examples, if τ is large, by taking advantage of the difference in the time constants of the reactions, there first appears fast response and slow relaxation. Such change is indeed characteristic of adaptive dynamics in biology.

Pioneering studies on such adaptive systems have been put forward by Koshland et al. (Koshland et al., 1982). One of the most thoroughly investigated is

---

[1] Of course, we need to check whether this steady state is stable. There are many examples to satisfy the stability, as will be discussed later.

chemotaxis, in which bacteria, *E. coli*, move in the direction of higher concentrations of certain chemical components. The movement of *E. coli* consists of a run and tumble. In the former, the flagellum rotates in one direction and the cell moves in one direction, and in the latter, the flagellar rotation does not go in one direction, when the cell stops motion and changes its direction. The frequency of this change in direction is adaptive to the density of the external environment: When the external concentration changes, the frequency of tumbles first increases, and later returns to the original level. This transient response is what gives rise to chemotaxis in response to concentration (Berg, 1975). Asakura and Honda explained this adaptation in terms of a stepwise phosphorylation response and proposed a basic model of adaptation (Asakura & Honda, 1984). Simplified models have also been extensively investigated (Barkai & Leibler, 1997).

Is such adaptation, then, reduced to a simple reaction circuit that gives rise to it? Can it be attributed to only a circuit given by few components? Does only a small fraction among thousands of components in a cell exhibit such adaptive response? Many researchers have examined transient responses in gene expression across thousands of genes. For example, Braun et al. have examined how the expression levels of each mRNA species i (the average over many cells of each mRNA concentration) change over time when environmental conditions are changed (Stern et al. 2007). Specifically, they measured the expression level $x_R(i)$ after a short response time following an environmental change, and the change in expression $x_S(i)$ after a sufficient amount of time has elapsed to reach a steady state. Then they expressed the log change of $x_R(i)$ and $x_S(i)$ from the original expression level $x_O(i)$ as a plot $(log(x_R(i)/x_O(i)), log(x_S(i)/x_O(i)))$ (Fig. 5.1). If the latter (the value of the y axis) is 0, we have perfect adaptation (i.e., $x_S(i) = x_O(i)$), if the slope is 1, we have no adaptation, and the transient change is fixed (i.e., $x_S(i) = x_R(i)$. If it is between 0 and 1, we have partial adaptation. The figure shows that many genes are scattered around a common slope of about 0.7. Rather than only a few genes changing and taking charge of adaptation, it appears that thousands of genes collectively change their expression levels with partial adaptation. In other words, it seems that not only few specific components are changing, but the entire large-degree-of-freedom component is adapting.

Recall that the environmental condition remains changed, so it is not trivial that so many components would change then return toward the original values (albeit partially). As a high-dimensional dynamical system, such common tendency over all degrees of freedom would be quite rare. In other words, there appears to be a conservative tendency for the cell to retain as much of its original state as possible. This may be because such cellular states that maintain themselves and continue to survive are rare, and hence there is a pressure to maintain them once they are generated.

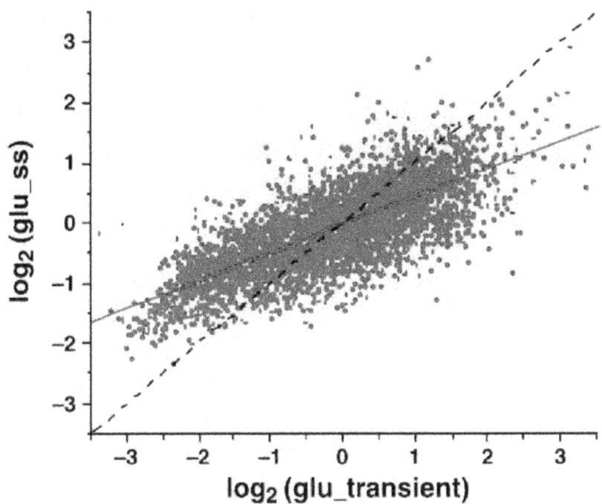

Figure 5.1 The relationship between the transient response and the final change when yeast cells are placed in a new environment. The degree of change in each mRNA expression level was measured when the medium for yeast cells were replaced from galactose to glucose. The logarithm of $x_R(i)/x_O(i)$, the ratio of change from the original expression to that after 4 hours in the new medium, is plotted on the horizontal axis, and the logarithm of $x_S(i)/x_O(i)$, the ratio of change after enough time has passed, is plotted on the vertical axis. Each point corresponds to the expression change of each different mRNA. Reproduced from Stern et al. (2007) with the courtesy of Erez Braun.

In this chapter, we will examine the adaptive system as a "general tendency to return to the original state," in order to understand one of the universal properties of living organisms.

5.2 Adaptation Model with Few Components: Phenomenology (A)

First, let us confirm that perfect adaptation can appear even in a simple intracellular reaction system (Inoue & Kaneko, 2011). In this case, X_0 is synthesized from an external resource S with a certain rate, X_1 is synthesized from X_0 using X_1 as a catalyst, and X_0 and X_1 are decomposed at a certain rate. In summary, the reaction system is as follows.

$$S \rightarrow X_0, X_0 + X_1 \rightarrow 2X_1, X_0 \rightarrow \phi, X_1 \rightarrow \phi.$$

Now denote the concentrations of S, X_0, and X_1, by s, x_0, and x_1 respectively. Then, with an appropriate scaling of variables, the rate equation for the concentration is given by

$$\frac{dx_0}{dt} = S - x_0 x_1 - x_0, \quad \frac{dx_1}{dt} = \frac{x_0 x_1 - x_1}{\tau}. \tag{5.5}$$

This equation has a stable fixed point $x_0^* = 1, x_1^* = S - 1$ for $S > 1$. In other words, x_0^* is independent of the external concentration S. Now suppose that the external S

concentration increases from S_0 to S. Since the first steady state is $(1, S_0 - 1)$, and at that moment, $\frac{dx_0}{dt} = S - S_0$, x_0 increases once and later returns to its original concentration. Thus, the component X_0 shows adaptation, while the change in external concentration is buffered by the increase in the concentration of X_1. This model is a simple autocatalytic reaction to synthesize the catalyst X_1 from the nutrient via X_0 without any elaborate mechanism. Many other examples for adaption are possible, such as replacing S with Sx_1 in Eq. (5.5) so that this nutrient uptake occurs via X_1.

Furthermore, whether in a linear chain $S \rightarrow X_0 \rightarrow X_1 \rightarrow \cdots \rightarrow X_{N-1}$ and the parallel case $S \rightarrow X_0 \rightarrow X_1^1, X_1^2, X_1^3, \cdots X_1^k$, or even parallel in intermediate paths, X_0 shows adaptation. In short, adaptation behavior is a property that is easily realized in catalytic reaction systems.

In most biological adaptation, the response occurs quickly, and the adaptation process toward the original state is slower, in other words, τ is much larger than unity. In such case, in the above autocatalytic system, not only does adaptation occur generally, but also the response is almost uniquely determined by the ratio of concentration change in the environment.

Consider in the present case, when S is changed as $S_0 \rightarrow pS_0 (p > 0)$. Then, the peak value of the response until X_0 before returning to the original value depends only on p and not on the absolute value of S_0. In other words, X_0 detects the fold change. Such a response was originally discovered in psychology and is known as the Weber–Fechner law. For example, when we perceive brightness, how dazzling we feel depends not on the absolute value of brightness, but the ratio of the novel brightness to the original level: when a light is turned on in a dark room, we are dazzled, but when the light is added to the same amount in a bright room, we feel little difference. Our response is given by the ratio of the stimuli before and after.

An example of this response in the current model (5.5) is shown in Fig. 5.2(a). Theoretically, this is easier to understand if we consider the adiabatic limit ($\tau \gg 1$). In this case, while x_0 quickly responds to a change in S, x_1 remains unchanged, at the original value $x_1 = S_0 - 1$. We then find the peak value at which x_0 remains unchanged while x_1 remains fixed. One can solve the peak condition $(dx_0/dt)_{x_0 = x_0^{peak}} = 0$ by taking $x_1 = S_0 - 1$ fixed. Then the value at the peak is $x_0^{peak} = p$, which depends only on the ratio p and not on S_0. In other words, it is appropriate to look at the response in terms of the logarithm of the input change. In the present case, this is derived due to multiplicative nature of (auto)catalytic reaction, and this point is common to the log-normal distribution of gene expression changes and fluctuations discussed in Chapters 2 and 3.

Note that in this model, the time to return to the original state is given by τ and does not depend on S, whereas the time for the peak to occur is proportional to $1/S$ and becomes faster with S. In contrast, in a stronger postulate for fold-change

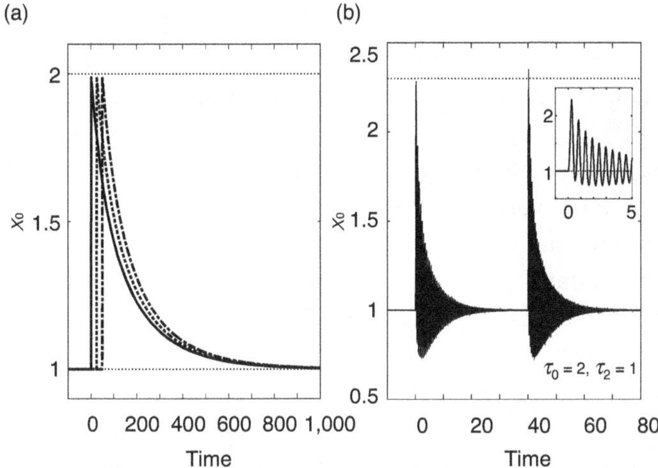

Figure 5.2 (a) Time series of x_0 when it is changed from S_0 to $2S_0$ at external nutrient concentration $t = 0$ in the model of Eq. (5.5). $\tau = 200$. Overlaying $S_0 = 100$ (solid line), 200 (dotted line) 400 (unidotted line). (b) Results when $S = 100 \rightarrow 200$ at $t = 0$ and $S = 200 \rightarrow 400$ at $t = 40$ in the 3-step model. The time constant for $S \rightarrow X_0$ are τ_0, that for $X_0 \rightarrow X_1$ is τ_1, and that for $X_1 \rightarrow X_2$ is τ_2, with $\tau_0 = 2, \tau_1 = 1, \tau_2 = 1$, (Insert is the magnification). Based on Inoue & Kaneko (2011). (Courtesy of Masayo Inoue).

detection, even the time course of the response itself, including the response time, depends not on the absolute value of S but only on the ratio of S changes, as has recently been found in cellular responses, and its model has been proposed (Shoval et al., 2010). In fact, it is obtained by appropriately choosing parameter values for the feed-forward network in reactions (Goentoro et al., 2009). The fold-change detection with this stronger sense will be discussed in Section 5.3.

The response of Weber-law type is valid even when catalytic reactions are connected in a series or parallel, under appropriate conditions, for instance under sufficiently large τ. In this case, oscillatory relaxation can occur. For example, if we assume that the reaction in the above network is catalyzed by X_2 in three steps, connecting $S \rightarrow X_0, X_0 \rightarrow X_1, X_1 \rightarrow X_2$, then by increasing (decreasing) S, X_0 first increases, and then it returns to the original value as in the adaptation discussed. In this case, however, depending on the rate (time constant) of the reaction, the return to the original value is accompanied by damped oscillations. Even in such a case, the peak of the response depends only on the ratio (Fig. 5.2(b)).

5.3 Microscopic Large-Degree-of-Freedom Model of Adaptation (B)

5.3.1 Adaptation in the Catalytic Reaction Network Model

In Section 5.2, we saw that adaptation with fold-change detection can easily occur in a catalytic reaction system with few components. Then, would such a response also appear in a reproducing cell model with multi-component catalytic reactions? For this purpose, we use a slightly modified version (Furusawa & Kaneko, 2012b) of the catalytic reaction network model discussed in Chapters 2 and 3 (Furusawa & Kaneko, 2003).

In brief, the model consists of catalytic reactions as already mentioned: Assuming that there is a network of catalytic reactions such as $x_i + x_j \rightarrow x_\ell + x_j$ (where x_i is the ith type of catalyst) among catalytic components and that nutrients exist at a certain concentration in the external environment, they penetrate through the membrane and enter the "cell." This "cell" takes in nutrient molecules from the environment and converts them into other molecules by the intracellular catalytic reaction network, and if this reaction proceeds successfully, the total number of molecules in the cell will increase as a result. When the total number of molecules in the cell increases, the cell divides into two, forming daughter cells of approximately half of the cell. As mentioned in Chapter 2, in this "ideal cell model," cell growth is maximal at a certain value of nutrient influx rate, and the cell growth does not continue when the influx rate is beyond this value. Around this optimal growth rate, "cells" of approximately the same composition are replicated in a steady state. There, we found a relationship in which the component abundances fall to the -1 power of the rank when arranged in order of increasing abundance. Such a "power-law" relationship is seen in statistical physics at critical points (where transitions occur). As already mentioned, when the amounts of many kinds of proteins in the cell are arranged in increasing order of abundance, this law holds true even in the present cell (see [Chapter 6 of Kaneko 2006]).

Now, in this system, by choosing a moderate value for the rate of nutrient influx, growth was optimized, and cells of approximately the same composition were produced. If the influx was too fast, growth would not be sustained since each catalyst needed to make each catalyst would not be synthesized in time. This is because the nutrient would occupy too much of the cell. On the other hand, as discussed in Chapters 1 and 2, a characteristic of living systems is that they autonomously form a state of macro-micro consistency. Hence, we added a minor change to this model. In real cells, the influx of nutrients is not simply diffusion from the outside, but is assumed to be achieved by active transport. In fact, cells use energy through channels in their membranes to take up components from the outside. Although several components are involved in this process, for simplicity, we assume that the uptake of nutrient i is carried out by the corresponding transporter component $m(i)$ and that its uptake rate increases with its concentration $x(m(i))$. Specifically, let the concentration of the external nutrient component be S and the uptake be given by $x(m(i))^\alpha S$ (for example, $\alpha = 2$; still, $\alpha = 1$ or larger than 2 will work).

In this simple model, the nutrient uptake rate F_0 is proportional to the cell growth rate. Figure 5.3 shows how F_0 changes with external nutrient

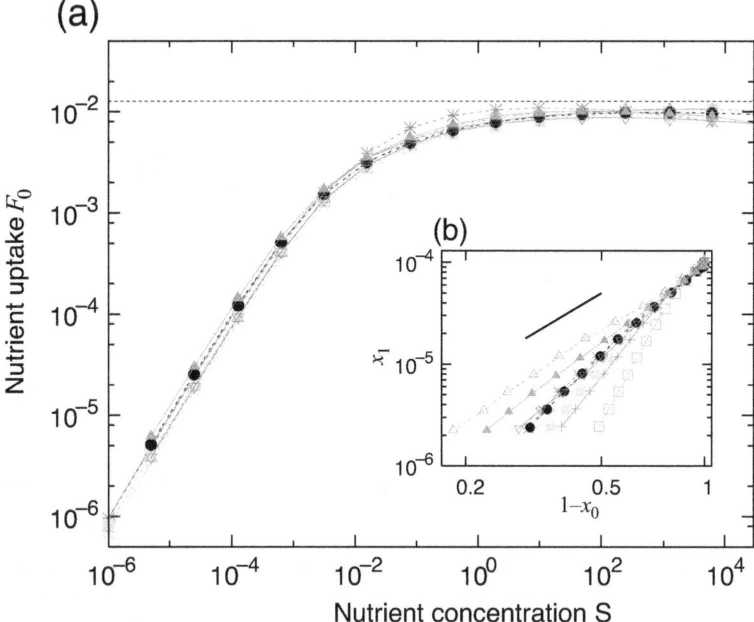

Figure 5.3 Nutrient uptake F_0 (that agrees with cell growth rate in the current model) plotted as a function of external nutrient concentration. Here, the nutrient influx at the critical state and transition from growth to non-growth in the earlier model, where nutrients enter simply by diffusion rather than active transport, is around 1.1×10^{-2} for F_0 in this figure, and that value is plotted as a horizontal dotted line for reference. The inflow is maintained around that front regardless of the external nutrient concentration. Overlaid results for different networks created with 10^4 number of components and reaction paths at 2.5 percent from each component. Based on Furusawa & Kaneko (2012a). (Courtesy of Chikara Furusawa).

concentration S. It shows that F_0 increases when S is small, but the growth rate remains almost constant above a certain level. On the other hand, in the case of simple diffusion, as S is increased, beyond this points, cells cannot grow anymore. In other words, the optimal growth rate is maintained by the introduction of transporters. This is because the decrease in transporter prevents an excess of nutrient flow. As external S increases, uptake increases, and internal nutrient concentrations increase instantly. Then the relative concentration of transporters drops, reducing the uptake, and nutrient concentrations decrease. As a result, negative feedback process works, and the nutrient concentration is kept at a certain level. This level is just around the critical point, where the growth rate is optimal, because above this level, there would be an excess of nutrients. There, a power-law relationship between the abundances of components holds. As already discussed, in actual cells, the amount of each mRNA follows a power-law distribution with a power of -1, which is consistent with this result (Fig. 5.4).

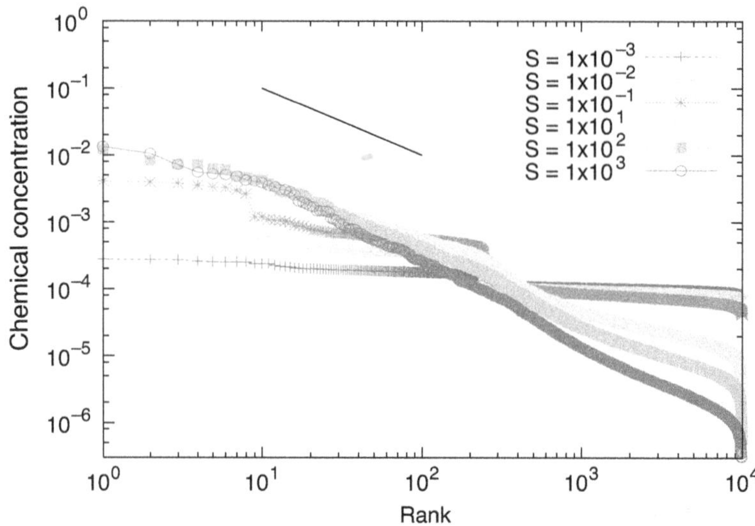

Figure 5.4 Relationship between rank and abundances. The abundances of each component are plotted in their order. Log–log plots. Overlaid results for different external nutrient S (from 10^{-3} to 10^{3}). As the amount of S is large and the plot asymptotically approaches approximately, abundances $\propto$ rank^{-1} as in the critical state. The solid line shows the -1 power for reference. Based on Furusawa & Kaneko (2012a). (Courtesy of Chikara Furusawa).

Suppose, then, that after reaching the steady state of each of these components, the external nutrient concentration is increased. The nutrient concentration in the cell would first increase, and then return to the original concentration due to the feedback from the transporter component(s). Therefore, the growth rate and nutrient concentration show perfect adaptation. Figure 5.5 plots the change in growth rate when the external nutrient concentration is changed from S_0 to S_0'. After increasing once, the rate returns to the original level.

At this time, the concentration of transporters decreases from its original value, so they do not show perfect adaptation. However, the initially large decrease of them is followed by an increase, indicating partial adaptation. To see how the other components have changed, the logarithmic change ($log(x^R(i)/x^O(i))$, $log(x^S(i)/x^O(i))$) of the concentration of each component $x(i)$ from the original (O) to the final steady state (S) through the response concentration (R) in the new environment (external nutrient concentration) is plotted in Fig. 5.6. This shows that almost all components are partially adapted, and the degree of adaptation is almost the same (about 2/3). Interestingly, this result is in good agreement with Fig. 5.1, which shows the adaptive response of expression levels across genes. The observed common partial adaptation over many components, thus, can be regarded as a universal property of systems in which many components can grow steadily

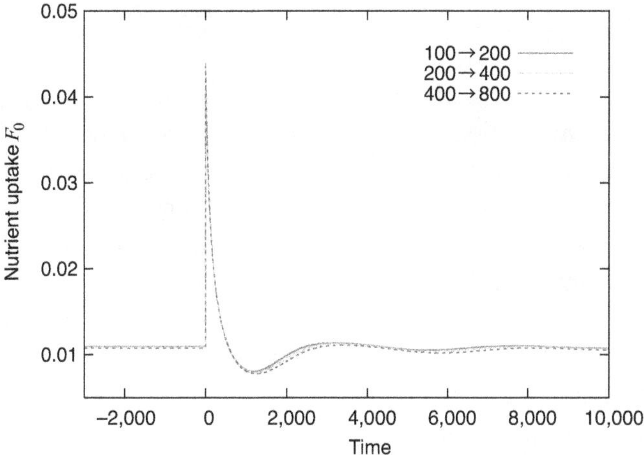

Figure 5.5 Time variation of nutrient uptake F_0 when the external concentration is doubled from 100 to 200, 200 to 400, 400 to 800, and so on. The same time variation is followed. Based on Furusawa & Kaneko (2012b). (Courtesy of Chikara Furusawa).

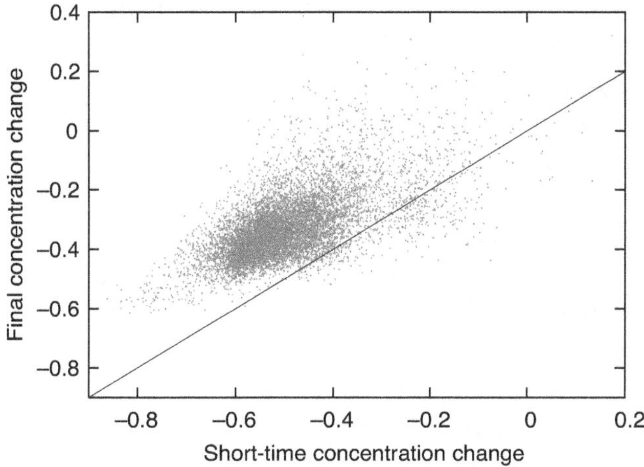

Figure 5.6 Short-time response and final change in each component when external nutrition is changed from 100 to 200. $((log(x^R(i)/x^O(i)), log(x^S(i)/x^O(i))))$ plotted across all components. Here $x^O(i)$ is the original steady-state concentration of each component i before the nutrient change, $x^R(i)$ is the concentration at 1,500 time units (see Fig. 5.6) in response after the nutrient change, and $x^S(i)$ is the concentration after a sufficient time after the nutrient change to return to steady state. A line with a slope of 1 is included for reference. The slope of the simulation results is about 0.7. Based on Furusawa & Kaneko (2012a). (Courtesy of Chikara Furusawa).

in relation to each other. (See also Chapters 2 and 9 for such common proportional changes across components).

As we have seen here, a common property emerges when the responses are viewed in logarithmic ratio. This is originally derived from the multiplicative nature of the catalytic reaction. Can we then see a logarithmic property, or a fold-change detection, against changes in external concentration? In Fig. 5.5, the change of the growth rate in time is plotted, after the external nutrient concentration is changed from S_0 to $S_0' = 10S_0$. In the figure, the time courses of the growth rate change are overlaid for $S_0 = 100$, 200, and 400. They show the same time course. Here, if we take a different multiplication factor from 10, the peaks in the middle would change accordingly, but for given multiplication factor, the time course is identical independently of S_0. In other words, the response depends only on the fold change of the external concentration change, but not on the original concentration. Thus, the response with the fold-change detection discussed earlier also holds in this system.

In summary, this *ideal cell model* of a catalytic reaction network with the active nutrient uptake shows (1) adaptation to a state of power-law abundance distribution of components (critical state) under sufficient nutrient supply, (2) perfect adaptation in the growth rate toward an optimal state, with a fold-change detection (3) partial adaptation over many components with approximately same ratio in the changes of response and adapted (final) values. (4) approximately log-normal distributions of the cell-to-cell fluctuations of each component (which we did not show explicitly here). Note that these four features are also observed universally in the present growing cells.

Can we extract the description with a small number of degrees of freedom as described in Section 5.2 from the present model for adaptation with many components? Such a description by a few macro degrees of freedom is a fundamental theme in statistical mechanics, whose application to living systems is a key issue in the theory of universal biology. For the current model, this question can be partially answered by mimicking mean-field calculation in the statistical physics. However, since this is mathematically detailed and is so far specific to the present model, we will discuss it in the Appendix at the end of this chapter. Formulating a general theory independent of models remains a future problem.

5.3.2 Adaptation in Gene Regulatory Networks

In Chapter 4, we considered gene expression changes by introducing gene regulatory networks. Adaptation occurs in the gene expression process, and in fact, a model of such gene-regulatory network for adaptation will be important. First, let us consider a simple gene regulation network with few components that can

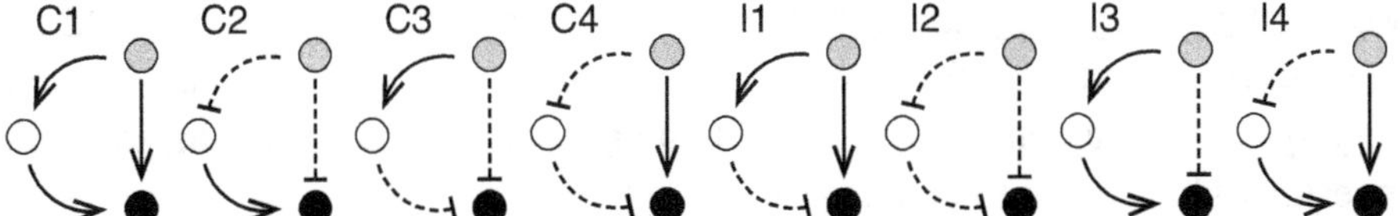

Figure 5.7 Three-gene feed-forward control network with one input and one output. The left four are "Coherent" networks with direct and indirect paths of the same sign, and the right four are "Incoherent" networks with different signs. The right shows adaptive behavior. (a) Positive direct path and positive–negative indirect control path (b) Negative direct path and positive–positive indirect control path. Based on Inoue & Kaneko (2013). (Courtesy of Masayo Inoue).

achieve the property discussed in Section 5.1: A set of variable(s) show perfect (or partial) adaptation, whereas the other set of variables absorbs the change applied externally. For it, we consider three-gene network consisting of a positive (activation) path and a negative (inhibitory) path, as shown in Fig. 5.7. Here, we focus on a network consisting of one-way network from gene 1 to 3 and paths from 1 to 2 and 2 to 3. This is called a feed-forward network because the information flows in one direction from upstream 1 to downstream. In general, there can be positive (activation) and negative (inhibitory) paths, so that there are 2^3 possibilities, as shown in Fig. 5.7.

If one of the two paths from 1 to 3 via 2 is positive and the other is negative, the activation of gene 1 suppresses that of 3 (as a negative pathway), whereas if both of them are positive or negative, then the expression of gene 1 activates the expression of 3. On the other hand, the influence of the direct path from 1 to 3 can also be positive or negative. Hence, there are cases where the direct path and the indirect path via 2 have the same sign (termed as coherent) (left half of Fig. 5.7) and the opposite sign (termed as incoherent) (right half).

Consider the incoherent case where the two pathways have opposite signs. When an input is added to the upstream gene 1, an adaptive change can be expected in the expression of the downstream output gene 3, as the change from the direct path $1 \rightarrow 3$ will be later suppressed by the pathway via 2. Take, for example, I1 (or I4) in the figure. If each path takes the same amount of time, the time required for the expression of the response of gene 3 by the direct path would be shorter compared with that of the indirect path. Then, when an input is received, the expression of gene 3 first increases through the direct positive path. Then the effect of the negative path would come in to cancel it out, and the expression would be directed to its original level. If the positive and negative paths are eventually balanced so that the difference is not beyond the threshold for the expression, the expression of the output gene 3 will return to the original level, indicating perfect adaptation, whereas if the difference goes beyond the threshold, partial adaptation will occur. Similarly, in the case of I3 (or I2), the direct path is negative, while the indirect (two-step) pathway is positive. In this case, the expression of gene 3 would drop once and then tend to return toward the original level. Such a three-gene regulatory system has extensively been studied as a network motif for the adaptation processes (Alon, 2006; Li et al., 2004).

This feed-forward network is considered to correspond to the simple example in Section 5.1. As described above, the expression of the output gene at the downstream end undergoes a transient change due to a direct path and then returns toward the original level by the influence of an incoherent indirect path.

Even though this simple motif will be important for the adaptation mechanism, can it be sufficient to fully understand the cellular adaptation? Can the cellular adaptation of gene expressions be fully understood by such a small number of components in a network motif? As shown in Fig. 5.1, when the external environmental conditions are changed, many components show adaptive responses. In contrast, according to the above motif, the expression of one gene shows non-adaption at all (just monotonic increase as in the gene 2 in the motif) to achieve the adaptation of the other. The overall tendency of partial adaptation of expressions across most genes is then hardly explained. The experimental result suggests that many genes are regulated by each other to make collective adaption behavior. This kind of collective (or cooperative) adaptation across many genes remains to be elusive.

We have carried out an evolution simulation to search for gene regulation networks (GRNs) that show adaptive behavior of an output gene. We took a gene regulation network of the form discussed in Chapter 4, and specified an output gene and an input gene. The behavior of gene expression dynamics depends on the GRN. Here we assign the fitness by using the dynamics of the output gene, so that it shows an adaptive behavior. Then we evolved the network structure by adding or deleting positive (activation) or negative (repression) paths, to select a GRN with the higher fitness. Specifically, we consider a network of N genes that activate and repress each other (see Inoue & Kaneko, 2013). The coupling from j to i is assumed to be J_{ij}; if it is positive, j activates i, if it is negative, j represses i, and if 0, there is no coupling. Some genes have external inputs that activate their expression. In this model, the expression of each gene, i.e., the corresponding protein, is produced, as a result of its mutual regulation, while it is degraded at a certain rate (see the appendix in Chapter 4 for the gene regulatory network model).

Among these N genes, there is one predetermined gene whose expression level defines the output. Suppose that this gene is not expressed initially, that is, its protein is not synthesized, and that the expression of this output gene changes as a result of the input. Let M be the maximum expression level during this process, and let $R(\geq 0)$ be the stationary expression level that is settled after sufficient time. The larger this M is, the stronger the response during the process, and the closer

R is to 0, the closer it approaches the original level after the response. Hence, we define fitness by $M - R$. If $M - R$ is larger, the degree of response-adaptation we are concerned with is larger. Then, we mutated the gene regulation network J_{ij} slightly per generation, and computed gene expression dynamics, and selected those networks with larger $M - R$. This allows the gene network with the larger response-adaptation to evolve.

In this case, no restriction is imposed on genes other than the input and output genes. Hence, adaptive process could be achieved just by a network of three genes as shown in Fig. 5.7, while keeping the expression dynamics of other many genes nonadaptive, that is, non-changing or monotonically changing. However, it is found that the expression levels of many genes rise and fall, or fall and rise, as is consistent with the behavior of (partial) adaptation, when the number of genes is larger than 5. In other words, after a transient response, there is a tendency that expression levels of most genes show the adaptive behavior (Fig. 5.8), in a similar manner with the experimental results shown in Fig. 5.1.

To show this behavior quantitatively, we take fitted networks, that is, those with which $M - R$ of the output genes has increased, and examined the proportion of genes whose expression changes show (partial) adaptation, that is, moves toward the original level (even if not complete) after a response. Figure 5.9 plots the proportion showing this "adaptive" behavior against the number N of genes in the network. It can be seen that the fraction of genes that exhibit adaptive behavior

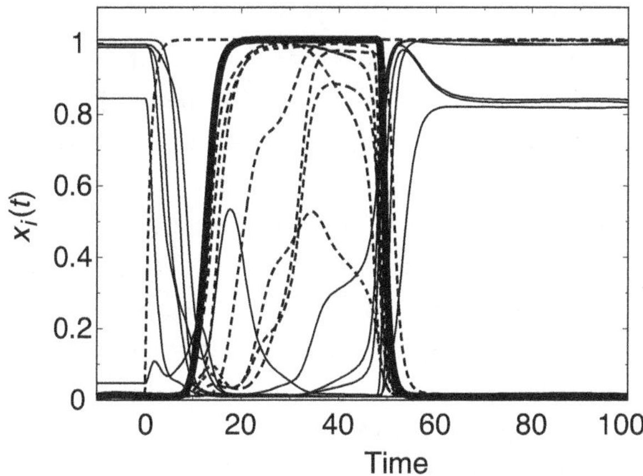

Figure 5.8 Time series of each gene expression level. Time series are plotted from the time of input until a steady state is reached. The output gene shows adaptive behavior (bold line) because the network is used after sufficient evolution. Here, expression of many other genes also tend to rise (and fall) and tend to return toward the original level, implying (partial) adaptation. Based on Inoue & Kaneko (2013). (Courtesy of Masayo Inoue).

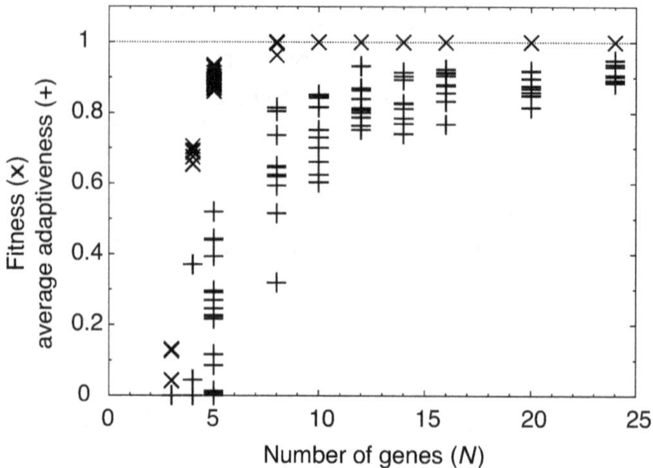

Figure 5.9 The proportion of genes that undergo adaptive change (+) and the degree of adaptation of the output gene $(M - R)$ (×). Plotted as a function of N number of genes. In $N \leq 5$, all the networks were checked, and only those with high adaptability were selected and plotted. For $N \geq 8$, evolutionary simulations were performed on a 10 samples, and the values were plotted after a sufficient (30,000) number of generations. For details, see Inoue & Kaneko (2013). (Courtesy of Masayo Inoue).

increases as N is increased beyond 5, whereas the average $M - R$ reaches maximal level around $N \sim 10$. When N is larger than 10, most of the genes show partial adaptation; the expression levels show large transient response and go back to the original state after responding.

Next, we took partial networks (motifs) consisting of three genes from the whole GRN, and examined which types of feed-forward network motifs in Fig. 5.7 are more included therein, and computed the frequency of each motif (Fig. 5.10). Here, if the simple adaptation mechanism examined in the three genes is employed, then it should be an incoherent network (with a different sign between the two pathways, type I) of Fig. 5.7, in particular I1 and I3. We found that for $N = 3$, I1 and I3 are indeed dominant, and also for $N = 4$, such motifs are frequent. However, as N increases, the fraction of such incoherent (i.e., +− signed) networks decreases, and on the contrary, the degree of networks with the same sign, C1 and C4 in the figure, increases. These network motifs exhibit behavior that is in synchronization with the responses of other genes, because coherent (positive) network motifs transmit positive correlation. In other words, many genes cooperate to produce an adaptive response as other genes produce an adaptive response, and the individual genes respond in alignment with each other. It is a self-consistent behavior, in which one's own adaptive behavior is exhibited because most of others exhibit the adaptive behavior, which in turn influences others to exhibit the

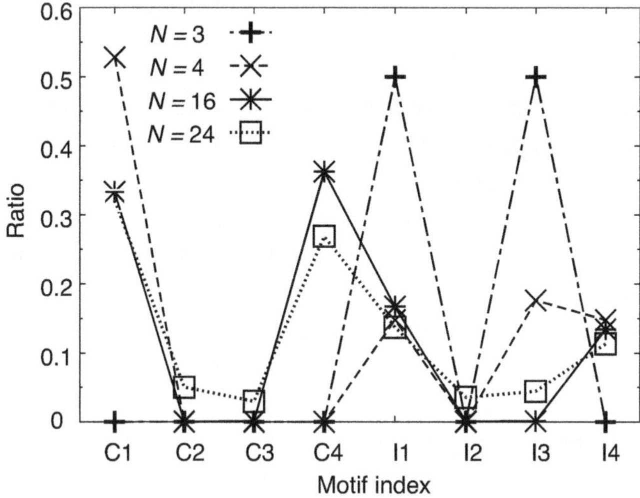

Figure 5.10 Percentage of each network motif. We took a gene network with sufficiently high fitness for adaptive behavior of the output gene, and computed the fraction of each of the three-gene feed-forward network motifs in Fig. 5.7 that are included in the network. The frequency of each motif was determined across the sample of networks with high $M - R$ as shown in Fig. 5.9. $N = 3, 4, 16, 24$. Based on Inoue & Kaneko (2013). (Courtesy of Masayo Inoue).

adaptive behavior. This is how the collective adaptation, as in Fig. 5.1. is achieved here.

In this section, we found cooperative adaptations by many components using catalytic reaction networks and gene regulatory networks. The cooperative adaptation behavior is achieved by the interaction of many degrees of freedom, in contrast to the adaptation by few degrees of freedom. Adaptation of each component is facilitated by the adaptive behavior of all other components, leading to a collective adaptation. In other words, adaptation satisfying macro-micro consistency. Many degrees of freedom (concentration of components or gene expression levels), then, change in coordination, with strong correlation with each other, as observed experimentally in Fig. 5.1.

Such collective dynamics in gene expression dynamics are also observed when the networks with sloppy units (i.e., those with lower Hill coefficients) are evolved to show on-off type response to inputs (Inoue & Kaneko, 2021). Such collective dynamics suggest that the changes in the dynamical system of many degrees of freedom are restricted to a fewer-dimensional manifold (state space), even though the expression dynamics of many genes are involved. This implies the dimensional reduction of the many-degrees-of-freedom system as will be discussed in Chapter 9.

5.4 Homeostasis of Biological Clock (D)

5.4.1 Robustness of the Period of Circadian Rhythm (D, E)

Even though there are many variables involved in the adaptation we have seen so far, there is a feedback process in a broad sense. Whereas the response to external change is direct and fast, the feedback is a slower process, which affects the entire system to make it stable. Here, there are different time scales: fast response and slow adaptation. This section discusses the relationship between this time scale and adaptation concerning the biological rhythms.

As mentioned in Chapter 3, chemical reactions in living organisms require high activation energy. Enzymes are required for the reaction to proceed. Although the equilibrium ratio of chemical concentrations is not dependent on the abundances of enzymes, the nonequilibrium state depends on them, as the rate of reaction depends on the concentration of enzymes. Because the amount of enzyme is changeable depending on the cellular state, the internal time scale of the cell, such as the relaxation time scale for adaptation or the period of biological rhythms, is regulated by the amount of enzyme.

In this section, we will discuss the circadian clock as an example of an adaptive rhythmic phenomenon. A circadian clock is a rhythm of approximately 24 hours in the organism. Our body clock is a typical example, but even some microorganisms show oscillation of gene expression or activity with the period of about 24 hours. Cyanobacteria, that carry out photosynthesis, are a typical example. Furthermore, Takao Kondo et al. made a remarkable experiment by using proteins termed as Kai A, B, and C (and ATP) extracted from cyanobacteria. When they were put in a test tube, their phosphorylation levels oscillated with a period of about 24 hours (Nakajima et al., 2005).

In addition, the period of this oscillation remained almost constant even when the temperature was changed. This "temperature compensation" is surprising, since the speed of a chemical reaction should increase with temperature T in the form $exp(-\Delta E/(kT))$ if its activation energy is ΔE, where k is the Boltzmann's constant. In fact, such a circadian clock cycle independent of temperature is common in all living organisms, where elaborate mechanisms are searched for, based on complex regulations with gene expression. However, since Kondo's *in vitro* experiment contains only three proteins (indeed essentially two, since KaiB plays only a supplementary role) and ATP as an energy source, it suggests that the temperature compensation may be possible without such an elaborate mechanism.

With this experiment in mind, a general mechanism for the temperature compensation of period was proposed (Hatakeyama & Kaneko, 2012). It is based on the time-scale control by called enzyme-limited competition, focusing on the aforementioned rate change due to the limitation of enzyme concentration.

Specifically, consider a situation in which the different substrate species compete for the enzymes necessary for a reaction. In this case, while an enzyme is bound to a substrate, it cannot be used by other substrates. The amount of free enzyme is then reduced. In this way, the amount of free enzyme changes depending on the concentration of the substrate, which may allow for autonomous adjustment of the time scale.

As an example of the application of this concept, let us take the circadian oscillation of the phosphorylation level of the Kai protein. Here, the KaiC protein is a hexamer consisting of six subunits, each of which has a phosphate group, so the six-step phosphorylation process is needed.[2] This phosphorylation process is a key factor for the circadian oscillation. In the case of the current circadian rhythm, the degree of phosphorylation (averaged over molecules) oscillates over 24 hours. Based on this experiment, a model has been introduced that takes into account of the reaction properties of the KaiC protein (van Zon et al., 2007). In this model, KaiC protein has phosphorylation levels ranging from 0 to 6, as well as active and inactive states. In the active state, phosphate groups are added to the protein one by one with the aid of an enzyme (KaiA), leading to progressive phosphorylation; once phosphorylated to six levels, the protein structure changes into the inactive form, where it is sequentially dephosphorylated (without enzyme assistance), and once the phosphorylation level goes down to 0, it is then changed into an activate form. Thus, as shown in Fig. 5.11, the active/inactive and phosphorylated states of the protein change sequentially, making a cyclic process. When this process is represented by a dynamical system and simulated, oscillations in the phosphorylation level are observed as in the experiment.

In this Kai system (and accordingly in the model), the enzyme KaiA is necessary for each phosphorylation step. While the enzyme is bound to the substrate (each subunit of KaiC), it cannot bind to other proteins. In other words, as the proportion of the complex with which enzymes are bound increases, the amount of free enzymes that are not bound to KaiC decreases, so that the abundance of available enzyme KaiA decreases. This "competition effect" of enzymes is taken into account of in the simulation of the model.

To focus on the temperature dependence, let us note that in this system, the rate of reaction for each step is given by $exp(-\Delta E/(kT))$ with ΔE as the energy difference between the active and inactivated states, and thus the fundamental reaction speed, increases as the temperature T is increased, as in the standard chemical reaction rate. However, the simulation results show that the period of the rhythm

---

[2] This is one of the most common posttranslational modifications of proteins in biological systems. The addition of a phosphate group changes the state of the protein.

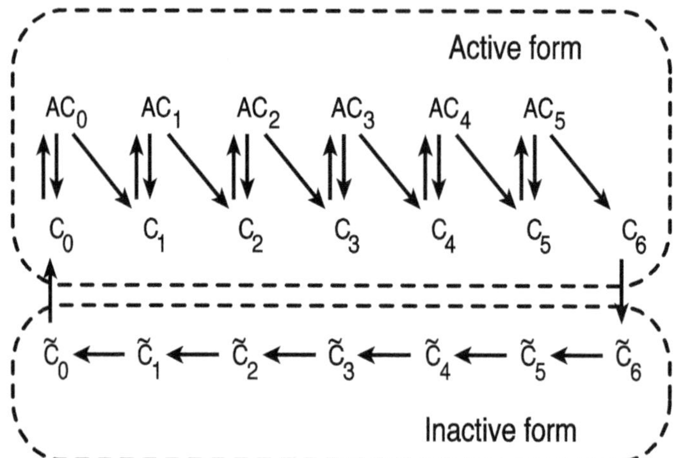

Figure 5.11 Schematic diagram of phosphorylation of Kai C proteins with the aid of enzyme Kai A. Based on Hatakeyama & Kaneko (2012). (Courtesy of Tetsuhiro Hatakeyama).

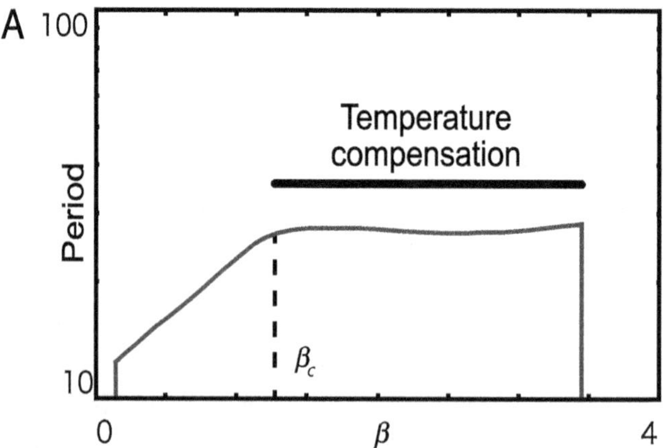

Figure 5.12 Kai's phosphorylation reaction model. (Figure 6.11) shows the periodic oscillations in phosphorylation levels. The temperature dependence of this periodicity is plotted, where β is the inverse temperature $1/(kT)$. Based on Hatakeyama & Kaneko (2012). (Courtesy of Tetsuhiro Hatakeyama).

remains almost unchanged, as shown in the graph in Fig. 5.12. Robustness of the period to temperature is thus achieved.

Then, how is this "temperature compensation" realized? The key point here is that the reaction rate depends on the amount of "available" enzymes that are not bound to the substrate. Each reaction that proceeds with phosphorylation uses the same enzyme. As the temperature increases and the amount of active protein increases, so does the proportion of enzyme bound to the protein. The amount of free enzyme is then reduced. Furthermore, the amount of enzyme available at each stage of phosphorylation is different because the affinity (chemical affinity) of the enzyme to the protein is different at each stage of phosphorylation.

When the energy difference between the active and inactive states is ΔE, the ratio of molecules in the activated state (the ratio of C to $\tilde{C}$ in the figure) changes by $exp(-\Delta E/(kT))$. Proteins in the activated state can bind to the enzyme, which reduces the amount of free enzyme. Therefore, when the temperature is increased, the fraction of activated molecules that bind to the enzyme increases as $exp(-\Delta E/(kT))$. As a result, if the total amount of enzyme is not abundantly supplied, the amount of free enzyme decreases by approximately $exp(\Delta E/(kT))$. Note that the individual reaction rates themselves increase as $exp(-\Delta E/(kT))$, so that the decrease in the amount of free enzyme just cancels out the increase in reaction rate. Hence, the period of the oscillation does not depend on temperature. This is a sketchy explanation; for detailed derivation of the period robustness see (Hatakeyama & Kaneko, 2012).

To summarize, as the temperature increases, the fraction of active state molecules that bind with enzyme increases, resulting in a shortage of free enzymes, and then the decrease of their abundances compensates the increase in reaction rate, leading to the robustness of the period of oscillation.

In the present case, for the temperature-compensation mechanism to work, there must be active and inactive states, and there must be several modification reactions (phosphorylation in this case) on the molecule, whose affinities differ by each step of modification, and thus each step competes for the available enzyme, whereas the overall enzyme concentrations are not too high. In fact, proteins in the cell that play an important function are often multimeric, consisting of several units, and have structural changes with different activities. Hence, the above conditions are often satisfied. Furthermore, the *in vitro* experiment by Kondo's group suggests that if the concentration of enzyme KaiA is high, the temperature compensation of period is poor (see supplement of Ito-Miwa et al., 2020). This also fits with the above picture of enzyme-limited competition.

This mechanism is not limited to robustness against temperature changes, but also applies to changes in other environmental conditions that affect the reaction rate, such as external nutrient concentrations. In fact, it has been shown that nutrient compensation is realized by the same mechanism as temperature compensation, that is, the period of the oscillation remains unchanged even when the external ATP concentration is changed (Hatakeyama & Kaneko, 2014b).

If the same enzyme is used in each step of a series of chemical reactions, then enzyme-limited compensation would be a general mechanism for robust adaptation to external environmental changes. In this case, when the temperature is changed, it first causes a change in the basic reaction rate due to the form of

$exp(\Delta E/(kT))$. This results in a transient change in the fraction active states. If the change is such that the activity of the reaction increases, the amount of free enzyme available is reduced, thus lowering the reaction rate and decreasing activity (and vice versa). As a result, imposed external changes are compensated for. Such time-scale regulation using enzyme levels as a buffer will provide a general principle for the stability of biological functions.

5.4.2 Reciprocity between the Robustness of the Oscillation Period and the Plasticity of the Oscillation Phase

In Section 5.4.1, an increase in temperature causes the reaction rate to rise once, after which the amount of free enzyme decreases and the rate returns to normal. In other words, changing the external conditions causes a transient change in the cycle period, followed by a change in the amount of free enzyme to counteract the transient change. The former is a transient change, and the latter persists as long as the external conditions are maintained. In other words, the variable of free enzyme abundance buffers the environmental change, and another variable, the period, returns to original. This is precisely the adaptation described in Sections 1 and 2 of this chapter. Thus, behind the homeostasis of biological rhythms is inherent change in the abundances of available free enzyme.

Here, once the temperature is increased, the oscillation speed is increased, and the phase of the oscillation will therefore advance faster at that stage as well. Therefore, the phase is expected to change more easily (plastic), alongside robustness of the period. To discuss the plasticity of the phase, let us introduce of the periodic change in the temperature with a 24-hour period in the current model. Initially, the external temperature oscillation and the phosphorylation oscillation of the biological clock can be out of phase, and the difference between them is arbitrary for each initial condition on the clock side. However, as this external temperature change continues, the phases of the external temperature cycle and internal circadian rhythm will be aligned, and the external rhythm and the internal clock will match. This entrainment is confirmed both in the above theoretical model as well as in actual experiments (Yoshida et al., 2009). (In the example of circadian rhythm in our body, this corresponds to the resolution of jet lag after a few days abroad).

In other words, the period T is homeostatic with respect to external temperature changes, whereas the phase ϕ is more changeable. We investigated this relationship between the period robustness and phase plasticity in various models of biological clocks (Hatakeyama & Kaneko, 2015).

To do so, we consider the case where the robustness of the period T_p is not necessarily perfect, and the period is shifted by ΔT_p when the temperature is externally changed. Here $\Delta T_p = 0$ if the compensation is perfect as in the example of Section 5.4.1. However, even in the model therein, depending on the value of the parameter, compensation may not be perfect. For example, if the activation energy ΔE is decreased or the amount of total enzyme (KaiA) is increased, the temperature compensation is not perfect and the period change ΔT_p remains as the temperature is changed. On the other hand, we define the degree of change in the phase $\Delta\phi$ of the circadian rhythm when the temperature is changed externally and computed how it depends on the parameter (Hatakeyama & Kaneko, 2015). Then, changing the parameter values of the model, we computed ΔT_p and $\Delta\phi$. Accordingly, we found the relationship

$$a(\Delta T_p/T_p) + \Delta\phi = constant. \tag{5.6}$$

(see Fig. 5.14). (where a is a positive constant). In other words, as temperature compensation is better (ΔT_p approaches 0), the phase change $\Delta\phi$ is larger, and as compensation is worse, the degree of phase change is smaller. This reciprocal relationship between the period robustness and phase plasticity has been confirmed to hold for other models of circadian clocks. Using dynamical systems theory, this universal relationship is derived for clocks in general, by introducing adaptive process to the dynamics to generate oscillation (limit-cycle) (see [Hatakeyama & Kaneko, 2015] for details).

In fact, this reciprocal relationship can be understood as the feed-forward motif discussed in Section 5.3. A change in the amount of free enzyme, which serves as a buffer for the external change, would cancel out the change in the period. During the change, the phase of the oscillation varies significantly. Therefore, in exchange for adaptation of the oscillation period, the phase is more likely to change. It is a reciprocal relationship between the robustness of the period and the plasticity of the phase.

To understand this relationship, let us draw the concentration of the buffer molecule on the horizontal axis and the concentration of molecules that are otherwise independent of environmental changes on the vertical axis. The original oscillation trajectory (limit cycle) is schematically depicted as a closed orbit in the two-dimensional space as in Fig. 5.14. If the environment changes, the trajectory is changed. When the period is robust (as in temperature compensation of the period), the environmentally induced change is absorbed by the buffer component, and accordingly the amplitude of the oscillation also changes. Let Δx^* be the change at this time. On the other hand, if the compensation of the period is not perfect, the orbit will lie between the original orbit and that for the perfect adaptation described above. Let Δx be the deviation from the original orbit in this case. Then the change in period is smaller as Δx approaches Δx^* and should become zero at

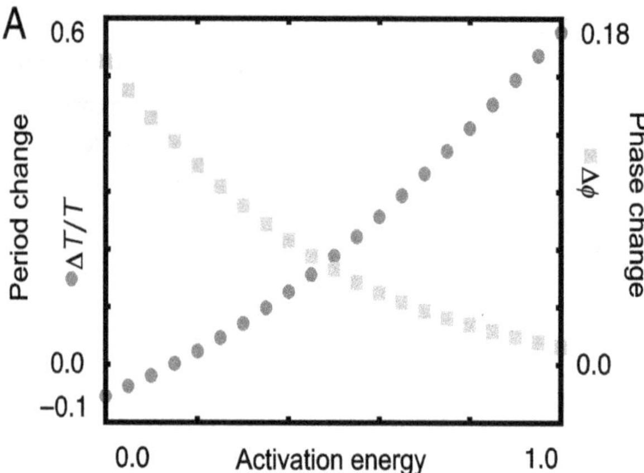

Figure 5.13 Reciprocity between the robustness of the period and the plasticity of the phase in the circadian rhythm model of Kai phosphorylation in Section 5.4.1; the degree of the change of the period against temperature, and the plasticity of the phase are plotted as a function of activation energy ΔE. The details of the computational method are based on Hatakeyama & Kaneko (2015). (Courtesy of Tetsuhiro Hatakeyama).

Δx reaches Δx^*. Assuming that the change is not so large, and adopting the linear relationship[3] the change of period will be proportional to $\Delta x^* - \Delta x$.

On the other hand, if there is deviation from the original orbit, the phase can be changed accordingly. If we make a linear approximation again, we can assume that the phase change is proportional to Δx. This Δx depends on a parameter that expresses the degree of robustness of the system.[4] Then, by eliminating Δx from the above two proportionality relations, we can express $a(\Delta T_p/T_p) + \Delta\phi$ only by Δx^*. Since x^* is in a state of perfect compensation, Δx^* does not depend on the previous external parameter. Hence we obtain the reciprocity relation Eq. (5.6). From this discussion, we expect that the relationship (5.6) between the robustness of the period and the plasticity of the phase does not depend on the specific mechanism for the oscillation and adaptation, but is universally valid for biological rhythms that exhibit adaptation. In fact, this relationship has been confirmed in different models (Hatakeyama & Kaneko, 2015, 2017).

What is noteworthy here is that the seemingly contradictory relationship between robustness and plasticity, that is, insensitivity to change and changeability, which are essential to living organisms, has been compatible, by assigning different variables for each, that is, period of the cycle and phase of oscillation,

---

[3] As seen in Eq. (5.6), the linear relationship holds rather well. As seen in Chapters 2 and 9, the region where the linear relationship is valid expands in living systems that have acquired robustness through evolution. The above result may also be related to such *deep linearity*.

[4] It corresponds to ΔE in the Kai C circadian model.

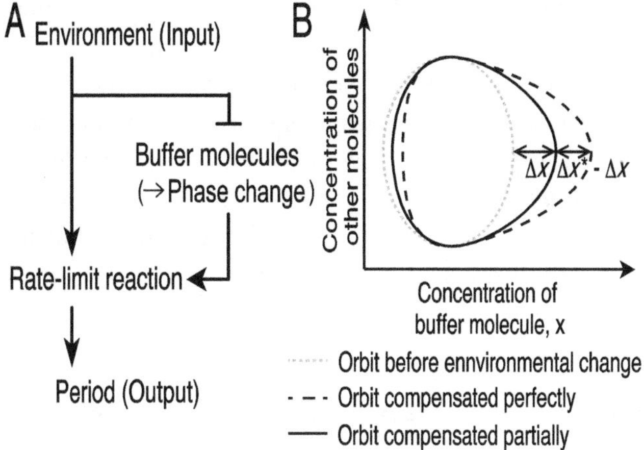

Figure 5.14 Explanation of the relationship between changes in oscillation orbits and buffer molecular concentrations due to environmental changes, and the resulting robustness of the period and plasticity of the phase. (Courtesy of Tetsuhiro Hatakeyama).

resulting in a reciprocal relationship between the robustness of one variable and plasticity of the other. This reciprocity between insensitivity and feasibility to change can also be seen as a property that links hierarchies; changeability of stem cell state and robustness cell-type distribution (Chapter 7) and evolutionary plasticity and robustness of a fitted state in the adaptive evolution (Chapter 9).

In the Kai protein model, when each protein molecule exhibits a temporal oscillation of the phosphorylation level, oscillation of each KaiC molecule is required to be entrained in order for the population of kaiC molecules to exhibit a macroscopic biological clock (Fig. 5.13). If the phases of each oscillation are desynchronized, the oscillation would be averaged out and disappear. If each of the microscopic clocks are entrained with each other, then a uniform macroscopic oscillation is generated. In other words, the plasticity of phase (i.e., entrainability between oscillators) is necessary for a biological clock to function collectively as a group of units. From this perspective, the robustness of the period can be considered as a consequence of the collective entrainment of microscopic clocks, or in other words, a consequence of the macro-micro consistency of the clock.

5.5 Collapse of Consistency in Adaptation

Adaptation is thought to result from the requirement that the original cellular state keeps macro-micro consistency between different time scales. In what cases, then, does adaptation to the original state collapse? When a cell is too long put to a new condition, and too much habituated to it, the cell may lose the memory of the original state and then the return to, and maintenance of, the original state would be lost. Conversely speaking, if the cell's adaptation process itself may have adapted and no longer operates, the new environmental condition is memorized as a new default state. In this way, memory can be viewed as *a state in which the adaptive process has adapted and can lose the return to the original*. The construction of a theory based on this view is a problem for the future, but in any case, the issues of memory with distinct time scales are important and will be discussed in Chapter 6.

5.6 Summary: Universal Properties and Laws Discussed in This Chapter

- In response to environmental changes, many state variables (e.g., gene expression levels) go up or down to respond and then tend to be back to the original. It is called perfect adaptation if they return to the original completely, and partial adaptation if incompletely. In systems with a few degrees of freedom, this can be represented by introducing some variables that buffer external changes (A, E).
- A *critical* state in which the abundances of components is inversely proportional to its rank is maintained in an *ideal* cell model with many degrees of freedom. In the model, nutrients are actively transported by transporters, which are then transformed by mutual catalysis, by simplifying the processes in real cells. This critical state is maintained even when the environmental condition is changed. Here, the transporter components buffer the response generated by the environmental change, so that the original macroscopic state is homeostatic (B, C, E).
- Many components, both in the present cells and in the above ideal cell model, show partial adaptation, where the ratio of transient response to the final remnant change is nearly the same across all components, suggesting the collective adaptive process (B, C, E).
- In both the ideal cell model above and in actual cells, response changes to the environment show "fold-change detection," in which the response ratio by environmental changes depends only on how many fold the environmental change is to the original value (A, E).
- In the gene expression regulatory network model, when the number of components increases, collective adaptation is observed, in which many gene expressions partially adapt as other components partially adapt. Such collective response of gene expression has also been found in experiments (B, E).
- In the biological circadian clocks, the period of oscillation remains almost unchanged against the change in temperature (temperature compensation) or other environmental conditions (E).

- When several reactions involved in the cyclic change use a common enzyme, there appears an enzyme-limited competition, if the enzyme abundances are not sufficient. This competition among substrates explains the temperature compensation mentioned above. Here, the change in the basic reaction rate with temperature is compensated by the change in the amount of enzyme available (D).
- Reciprocity between the period and plasticity of biological clocks: The period T_p is more robust to environmental changes, as the oscillation phase ϕ is more easily to change. Quantitatively, the change in period ΔT_p and phase change $\Delta\phi$ with respect to changes such as temperature follows the relationship $a(\Delta T_p/T_p) + \Delta\phi = const.$ (D).

5.7 Appendix: Mathematical Analysis

Let us approximate the reaction network by a hierarchy of chemical reaction processes sequentially from a nutrient L_0 to $L_0 \rightarrow L_1 \rightarrow L_2 \rightarrow \cdots L_k$ where the number of all component types is N.[5] Let m_i be the average concentration at each L_i level, (i.e., we take mean-field approximation to neglect the difference in abundances among the components within the same layer). The catalyst concentration of each reaction is approximated commonly by the average of the catalyst concentrations, which is $(1 - m_0)/N$. Then the change in concentration at each level is given by

$$dm_0/dt = Sm_k^\alpha - (M/N)(1 - m_0)m_0 - Sm_k^\alpha m_0, \tag{5.7}$$

$$dm_j/dt = (M/N)((1 - m_0)m_{j-1} - (1 - m_0)m_j) - Sm_k^\alpha m_j, \tag{5.8}$$

where $j = 1, \cdots, k$ and $M = \rho N$ is the average number of reaction paths from each component. The last term in Eq. (5.8) is the dilution due to cell volume growth (the model does not consider decomposition or flow to the outside of the cell, so that the growth occurs only as nutrients flow in). Let $dm_0/dt = 0$, to obtain the steady-state concentration of nutrients

$$F_0 = Sm_k^\alpha = \rho m_0, \tag{5.9}$$

On the other hand, $m_j = m_{j-1}(1 - m_0)$ since $dm_j/dt = 0$ for the concentration at each level. Then it follows

$$m_k = m_0(1 - m_0)^k, \tag{5.10}$$

From these equations we obtain F_0, m_k, m_0 as a function of the external nutrient concentration S. In particular. If S is large, $m_k \propto (1 - m_0)^k$ and $F_0 \sim \rho$, which does not depend on S. Also, noting that the types of components at each level of the

---

[5] This appendix is technical. It can be (or recommended to be) skipped, unless one is especially interested in dynamical systems or statistical physics.

hierarchy increase in power, when S is large, the -1 power law between that rank of component concentrations is also derived.

Finally, the fold-change detection in response can also be derived. For simplicity, let us look at the response at two levels, $k = 2$. First, we find the stationary solution (m_0, m_1, m_2 (but $\sum_{j=0}^{2} m_j = 1$ so that two degrees of freedom)) under an approximation assuming that S is large. (Specifically, $m_0 \approx 1 - (\rho/S)^{\frac{1}{2\alpha}}$, $m_2 \approx (\rho/S)^{\frac{1}{\alpha}} (1 - \rho/\alpha(\rho/S)^{\frac{1}{2\alpha}})$).

Now, in order to see the relaxation of M_0, M_1 to stationary solutions, the eigenvalues of the Jacobi matrix around this fixed point are obtained, as $\alpha \rho (1 \pm \frac{1}{2}\sqrt{1 - 8/\alpha})$, which are independent of S. On the other hand, when S changes from S_0 to S'_0, the change in nutrients is given by $1/F_0(t) - (S_0/S'_0)/F_0(0) = t$ when t is small and S is large, which depends only on S'_0/S_0 and not on S_0 alone. Thus, the time evolution depends only on the fold ratio of the S change.

6

Cellular Memory

6.1 Introduction

So far, in each chapter we have discussed the consistency of different time-scale phenomena. The most important of the slow time-scale phenomena in living organisms is memory in its broadest sense. In general, cells "remember" their history. Such memory exists in cells, and does not need the nervous system.

Let us discuss a classic example. A *Paramecium* is incubated at a certain temperature over a long time and then is placed in an environment under a temperature gradient. The cell then moves to the temperature at which it was originally cultured (Jennings, 1906; Nakaoa et al., 1982). During this movement, an adaptive process occurs within the cell, as discussed in Chapter 5. When the cell is moved to a different temperature, its internal state changes once, and then returns to its original state. This transient change in state alters the frequency to change the orientation of motion and moves toward the original temperature, as discussed in Section 5.1. At this stage, it can be said that this *paramecium* remembers the temperature at which it was cultured. However, this "memory" of the original temperature lasts only for a certain time span. If it is then cultured at a new temperature for a longer time span, it will then remember the new temperature as the environment to which it should return (Oosawa, 2001). In other words, the process of returning to the original temperature in the adaptation process itself can be regarded as adaptation, whereas when this adaptation process itself was adapted and lost, the memorization of the new temperature was shaped.

In recent years, extensive studies have shown that the state of cells is remembered even when genes are not changed. First, in multicellular organisms, the differentiated cell state is maintained even after cell division. This is often referred to as epigenetics, but of course, by assigning the name does not mean that we understand it.[1] As will be discussed in Chapter 7, the differentiated state first differs in the composition of protein expression, which is then further consolidated by molecular modifications, such as the addition of methyl groups to DNA molecules, histone modification by proteins that bind to DNA, and change in chromatin structure. This epigenetic modification makes the expression of genes more feasible or harder. In short, the epigenetic modification solidifies the changed degree of expression levels, leading to memory over cell-division generations. However, even if memory is maintained by the modification of DNA molecule, it is still a molecular change, and how the time scale for it can be quite long needs to be solved. We do not understand the cellular memory until we understand how molecular processes can shape long-term memory.

Such cellular (nongenetic) memory has been reported even in unicellular organisms. When a unicellular organism divides, the cytoplasm is divided in half to form descendant cells, and the composition of proteins is directly transmitted to daughter cells. Therefore, if the amount of a certain protein is high or low, this tendency should remain to some extent even after the cell division. Hence, it appears that the cellular state can be memorized. However, in order to sustain the memory over cell divisions, a cell that initially happens to contain a higher amount of a certain protein must synthesize more of that protein than average. For this, a special mechanism to synthesize more of the protein is needed, when the amount of that protein is high. Otherwise, the deviation of the concentration from the average would be reduced approximately to $1/2$ per division. In other words, the half-life time is one cell division, and if we divide 10 times, the difference will be about $2^{-10} \sim 10^{-3}$. Thus, such cytoplasmic "memory" will soon disappear. If there is a memory of cellular state that remains for more than 10 generations, there must be some other mechanism.

In fact, there are known cases in which the phenotypic changes (e.g., changes in protein levels) caused by environmental changes remain for several or dozens of generations without any changes in genes (DNA sequences). Classically, Sonneborn et al. showed that reversing the orientation of cilia on the membranes of a *Paramecium* is maintained after division (Beisson & Sonneborn, 1965). On the other hand, Siegal et al. have shown that phenotypes with different growth rates are transmitted for about 40 generations in yeast (Levy et al., 2012). Some other examples of such nongenetic inheritance are discovered. In fact, Furusawa et al. have shown in environmental adaptation experiments in *E. coli* that phenotypic changes may be transmitted without genetic changes for more than several dozen

---

[1] The term *epigenetic* comes from Waddington (1957), but he did not necessarily attribute it to molecular modifications. He did not necessarily think the epigenetics in terms of specific molecular process, but probably searched for dynamical-systems representation. See also Chapter 7 of the present volume.

generations. In another example of multicellular organisms, Soen et al. reported that resistance to drugs in *Drosophila* is transmitted to the next generation (Stern et al., 2012).

Furthermore, even if the memory may not be extended over generations, phenomena in which cells maintain a certain state for a long period of time can also be observed in cell dormancy, spores, and seeds, as discussed in Chapter 3, and these can also be considered as cellular memory in the broad sense of the term.

The molecular mechanisms underlying such long-term "memory" states of cells as described above are not yet well explored, and many of them require further experimental scrutiny. However, theoretically speaking, it is an essential question in universal biology to ask whether changes in the cellular state can be maintained for many orders of magnitude longer than the timescale for chemical reactions. This is an important issue for establishing the theory of life.

The above discussion on the possibility of memory storage at the cellular level will be related to the long-term changes in microscopic molecular states, such as modifications of DNA molecules, whereas such changes are also observed in the memory of the neural system. Memory in the brain is embedded in neural networks, and at a molecular level, it is thought to be embedded in the state of synapses between neurons. How, then, is long-term memory maintained within the synapses responsible for memory? This is termed as Long Term Potentiation (LTP) of synapses. This LTP is mediated by changes in the state of molecules such as Ca2+/calmodulin-dependent kinase II (CaMKII) and protein kinase C (Protein Kinase C) in the early phase (without protein synthesis) and by cAMP-responses with protein synthesis in the late phase. It is thought to be embedded in changes in the state of proteins such as cAMP-responsive element-binding protein (CREB), which accompanies protein synthesis in the later stages (Lisman et al., 2002; Silva et al., 1998).

The conventional view of such memory is that of multiple stability. For example, a well-known example is that there are two stable states of a magnetic material in which the magnetization is either upward or downward, and that memory is created by using and maintaining these upward and downward states. In this case, the embedding and erasure of memories are performed by applying an external magnetic field, which is a prototype of computer memory. In fact, such bistable states can also be created by controlling gene expression. For example, as discussed in Chapter 4, a toggle switch in which two genes (from which proteins are synthesized by transcription) repress each other's expression has two states depending on which protein is being expressed. Furthermore, as discussed in that chapter, in more complex gene regulatory networks consisting of many genes, there can be many attractors. Such multiple stable states have also been extensively investigated in neural networks in which neurons activate or inhibit each other's firing.

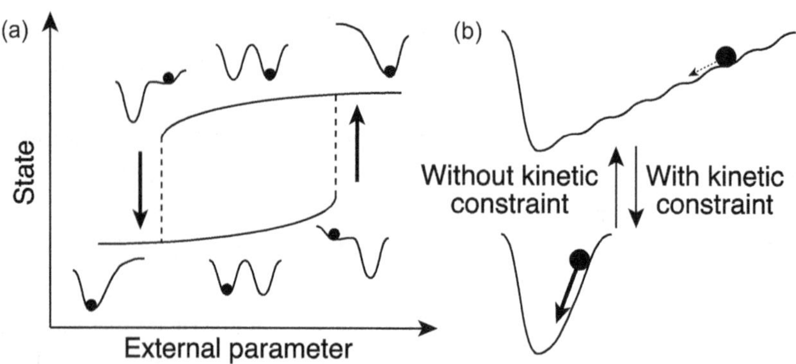

Figure 6.1 Two ideas about memory: (a) multiple stability and its erasure (b) kinetic memory. Courtesy of Tetsuhiro Hatakeyama.

Since these are memories in two or more discrete states, they can also be called digital memories.

However, there are several questions as to whether the cellular memory discussed so far can always be explained by conventional multi-stability. First, in the case of the memory of incubation temperatures, say 36.5°C, 36.7°C, 37°C. It would be more natural to think that the history of the former temperatures remains continuously (within an appropriate precision). In other words, we can regard it as an analog memory rather than a digital one.

Secondly, the fact that such memories are lost after a certain time span makes it somewhat questionable whether this is consistent with the picture of memories being kept in multi-stable states. If the memory is in a multi-stable state, it would require a large external input to reset the memory, in order to destabilize the original state: The operation to raise the valley is needed in order for a state to flow out of it (see Fig. 6.1). (In the case of magnetic memory, this corresponds to applying a large external magnetic field.) On the other hand for some cellular memory, it seems more likely that the history of past states is gradually lost over time rather than being lost by such a resetting operation.

In the case of other cellular memories lasting several to several dozen generations, they also seem to deteriorate gradually rather than being lost at a certain stage abruptly as in the resetting operation. In other words, rather than being directed into a different stable state, the memory here seems to be generated by the formation of a transient state that is maintained for a very long time relative to the time scale of the original chemical reaction, so that it is not easily relaxed to the true equilibrium state. In this sense, it is more like considered as *kinetic memory* rather than static memory by multi-stability.

Since such kinetic memory is not as stable as the static memory by multi-stability, it may be considered inappropriate as a long-term (ever-lasting) memory.

On the other hand, because it is kinetic, memory can be easily written and erased, without much thermodynamic cost. In fact, memory in living systems does not necessarily last forever. Short-term and long-term memory exist therein, whereas sometimes they are not clearly separated, and the timescales of memories are sometimes intermediate between the two. Also, the time scales themselves depend on the situation. Memory, then, can be kinetically written to, and read from. Therefore, it would be of significance to explore a possible logic for kinetic memory processes that is different from multiple stability.

The physics of glass has been regarded as a useful reference for considering this issue. Many experiments have shown that the glass we see is not in perfect thermal equilibrium and that its behavior changes very slowly with the time scale of days or years (Debenedetti & Stillinger, 2001; Berthier & Biroli, 2011; Cavagna, 2009). Over the past few decades, statistical-physics research has developed to understand the glass, and our understanding has deepened, even though, it has not yet been fully resolved. The two main concepts that have developed in this context are structural multi-stability and kinetic constraint.

The former was advanced in studies originating from magnetic materials, called spin glass (Mezard et al., 1987; Nishimori, 1999). In ferromagnetic materials, interactions to align the direction of spins (microscopic magnets, taking up and down directions) lead to the total alignment to generate a macroscopic magnet. In contrast, different types of interaction reverse the directions of spin (up versus down). Then, an abstract model was introduced with an upward and downward state (spin) and interactions that align or reverse the orientation of the two spins.[2] If there exists only the interaction so that aligned spins have a lower energy, then a ferromagnetic state is realized at low temperatures, where all the spins have the same orientation. (We call such interaction J_{ij} as +). In contrast, for the interaction $J_{ij} < 0$, the opposite spin configuration is favored energetically, and the antiferromagnetic state is realized.

Now we consider the case that both the positive and negative interactions between spins are involved. In fact, such material is known to exist then, the question of which configuration of spin orientations can lower the energy is no longer simple. For instance, let us consider the simple case of three spins, where the interaction between 1 and 2, 2 and 3 is +1, and the interaction between 3 and 1 is −1. Now, due to the positive interaction between 1 and 2, the spin 1 and 2 will be oriented in the same direction. If the third spin is also oriented in the same direction, the interaction between 2 and 3 lowers energy, but that between 3 and 1 cannot lower the energy. In contrast, if the spin 3 is oriented in the opposite direction, the interaction between 2 and 3 cannot lower the energy (Fig. 6.2(a)).

---

[2] For physicists, this corresponds to the Ising model with ferro- and anti-ferro couplings.

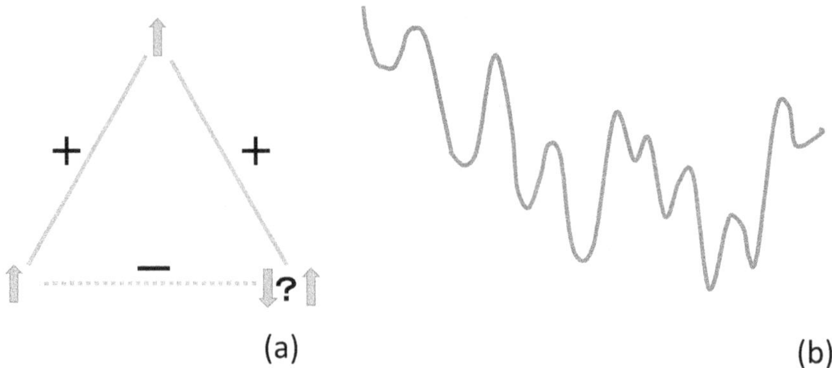

(a)   (b)

Figure 6.2 Schematic representation of spin glass. (a) If the interaction between the three spins is $+, +, -$, there can be no configuration that lowers the interaction energy between all spins due to frustration. (b) Rugged energy landscape. The height is the energy, and the horizontal axis is the axis of the configuration space that schematically represents successive changes in the configuration of the spins.

The four (up,up,down; up,up,up; down,down,up; down,down,down) spin configurations have the same "minimum" energy. There are twice as many cases of ferromagnetic interactions where all of them are aligned with the lowest energy. As the number of spins is further increased, more and more various spin configurations have the same (or similar) lower-energy states, so that more and more (meta)stable states with lower energy exist: Here a metastable state is the one in which the energy is not completely minimum, but is not decreased when any single spin orientation is individually flipped one by one from the original state. Since there are many such states, a schematic representation of the energy landscape is depicted as in Fig. 6.2(b). In this case, it may take a very long time to find the lowest energy state, while each metastable state stays for a very long time even if the spin direction is flipped due to thermal noise. Therefore, it will be possible to use these many metastable states as memory.

In fact, this spin statistical physics model (the so-called spin glass model) has been adopted to consider multiple memory states in a simplified neural network system (Hopfield, 1982; Mezard et al., 1987), in which case firing or non-firing of neurons corresponds to the up and down spin states. This is an example of memory as multi-stable states. Similarly, the gene expression network model discussed in Chapters 4 and 5 has multiple states in which each gene is expressed or non-expressed. Since gene expressions interact, many attractors are generated to form multi-stability. Depending on the nature of gene-gene interaction, they have properties similar to those of the spin glass.

On the other hand, an alternative picture on glass exists, in which an energy landscape with many metastable states is not postulated. Rather, the relaxation to a single (or few) stable states is slowed down by kinetic constraint (Ritort & Sollich 2003). Consider a case in which there exists a unique lowest energy state and there are no other metastable states, but it is somehow hard to relax toward it. For example, in the spin model, the spin-spin interactions are simply all + and the states in which all the spins have the same orientation (up or down) have the lowest energy. In a simple kinetic model, in which a spin flip-flops if the interaction energy with another neighboring spin is decreased by it, ordinary relaxation follows: the spins will be aligned sequentially, even if they are initially oriented randomly. This ordinary relaxation to the equilibrium state follows an exponential form $exp(-t/\tau)$ with time t.

In contrast, suppose that the relaxation is determined not only by the two interacting spins but depends also on some other spins: Only if they are in certain orientations, the spins can flip-flop (i.e., the other spins are the catalysts for the change). If the initial spins have different orientations, then it will be difficult to satisfy the conditions of the other (third or fourth) spin configurations, and the relaxation to a lower energy state will be hindered. In this case, the spin configuration stays around a certain state over some time, after which the energy slowly drops. Then, the relaxation to the equilibrium state does not follow the standard exponential form, but is much slower, say with the form of $-\log t$. Such slow relaxation is indeed observed in real glass.[3]

Now, in theorizing the two views on memory discussed in the beginning of this section, the two theories developed in the study of glass are instructive. Indeed, the multiple stable states in gene expression networks were shown by Kauffman (1969) in a Boolean network model, and later this model was analyzed by Derrida et al. (Derrida & Pomeau, 1986) by connecting it with the statistical physics of spin glass (with metastable states). The neural network study by McCulloch & Pitts (1943) and Amari (1978) was later shown to correspond to the spin glass model by Hopfield (1984). Such many metastable states, as in Fig. 6.2(b) are often represented to be located at the valleys in a rugged landscape (Mezard et al., 1987).

In contrast, the relationship between slow relaxation due to kinetic constraints and biological phenomena has been less explored. Note that many chemical reactions in the cell are enzymatic. In such a catalytic reaction system, the reaction from a component A to B depends on a catalyst C. If the concentration of catalyst C is low, the reaction rate is suppressed (see also Section 5.4), and then kinetic

---

[3] The full understanding of real glass will presumably be achieved by integrating both the pictures of former multiple metastable states and the latter kinetic constraint.

constraints as described above naturally arise. Furthermore, molecular modifications and protein reactions generally involve many steps of such catalytic reactions. Thus, the slow relaxation as in the kinetic constraint in glass can occur. In the following, we will discuss such examples and see how they can provide a theoretical framework for kinetic memory.

6.2 Kinetic Memory Theory with Enzyme Competitive Rates (B and D)

Consider the stepwise modification reaction of a molecule with many modification sites, as in Section 5.4.[4] This is a protein macromolecule attached to a small molecule such as phosphoric acid. Each of the modification reactions is generally catalyzed by enzymes. In this section, we show that the chemical reaction to return from the modified state to the original state is suppressed and the relaxation process is drastically slowed down due to the enzyme-competition, resulting in kinetic memory (Hatakeyama & Kaneko, 2014a).

For this purpose, we consider the case where modification reactions such as phosphorylation and methylation occur stepwise in a multimeric protein. (see Fig. 6.3). In this specific example, there is an $N(=6)$-modified protein, and there are reactions that sequentially demodify the states from the fully modified (e.g., phosphorylated) state to non-modified state, that is, $N \to N - 1 \to \cdots \to 0$. These reactions are catalyzed by a common enzyme. Since the modification changes the properties of the protein, the affinity of the protein for the enzyme in the reaction $m \to m - 1$ depends on m. This is because the phosphate or methyl group changes the shape of the protein, which in turn changes the feasibility of binding to the enzyme.

Let us use this model to study the relaxation process from the initial state in which the protein is sufficiently modified, and monitor the loss of the modification. First, if the amount of enzyme is sufficient, the fraction of modified molecules will decay exponentially. The decay time is given by the time constant of the reaction. This is natural since the modification is removed at a constant rate.

Next, if the amount of enzyme responsible for the reaction to change the modification becomes less than the amount of substrate (modified protein), the relaxation will not proceed as shown in Fig. 6.4. In this case, the fraction of modified molecules does not decrease exponentially, but decreases slowly, approximately logarithmically with time (Fig. 6.4). This is because each demodification reaction uses the common enzyme, and when its concentration is low, a demodification step is hindered if the enzyme is used in another step. Since this is a logarithmic relaxation, it will take orders of magnitudes longer, and the relaxation will be very slow. Figure 6.5 plots the relaxation time (here, the time it takes for the

---

[4] Since this chapter is all about time, (D) is not explicitly mentioned.

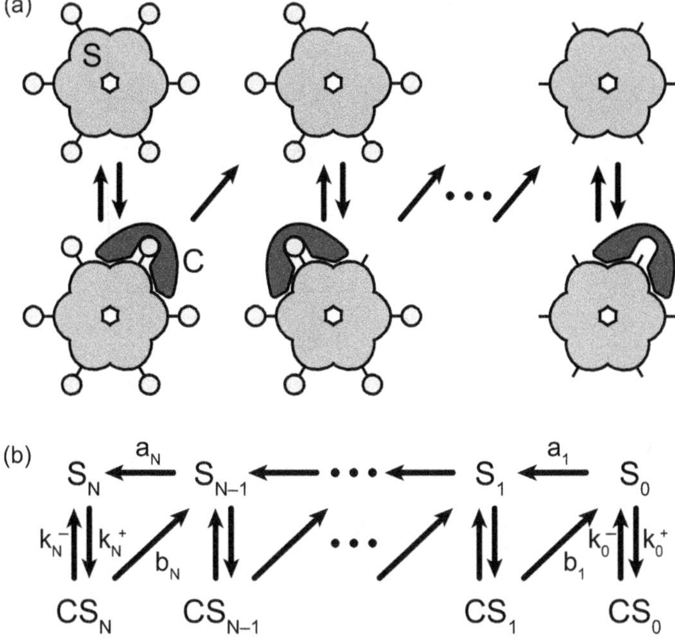

Figure 6.3 Schematic representation of the reaction (a) and model (b) of the chain modification model: There are N modification sites on the substrate, and phosphate groups (or methyl groups, etc.) are attached to them. The modification proceeds at rate a_i, and the demodification process proceeds at rate b_i by the catalyst. Based on Hatakeyama & Kaneko (2014a). (Courtesy of Tetsuhiro Hatakeyama).

degree of modification to get close enough to zero) as a function of the amount of enzyme. The figure clearly shows that the relaxation time increases rapidly when the amount of enzyme is less than that of the substrate. This means that modification states (such as phosphorylation) are maintained for a very long time compared to the individual reaction time scales.

In this case, the characteristic nature of relaxation is also distinguishable from that under sufficient enzyme abundance, where the population of each modified state changes successively. In the enzyme-limited case, the less modified state first increases and then relaxes, suppressing the relaxation of the highly modified states. The relaxation takes the form shown in Fig. 6.6. This can be understood as the enzyme-limited competition seen in Chapter 5. First, note that the less modified state has a higher affinity and is therefore more likely to bind to the enzyme. Now, because the amount of enzyme is less than that of substrate, there are not enough enzymes to bind to the highly modified molecule. Therefore, the highly modified molecule is difficult to demodify and remains so for a long time. In other words, the phosphorylation or methylation is maintained for a long time by enzyme-limited competition.

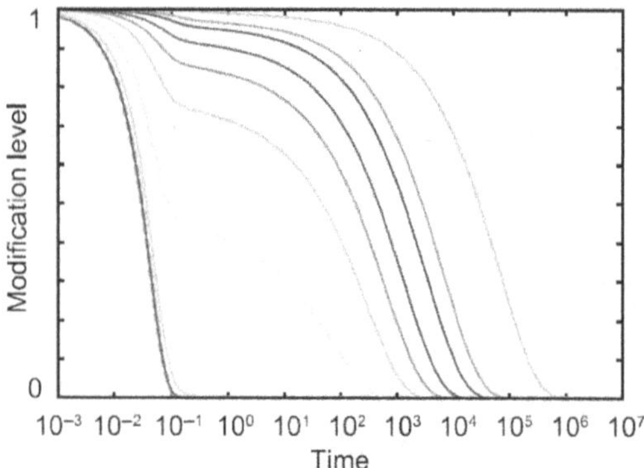

Figure 6.4 Slow, logarithmic relaxation in chain modification model. Initially with everything fully modified $[S_N] = [S]_{total}$, and $[S_i] = 0$ for $i \neq N$, we plot the relaxation in the degree of modification, where we set $N = 6$ and $S = 1$. The affinity (chemical affinity) K_i of the molecule at each modification level i is assumed to be $K_i = K_0 \gamma^i$, with $K_0 = 0.1 \times \gamma^{-6}$, $\gamma = 5$. Time variation of modification degree is plotted with the enzyme amount 10.0, 1.0, 0.5, 0.2, 0.1, 0.05, 0.02, 0.01, 0.001 from right to left (i.e., relaxation slows down in this order). If the amount of catalyst is greater than that of substrate, the relaxation is fast and exponential. If less, the relaxation slows down drastically. Based on Hatakeyama & Kaneko (2014a). (Courtesy of Tetsuhiro Hatakeyama).

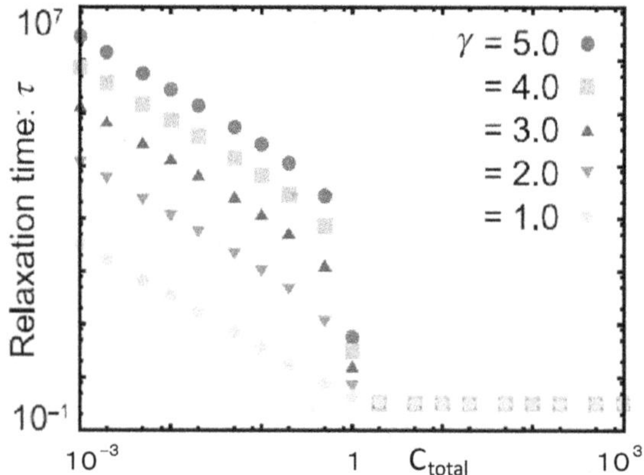

Figure 6.5 Dependence of the relaxation time τ on the amount of enzyme ($[C]_{total}$). Starting from the initial conditions in Fig. 6.4, the time for S_i to sufficiently approach the final state $[S_i]^*$ (specifically, the time for $|[S_i] - [S_i]^*| < 10^{-8}$) is plotted against $[C]_{total}$. The γ represents the degree to which affinity K depends on the modification level i. (When $\gamma = 1$, there is no dependence on i, and as γ increases, its i-dependence is more prominent. Specifically, as shown in Fig. 6.4, the affinity for i is $K_i = 0.1 \times \gamma^{i-6}$, and the results for $\gamma = 1.0, 2.0, 3.0, 4.0, 5.0$ are plotted. Based on Hatakeyama & Kaneko (2014a). (Courtesy of Tetsuhiro Hatakeyama).

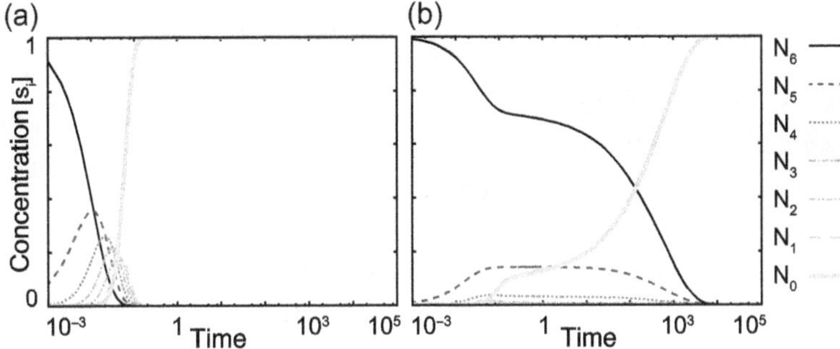

Figure 6.6 Time course of the concentration $[S_i]$ of protein modified up to stage i. Starting from the fully modified state, the temporal change of each $[S_i]$ is plotted. (a) For large enzyme amounts ($[C]_{total} = 2.0$): $[S_i]$ relaxes in order from the largest i. (b) When the amount of enzyme is small ($[C]_{total} = 0.1$): $[S_i]$ relaxes by increasing the concentration of proteins with a low degree of modification (small i) first. The parameters in Fig. 6.4 were used. Based on Hatakeyama & Kaneko (2014a). (Courtesy of Tetsuhiro Hatakeyama).

As seen in Figs. 6.4 and 6.5, this enzyme-limited competition occurs at the point where the amount of enzyme is less than that of the substrate. At this point, the relaxation time increases by orders of magnitude. Thus, a small change in the amount of enzyme around this point can control a shift between a state in which the memory is retained for a long time and the one that is erased in a short time.

Derivation of the logarithmic relaxation requires a more detailed analysis, but it can be roughly explained as follows. Note that the relaxation of the modified state at the k-step can be expressed as $exp(-t/\tau_k)$, whereas the relaxation time τ_k is written as $exp(E_k/(kT))$, and its affinity E_k is distributed. Then by summing the relaxation over all k, this slow logarithmic relaxation is derived approximately.

For the slow relaxation in this section to occur, it is necessary that:

(1) there exists a protein with multiple modification sites
(2) the reaction to change the (de)modification is carried out by an enzyme
(3) the enzymatic reaction has different affinity for each (de)modification
(4) the amount of enzyme is less than that of the substrate, resulting in enzyme-limited competition.

A detailed explanation of this mechanism with equations is given in the Appendix at the end of this chapter.

6.3 Glass-like Behavior in Chemical Reaction Networks (C and D)

In Section 6.2, we saw that a lack of enzyme amount causes a drastic slowing of relaxation and a kinetic memory. Here, we should note that the enzymes themselves are synthesized as a result of intracellular reaction within the cell. For example, if there is a feedback process so that the modified molecules inhibit the synthesis of enzyme, then once a modified state is formed, it would be expected to be fixed as a long-term memory. The memory would then be maintained as a system by a network of reactions.

Therefore, it is necessary to consider a system in which the amount of enzyme changes as a result of the reaction network. Here, we will first put aside for the moment the correspondence with reality and use the catalytic reaction network discussed in Chapters 3 and 5 to show that relaxation in this system slows down and that it has the properties of a glass in physics. This will give us a prototype of slow relaxation and memory in reaction network systems.

In this section, in order to make a clear comparison with the glass problem in physics, we use a system that is exposed to a constant temperature heat bath with no external material flow. In other words, since it is a closed system, relaxation to equilibrium is guaranteed at some point. In this sense, it may not be very suitable for the study of cells. This simplified setup is used here for the sake of correspondence with statistical physics.

(Nevertheless, the problem of maintaining a state in a closed system with no flow of matter or energy will also be important in living systems. Consider, for example, a plant seed, which does not have much exchange of matter with the outside world, but is not expected to reach equilibrium over a long period of time. As discussed in Chapter 3, if we consider the long-term persistence of a dormant state in bacteria and other organisms, even in a closed system without much flow of energy or matter from the outside, a state out of equilibrium can be maintained for a long period of time.)

In this section we will discuss whether such a nonequilibrium state can be maintained in a complex system of chemical reactions by relating it to a physical system such as glass, which does not easily fall into an equilibrium state.

We will adopt the catalytic reaction network used in Chapters 2 and 5. In this model, M types of catalysts (enzymes) $X_i (i = 1, \cdots, M)$ catalyze and react with each other (Awazu & Kaneko, 2009) (see Fig 6.7). In other words, changes in X_i and X_j are catalyzed by $X_{c(i,j)}$, as

$$X_i + X_c \underset{k_{j,i}}{\overset{k_{i,j}}{\rightleftharpoons}} X_j + X_c, \tag{6.1}$$

where $c(i,j)$ (which is neither i nor j) is the index of the catalyst for the reaction $i \leftrightarrow j$. It is chosen and predetermined from the components 1 to M. (Here it is assumed that the reaction network has other raw material components that are kept at a certain concentration and that these components are in a particle bath. The diversity of the raw material part of the network causes a reaction to occur from one component to several components). The reaction paths from each component are assumed to be only a few (as long as the reaction network is sparse, i.e., the number of paths from each component is much smaller than M, the result to be shown below does not change). The difference from the models discussed in Chapters 2 and 5 is that the reaction paths are no longer unidirectional. Here the reaction is bidirectional to satisfy detailed balance required to approach thermal equilibrium; this bidirectional reaction includes forward and backward paths to ensure relaxation to equilibrium. To do this, a chemical potential (energy) E_i is assigned to each molecule, and the reaction rates in the forward and backward directions satisfy detailed balance as,

$$k_{i,j}/k_{j,i} = \exp(-\beta(E_j - E_i)), \tag{6.2}$$

where β is the inverse temperature $1/(kT)$. In this case, according to statistical physics, the concentration of each component at equilibrium follows the Boltzmann distribution $x_i^{eq} \propto \exp(-\beta E_i)$.

The change in concentration of each component due to the reaction in Eq. (6.1) can be calculated by the rate equation as

$$\dot{x}_i = \sum_{j,c} Con(i,j; c) x_c (k_{j,i} x_j - k_{i,j} x_i), \tag{6.3}$$

where $Con(i,j; c) = Con(j,i; c)$ is 1 if there is a reaction path between i, j and 0 otherwise, and the reaction rate $k_{i,j}$ satisfies Eq. (6.2).[5] If we evolve this rate equation in time, $x^i(t)'s$ eventually approach the equilibrium state x_i^{eq} given by the Boltzmann distribution. Now, let us assume that the energy E_i is distributed between 0 and ε. Since the reaction rates in this model always appear in the form of βE_i pairs, $\beta\varepsilon$ is the fundamental parameter that determines the properties of the model.

Let us now look at the relaxation dynamics of x^i starting from initial conditions that are out of equilibrium. Specifically, let us start from the initial conditions where all components have the same concentration (in other words, where the temperature is (infinitely) high) and see how they relax to equilibrium. The degree of deviation from equilibrium $(C(t) = \{\sum_i (x_i(t) - x_i^{eq})(x_i(0) - x_i^{eq})\}/\{\sum_i (x_i(0) - x_i^{eq})^2\})$ is plotted as a function of time in Fig. 6.8.

As can be seen in the figure, if $\beta\varepsilon$ is sufficiently smaller than 1 (i.e., the energy distribution width is small or the temperature is high), there is a standard, quick exponential relaxation. However, if $\beta\varepsilon$ is larger than 1, the relaxation is much slower. Here, two salient properties are noted.

(1) Overall, the relaxation is logarithmic $(-\log(t))$ rather than exponential.
(2) There are several plateaus in the middle of the relaxation.

---

[5] For example, $k_{i,j} = \min\{1, \exp(-\beta(E_j - E_i))\}$.

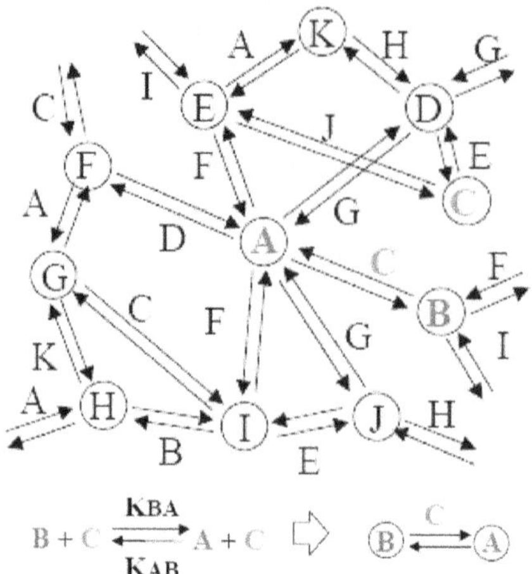

Figure 6.7 Catalytic reaction network model. Based on Awazu & Kaneko (2009).

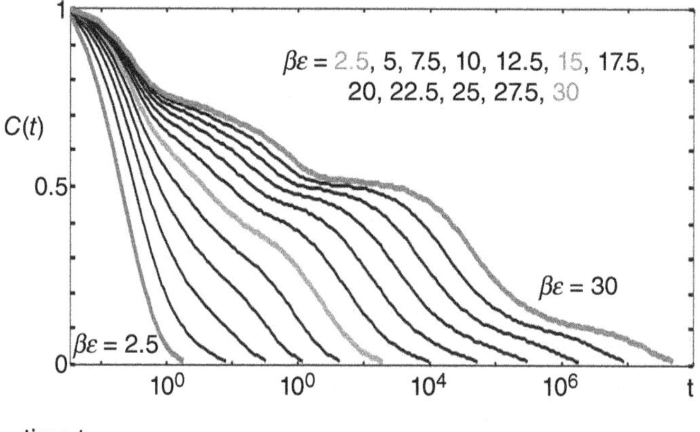

Figure 6.8 Example of time evolution of deviation from equilibrium. Time evolution of the degree of deviation from equilibrium $C(t)$. By using a random reaction network as in Fig. 6.7 and the time course of $C(t)$ is computed and plotted for different β values. Based on Awazu & Kaneko (2009).

In fact, such $log(t)$ relaxation with several plateaus are often found in glass studies. First, the $log(t)$ relaxation is derived as in Section 6.2. Since the relaxation times are distributed as $\tau(E) \sim exp(\beta E)$ depending on the energy of each molecule, the relaxation is given by a superposition of these $exp(-t/\tau(E))$, from which $log(t)$ type relaxation arises.[6] (Transformation to $z = texp(-\beta E)$ variable reveals the $log(t)$ dependence). For more details, see Awazu & Kaneko (2009).

___

[6]　The integration of $exp(-texp(-\beta E))$ over E leads to $log(t)$.

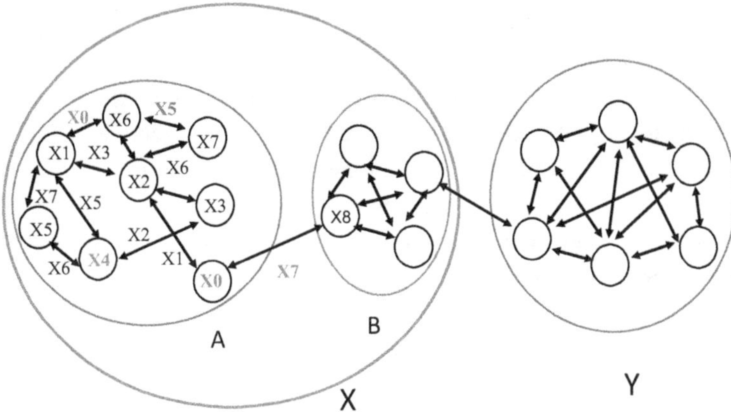

Figure 6.9 Schematic representation of the formation of locally equilibrated clusters.

More interesting is the origin of the plateaus. At each plateau, all components are divided into several clusters, within each of which equilibrium is approximately realized (see Fig. 6.9). This is not to say that there are no reactions that connect different "local equilibrium clusters" here. There are certain reactions in between, thereby satisfying the reach to unique equilibrium (as ergodicity is satisfied). However, the concentration of the catalytic component ($X7$ in the figure, for instance) responsible for such reaction between the clusters is quite low at the moment, so that the reaction does not progress well. Suppose, for example, that in schematic Fig. 6.9, local equilibrium is reached within each of clusters A and B approximately, but the component in A is relatively less abundant than in the equilibrium case by comparing with the component in B. Here, as the result of the reaction dynamics, the amount of catalyst $X7$ connecting between $X8$ that belongs to B with $X0$ that belongs to A is negatively correlated. Then, if the concentration in the members in B is relatively higher, the reaction to reduce them to approach the global equilibrium is suppressed, because catalyst $X7$ is less abundant. Thus, the **enzyme limiting** and the negative correlation between the amount of enzyme and substrate suppress the progress of the reaction toward equilibrium, and a plateau is formed. Of course, after sufficient time, the ratio of A to B approaches the equilibrium value, so that A and B form a single equilibrium cluster. If it spans all components, the thermal equilibrium state is achieved. If it still does not cover the whole components, the substrate-enzyme negative correlation by enzyme-limiting again suppresses the equilibration between larger clusters, say X and Y in Fig. 6.9. Then the next plateau is formed for a while. The relaxation plateau found in numerical calculations can be explained by this mechanism.

In a catalytic reaction network with a large number of components, the plateau formation described above is observed for large $\beta\varepsilon$, as long as the reaction paths in the network are not too dense. When $\beta\varepsilon$ is large, there is a large gap between the amounts of each component, and there are enzyme components with low amounts. If the number of components is large and the reaction paths are not dense (i.e., not all reacting directly with each other), then there will be some reactions catalyzed by enzymes with a low concentration, and some of these reactions will be negatively correlated with the amount of substrate. Then, the presence of plateau(s) and slow relaxation are universal, and thus the glassy relaxation is common in the catalytic reaction networks with many components.

In addition, even if there exists only few number of chemical species, networks showing such glassy relaxation can be easily constructed (Awazu & Kaneko 2009). In fact, even in a simple example with five components, plateaus and formation of local equilibrium clusters can exist. In the five-component catalytic reaction network shown in Fig. 6.9, we set the energy as $E_0 < E_1 < E_2 < E_3 < E_4$ and choose the initial conditions with the same concentration for all components. Hence, initially the concentrations of $X0, X1$ are lower than those in equilibrium, while those of $X3, X4$ are higher. In this case, the relaxation shows a two-step plateau as shown in the figure. Here, although the concentration of $X1$ is less than that at the equilibrium state, the reaction catalyzed by it proceeds quickly because of its large amount, and a local equilibrium cluster consisting of components 0, 2, and 4 is formed first. Subsequently, the reaction catalyzed by $X2$ proceeds and component 3 is added to this local-equilibrium cluster, and finally component 1 is added. Thus, a two-step plateau is formed. It is important to note that such plateau formation depends on the initial conditions. For example, in this example, if $X4$ is initially abundant, $X1$ can reach equilibrium with the others first.

In this case, there can be many different patterns of cluster formation and fusion, depending on the initial conditions. The relaxation path can be split due to fluctuations.[7] Slow relaxation is observed under many initial conditions, but different relaxation paths may be taken depending on the initial conditions. This diversity could be the capacity of "kinetic memory" that this network can have (Hatakeyama & Kaneko, 2019).

In this section we have seen that due to a negative correlation between the amount of a certain substrate and that of the enzyme needed for the reaction, the enzyme-limitation leads to maintenance of a nonequilibrium state for a long time. As discussed in Chapter 3, a living system is, in some sense, maintains and

---

[7] Since the arrangement of the plateau can be changed depending on how the initial excitation is performed. This has recently been investigated in a stochastic version of the model in Section 6.2 (Hatakeyama & Kaneko, 2018).

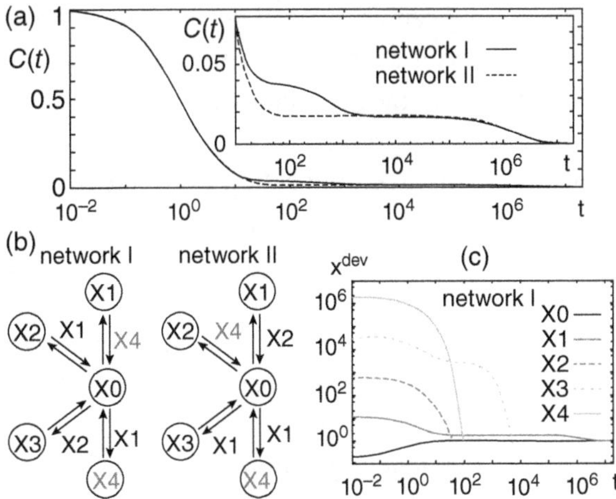

Figure 6.10 Slow relaxation (a) seen in the simple catalytic reaction network shown in the figure. (b) Deviations from equilibrium $C(t)$ are plotted versus time. The subfigure (c) plots $\log(X_i(t)/X_i^{eqb})$ how much each component X_i deviates from equilibrium. Locally equilibrated clusters are formed in two steps. Based on Awazu & Kaneko (2009).

reproduces a nonequilibrium state. If a state that is "harder to fall into thermal equilibrium" can be created and is reproduced before the system falls into equilibrium, it will be transferred to the next generation, repeatedly. This may help us understand how life is possible.

6.4 Epigenetic Fixation to Slower Time Scales (C and D)

So far, we have seen that slow time-scale phenomena are generated in biochemical reactions. In fact, different hierarchies exist in biological phenomena: the level of gene expression, that is, the scale of change in protein levels; slower epigenetic changes, for example, due to modifications of DNA molecules; and much slower (rarer) phenomena such as genetic evolution (see also Fig. 2.7). Memory, then, can be regarded as the process of embedding fast phenomena into slower processes within this hierarchy (Fig. 6.10). Kinetic memory, as we have seen in Sections 6.2 and 6.3, embeds a fast chemical reaction process into a slow-changing process.

In the case of neural systems, fast phenomena include the firing of neurons, and slower dynamics include changes in synapses connecting neurons. In neural networks, the Hebbian rule, which changes the strength of synapses depending on the correlation in firings of two neurons, was proposed as a mechanism to generate memory by embedding correlates of neural firing into synapses (Hebb, 1949). Although various extensions and modified versions have been proposed since then, Hebb's basic idea has been preserved. In the case of cellular epigenetic memory, then, how can changes in gene expression on a fast time scale be embedded in slower processes such as DNA methylation and histone modifications? Can we propose a general mechanism that corresponds to the Hebbian rule in this case?

In the case of neural circuits, the connection between two neurons exists that transmit firing of neuron to the other. If the change in the strength of the connections is slower than the firing dynamics, it is possible to embed memories in them, which makes the Hebbian rule possible.

On the other hand, in the gene expression system in cells, the relationship between expression of one protein and another is determined by a chemical reaction network. In contrast to direct physical connection by synapses in the neural network, the connection in the gene regulation network is via reaction due to collision of molecules. Then, it is unlikely that the connection strength, that is the reaction coefficient itself, can be changed by learning.[8] Therefore, to consider memory embedding in the epigenetic processes, we need to introduce a concept different from the Hebbian rule (Furusawa & Kaneko, 2013).

To this end, we adopt the gene regulation network model as discussed in Chapters 4 and 5, given by $dx_i(t)/dt = F(\sum_j J_{ij}x_j - \theta_i) - x_i$. As explained in Chapter 4, $F(x)$ is a nearly step function of x. As x goes larger in the positive side, it approaches 1, and as the magnitude of x is larger on the negative side, it approaches 0.[9] This threshold was treated as a constant parameter in the models of Chapters 4 and 5. Let us replace it by a variable $\theta_i(t)$, and assume it can slowly change in time.[10] For example, if a gene is masked by a histone modification and as a result is less likely to be expressed, $\theta_i(t)$, the threshold for the expression, is increased. Now, these epigenetic changes depend on the expression state. The change in expression gives feedback to the change in $\theta_i(t)$. Specifically, we consider a positive feedback process that makes the expression more feasible when it is expressed and harder when it is not expressed (Dodd et al., 2007). This can be realized, for example, by taking

$$d\theta_i(t)/dt = -a(x - \Theta_0) - \theta_i. \tag{6.4}$$

If x_i is below the threshold and the protein is not expressed, then the threshold increases toward Θ_0, and if x_i is above the threshold, the threshold is

---

[8] If we consider a long time scale including evolution, it may be possible to change due to the change in enzyme activity, though.

[9] If we write the response in Hill form with Hill coefficient n, the expression of $X^n/(X^n + K^n)$ (active) and $K^n/(X^n + K^n)$ (repressed) for the input X. In this case, the value of K corresponds to the threshold value and approaches a staircase function as n becomes larger.

[10] From the chemical reaction-based model up to Section 6.3, one may feel a gap in the following model, because the slow process in this section is abstracted as a threshold change for gene expression. If we consider that the epigenetic process is due to a molecular modification, this would indicate a slow change as described in Section 6.2. Connecting a molecular model in Section 6.2 with the present model is a future problem.

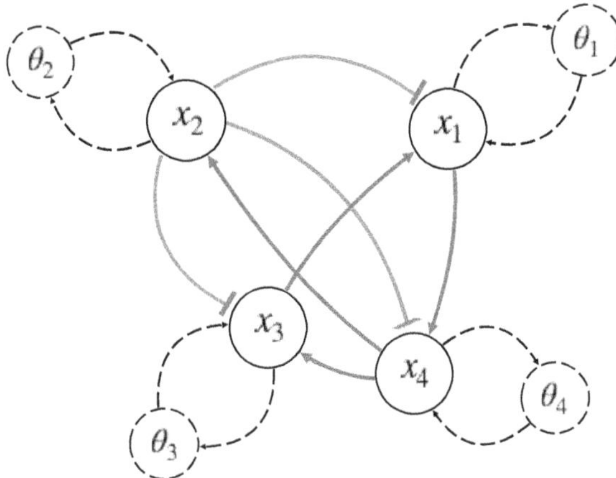

Figure 6.11 Schematic representation of a gene regulation network model with epigenetic feedback regulation. The dynamics of protein expression levels x_i are mutually regulated with activation (arrow) and inhibition (stop arrow), whereas the threshold θ_i for it changes with mutual positive feedback with x_i.

lowered. Here, a represents the strength of this epigenetic feedback. (see Fig. 6.11 for schematic representation). This embeds gene expression into the epigenetic modification.

This feedback reinforces the expression or suppression, thus stabilizing the expressed or suppressed state. Thus, compared to the original fixed-threshold gene expression dynamical system, novel stable states can be created and stabilized. This allows for both plasticity and stability by initially adopting a changeable state and then stabilizing it through epigenetic fixation.

Let us introduce this feedback mechanism into the attractor selection in Chapter 4. Specifically, the model in Section 4.4 is given by

$$dx_i/dt = v_g(F(\sum_j W_{ij}x_j - \theta_i) - x) + \eta_i. \tag{6.5}$$

Now this threshold θ_i changes according to Eq. 6.4. Let v_g be, for example, $v_g = v_{max}exp(-\sum_{i=1}^{N_{target}}(x_i - X_i^{target})^2)$ and assume that the growth rate increases if the expression is close to the target pattern. In Section 4.4, we discussed the mechanism by which an attractor is selected if there exists an attractor close to the target pattern and if the noise is of appropriate magnitude. The challenge then was that an adaptive attractor must exist in advance.

This was the case when the threshold θ_i was fixed. In contrast, as θ_i changes with the above feedback, it forms a new x_i expression pattern that can be stabilized by the change of θ_i. Once an expression pattern close to the target is formed, the attractor selection as described in Chapter 4 is expected to work to fix the adaptive state.

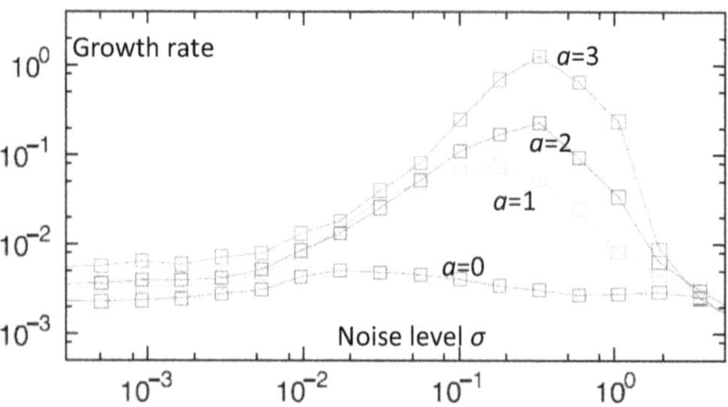

Figure 6.12 Adaptation to different environments through epigenetic feedback. Growth rate increase plotted as a function of noise. Results for feedback strengths $a = 0, 1, 2, 4$. Higher growth rates are stabilized as $a > 0$ increases if the noise is moderate. Based on Furusawa & Kaneko (2013). (Courtesy of Chikara Furusawa).

We simulated, this model. Initially, the expression pattern changes significantly due to noise. Eventually, as the expression pattern approaches the target, v_g becomes larger, and the effect of noise turns to be smaller and the pattern remains stable. When v_g is large, the time scale for θ change is larger than that for x, so that the feedback of θ works to fix this state. Thus, θ_i is fixed in a state where the expression pattern matches the target. As a result, it can adapt to many environments. Figure 6.12 plots how much the growth rate increases under noise for different feedback strength a. Compared to the case where θ is constant ($a = 0$), the growth rate v_g increases as a increases and feedback becomes stronger (Furusawa & Kaneko, 2013).

Further, by taking into account of oscillatory dynamics in gene expression, the exploration of adapted state and epigenetic fixation works more efficiently, as has recently been shown (Matsushita & Kaneko, 2023). Furthermore, gene expression dynamics with randomly connected regulation networks often show chaotic oscillations, if the number of genes is large (say more than 50). Then an adaptation mechanism based on fast (chaotic-)oscillation in gene expression dynamics coupled with a slower epigenetic fixation process is investigated: The chaotic oscillatory dynamics are relevant to the search for optimal states depending on the input, and once optimal states are approached, a slower epigenetic modification process fixes those states. As long as the search by using oscillatory (chaotic) dynamics adequately covers the state space, this adaptation mechanism works efficiently. The degree of chaos, or the region of the state space traveled by orbits, is correlated with the capacity of environments that the cell can adapt to. As the number of genes increases, the fraction of networks allowing for such dynamics also increases, supporting the generality of the proposed mechanism.

The present adaptation process can also be depicted by Waddington's epigenetic landscape. The initial chaotic-oscillation state is located in a shallow valley in the epigenetic landscape, during which the state travels over a large portion of the state space. As the slow variables change, deeper valleys are generated to which the state is attracted, leading to the adaptation to the environmental condition in concern.

Notably, the mechanism works robustly under strong stochasticity, whereas it does not require attractors that achieve optimization in advance. Hence, it would also support the generic adaptation of cells to unforeseen environmental conditions as suggested by Braun (2015). Furthermore, as the present mechanism adopts a simple setup for oscillatory dynamics with on-off type dynamics and a slower fixation process, it may be applied to neuroscience as well.

In this section, epigenetic memory is taken as an example to link between macroscopic states of cells (such as protein abundance composition) and microscopic states at the molecular level, such as histone modifications in DNA. In a sense, this section bridges Section 6.2 with Section 6.3 of the present chapter. In the model of Section 6.2, the amount of enzyme is given, but in the cell, the amount itself changes under the influence of other reactions, as in Section 6.3. This also makes it possible to stabilize memory by linking the molecular modification and the chemical composition in the cell, that is, by achieving macro-micro consistency. In this section, we introduced it as a feedback process between modification θ_i and protein expression level x_i, but it will be necessary to investigate specific reactions in the future.[11]

6.5 Biological Significance of Kinetic Memory and Experimental Verification (E)

Before discussing the possibility of experimental validation, let us once again discuss the cell-biological significance of kinetic memory in this chapter, compared to the standard picture of memory as multiple stability.

(1) Erasing memories is easier. In the case of memories stored in multi-stable states, erasure of memories generates heat. As the erasure process is faster, the heat generation is larger, and thus the process needs more cost. In contrast, kinetic memories can be easily erased by relieving enzyme limitation that slows down the process.

(2) Kinetic processing and memory embedding are easy. Since the kinetic memory is originally a dynamic process that progresses with a slow time scale, it can easily interact with other dynamic processes and is suitable for gradually

---

[11] For an epigenetic adaptation model that considers microscopic modification processes, see (Himeoka & Kaneko, 2019).

embedding fast changes in the external world into slow internal processes. Such embedding of faster external process into slower internal processes is essential for cell differentiation in Chapter 7 and for phenotypic evolution in Chapters 8 and 9.

(3) Continuous (analog) memory is possible, rather than digital memory such as binary on/off. This is a different type of memory from digital memory, and this form of memory will be important in living systems.

(4) Of course, there is a drawback: the gradual loss of memory (in $log(t)$ relaxation) may be a weak point. Instead, it would be possible to use this kinetic memory as a short-term memory to mediate transfer to the long-term memory of multi-stability.

At present, there has been little direct experimental verification of the kinetic memory described here. This lack of evidence of this picture may be due in part to preconceptions that memory must be represented by multistability and that relaxation be exponential decay (as measured by the half-life time assuming this exponential decay). One way to confirm the kinetic memory discussed here would be to find a logarithmic relaxation, which is slower than exponential, and another way would be to check whether the enzyme abundances can regulate the memory duration.

For example, in the neural system, long-term potentiation (LTP) is important for memory. As described in Section 6.1 of this chapter, early potentiation does not involve protein synthesis, where calmodulin kinase II (CaMKII) and other enzymes are thought to be important (Lisman & Goldring, 1988; Lisman et al., 2002). On the other hand, cAMP-responsive element-binding protein (CREB) and others are considered to be important in the later stages with protein synthesis (Silva et al., 1998).

It is interesting to note here that CaMKII and CREB are also multimeric proteins with many phosphorylation or modification sites, as schematically shown in Fig. 6.3. Synaptic plasticity in the neural system that is responsible for neural memory is thought to be related to the phosphorylation state of this hexameric CAMKII. Researchers examined if this phosphorylation state involves bistability, but so far no support for it has been obtained (Lisman et al., 2002). In this sense, it would be interesting to experimentally verify the results of Section 6.2 of this chapter in this system.

Furthermore, it has been suggested that the phosphorylation state of CaMKII can be controlled by known signaling systems involved in synaptic plasticity. In this sense, as described in Section 6.4, the intracellular state and the modification state of the related molecule may be mutually stabilized to form a long-term memory in which macro (i.e., cellular) and micro (i.e., molecular) states are consistent.

Thus, it will be important in the future to consider the basis of memory in the neural system from the standpoint of the consistency of intramolecular modifications and intracellular chemical composition, and to experimentally verify such modifications.

Of course, kinetic memory still has many issues to be elucidated theoretically and to be verified experimentally in the future. For example, in the case of multi-stability, the selection of two states gives a one-bit memory unit, and it has been clearly investigated how much memory capacity a given system can have. In contrast, the theory of kinetic memory capacity remains elusive.

6.6 Summary: Universal Properties Discussed in This Chapter

- Digital memory is understood as a different stable state, but in the cell, there is another memory form as a continuous (analogue) change (E).[12]
- Cells are capable of maintaining long-term memories beyond the genome, which are generally referred to as epigenetics. Some of this is thought to be due to modifications of molecules such as DNA (E). Hence, there exist fast phenotypic changes, and slower epigenetic modification with an intermediate-time scale, and much slower genetic change. These processes with different time scales interfere with each other, to embed fast changes into slower state changes (A).
- Referring to the kinetic constraints of glass theory, one can think of kinetic memories with slow relaxations as an alternative to the conventional memories of multi-stable states. They are characterized by slow logarithmic change with several plateaus that can be taken during the relaxation process (A, C).
- When the same enzyme catalyzes a stepwise reaction and if the amount of the enzyme is not sufficient, the reaction process can be hindered by enzyme-limited competition, resulting in kinetic memory in the sense described above (B).
- In a network of catalytic reactions, a negative correlation between the amount of substrate and enzyme can be created thereby allowing a slow relaxation process with many plateaus as described above, where multiple states can be maintained over a long period of time (C).
- It is possible to shape epigenetic memory by embedding intracellular states into molecular modifications that reinforce each other, thereby allowing adaptation to different environments that have not been experienced before. This allows both plastic change and robustness of the adapted state (B, C).

---

[12] Because all the discussions in this chapter are related to time, (D) is not specifically added.

6.7 Appendix: Chain Modification Model for Kinetic Memory

Consider a substrate S with N modification sites and a catalyst C, where S_i and CS_i are the substrate and substrate-catalyst complex with i sites modified, respectively. Let $[S]_{total}$ and $[C]_{total}$ be the summed total amount for each i, both of which are conserved. Then the change in each concentration, S_i and CS_i, is given by

$$S_i \xrightarrow{a_{i+1}} S_{i+1}, \tag{6.6}$$

$$S_i + C \underset{k_{j,i}^-}{\overset{k_{i,j}^+}{\rightleftharpoons}} CS_i \xrightarrow{b_i} S_{i-1} + C. \tag{6.7}$$

Now, let $[C]_{free}$ denote the amount of free enzyme (i.e., that is not making a complex). Then, the time evolution of the amount of each substrate is given by

$$\frac{d[S_0]}{dt} = -a_1[S_0] + b_1 \frac{[C]_{free}[S_1]}{K_0 + [C]_{free}} \tag{6.8}$$

$$\frac{d[S_i]}{dt} = -a_{i+1}[S_i] + b_{i+1} \frac{[C]_{free}[S_{i+1}]}{K_{i+1} + [C]_{free}} + a_i[S_{i-1}] - b_i \frac{[C]_{free}[S_i]}{K_i + [C]_{free}} \tag{6.9}$$

$$\frac{d[S_N]}{dt} = a_N[S_{N-1}] - b_N \frac{[C]_{free}[S_N]}{K_N + [C]_{free}} \tag{6.10}$$

$$[C]_{total} = \sum_{i=0}^{N} \frac{[C]_{free}[S_i]}{K_i + [C]_{free}} + [C]_{free}, \tag{6.11}$$

where $K_i(= k_i^- / k_i^+)$ is the dissociation constant of the S_i and complex C, which is assumed to increase with i, that is, $[S_i][C]$ is easier to be detached as i is larger. Specifically, we choose the form $K_i = K_0 \times \gamma^i$. From the above equations, we can calculate how the modification level ($\sum_{i=0}^{N} \frac{i[S_i]}{N[S]_{total}}$) relaxes toward equilibrium.

7

Cell Differentiation through Development

7.1 Introduction

As cells grow and divide, they form a cluster of cell aggregates. This inevitably depletes nutrients, eventually hindering the growth and maintenance of such cluster of cells. In this case, as discussed earlier, the expected path will be diversification of cells (see also Fig. 1.1 in Chapter 1). If identical cells compete for the same resources, the resources will soon be depleted. A possible remedy for this is the aggregation of cells with different genetic backgrounds that do not compete for the resource with each other, but rather help maintain one another. The ecosystem can be considered an example of such a mutualistic system. A biofilm consisting of a wide variety of bacteria is a small-scale example of this. In this example, for the maintenance of such system, the growth of each type of organism should be balanced. Otherwise, one type of organism (cell) drives others out.

Alternatively, consider the cell division process starting from a single cell. In this case, the cells share the same genetic background. Still, if they diversify in their phenotypic states, the competition will be relaxed. We can see such examples in a multicellular organism, where a population of cells take different cell types, even though they share common genes. Furthermore, if a single cell (few cells) is taken out of this population and if the proliferation leads to a similar cell-aggregate, we may expect a higher-level reproduction system, that is, the recursive proliferation of the cell aggregate. This is thought to be the origin of multicellular organisms.

Macro-micro consistency

In fact, in the developmental process of multicellular organisms, cells with complex chemical reaction processes differentiate into different states as the number of cells increases, and the fraction of each differentiated cell type is rather robust to perturbations. How, then, is this robust differentiation process possible?

181

First, replication and differentiation must be compatible. The former is a stable reproduction of the same cellular state, that is, the stability of the state, whereas, the latter is the change to other cellular states, referred to as the plasticity. Of course, the latter may require some extracellular stimuli, but in any case, both replication and differentiation occur together in almost the same environment. How is it possible for them to be compatible?

Stemness: compatibility between plasticity and stability

Multicellular organisms usually have cells that can form other types of cells, called stem cells. Stem cells can both self-renew and differentiate into other cell types. While stem cells can self-replicate, they also have the plasticity to differentiate into other types.

Alternatively, as the cells differentiate, they eventually reach a state that produces a determined cell type that can only create itself. This is a commitment to differentiation. Once a cell is committed, its cellular state is stable against cell division and replication. Hence, the questions to be asked are how is cellular plasticity lost and determined cell states emerge as the cell population grows?

Plasticity and pluripotency for differentiation

The range of possible differentiated cell types is dependent on the stem cell type. In other words, there is a difference in the range of cell types that stem cells can produce, that is, pluripotency for cell differentiation. At the top level, there is a totipotent cell that can produce all cell types in an individual. Next, pluripotent stem cells can produce some, but not all, cell types. For example, hematopoietic stem cells produce all types of blood cells, such as white and red blood cells, and platelets. Neural stem cells produce all the cells of the nervous system, such as neurons and glia. Here, hematopoietic stem cells do not generate neurons during normal development. The ability to differentiate into other types of cells declines with development.

In this sense, in a "normal developmental process," there is irreversibility in the degree of differentiation potentiality (Kaneko, 2006; Forgacs & Newman, 2005).[1]

Roughly speaking, the capacity to change into other cellular states (pluripotency or potency) should be related to plasticity. The "plasticity" of a cell can be considered the "changeability," that is, how the state of the cell changes in response

---

[1] When not in "normal" development, for example, under some external manipulation, cells may reverse the process of differentiation. This was reported earlier in plants, but in animals, Gurdon et al. (1975) first succeeded in reversing the differentiation in somatic cloning of frogs. They transplanted the nuclei of the committed (determined) somatic cells into another cell type, from which they regained pluripotency. Subsequently, somatic clones were created (through starvation) in mammals, as well as by Campbell and Wilmut et al. (Campbell et al., 1996). Then, Takahashi and Yamanaka succeeded in restoring pluripotency by activating the gene expression of some of the determined cells (fibroblasts). These cells were named as induced pluripotent stem (iPS) cells (Takahashi and Yamanaka, 2006), as will be discussed in Section 7.6.

to changes in the external environment. The larger the plasticity, the greater the pluripotency. Therefore, plasticity can be used as an indicator of differentiation. The differentiation process can be understood as a loss of plasticity. In this case, one needs to address the question of how this plasticity is represented in terms of gene expression in the cell. If we can answer this question, we will be able to understand what is required for the operation to reverse the differentiation process, that is, reprogramming.

Robustness in the developmental process

The developmental process is surprisingly robust. As we have seen, the cell types produced by complex developmental processes, their states, the number fraction of each cell type, and their spatial arrangements are robust, despite significant fluctuations in intracellular gene expression processes. This robustness is hierarchical. Each cell state remains within a certain range, despite some variance in the gene expression patterns, implying "cellular-level" robustness. As for the number of each cell type in a cell population, the ratio remains within a certain range against noise and disturbances during development. Thus, the developmental process is robust at both cellular and multicellular levels.

As mentioned already, the "attractor hypothesis" of cell types explains the robustness of each cell state. In the case of an attractor, the cellular state returns to its original state even if there is perturbation by noise. To explain the robustness of the cell population, such as the presence of each cell type in a tissue and the stability of the number ratio of each cell type, we must understand the hierarchical consistency between the single-cell and the cell-ensemble states. This requires consideration of the interplay between internal cellular dynamics and cell–cell interactions within the cell population.

7.2 Possibility in Macroscopic Phenomenological Theory (A and E)

Waddington (1957) schematically represented the developmental process by the motion of a ball along the landscape with branching valleys, as shown in Fig. 7.1. The ball falls down to the deeper valley, which corresponds to a stable cellular state. With developmental time, the valleys in the landscape successively branch. As a result, there are multiple valleys within which the ball can settle. Because these correspond to stable cellular fates, the landscape represents the differentiation of a cell into multiple states. Waddington coined the term "epigenetic landscape" for this figure. The horizontal axis represents the cellular state giving rise to the cellular type, and the depth axis is considered to be the development time. This picture has seen a revival recently and has gained much attention.

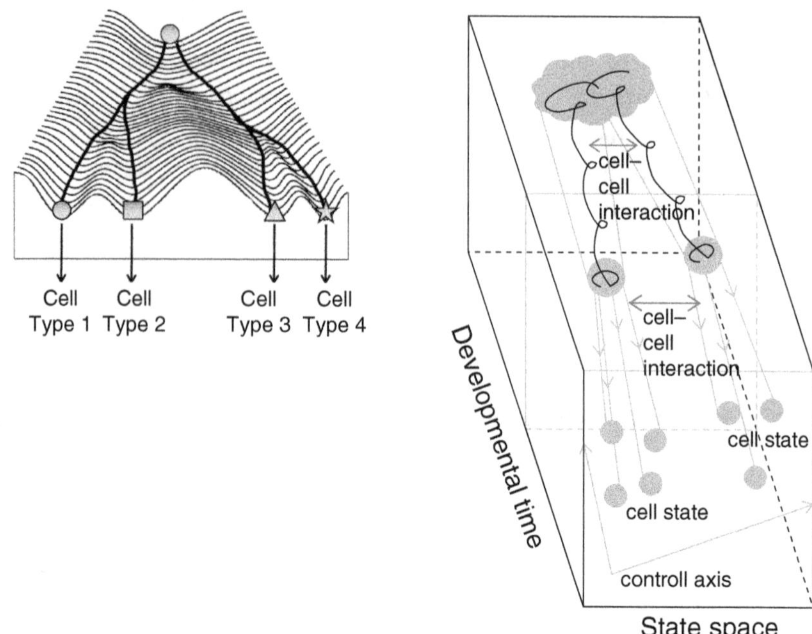

Figure 7.1 (a) Schematic representation of the epigenetic landscape and cell differentiation by Waddington. Valleys in the landscapes are branched and deeper with developmental time. They correspond to differentiated cell types (Waddington, 1957). (b) Waddington's complex phase-space diagram is redrawn under the present dynamic systems picture, as presented in the present chapter.

Here the horizontal axis represents the cellular state, possibly represented by the gene expression pattern. Then, as the stable state of each differentiated cell type settles at the bottom of the valley, the height of the landscape probably represents the plasticity (changeability) of the state. This may fit the attractor picture of a cellular state: The cellular state is attracted to the bottom of valley, at which the state reaches a fixed-point attractor, and no more change occurs. Here, however, the landscape itself is shaped along the depth axis representing the developmental course, in which the valleys are branched successively and are deepened in a process known as *canalization*.

Now, can we extract the landscape from the experimental data and understand it theoretically? Waddington sought to uncover the possible mathematical structure behind this diagram, as he called the phase-space diagram of development in his book (redrawn in Fig. 7.1(b)). It schematically shows how the state of cells is attracted to each cell type with developmental time. Waddington also organized a series of meetings, by collecting renowned mathematicians, including Rene Thom, to understand the landscape picture in terms of a field of mathematics, now known as a dynamical system. Furthermore, Waddington's book contains a figure entitled

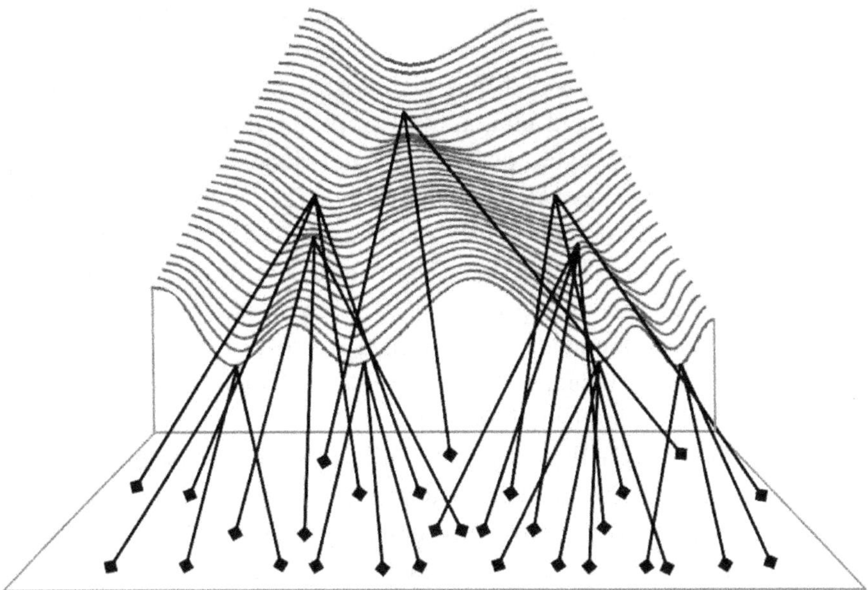

Figure 7.2 Schematic of landscape shaping by genes, represented by strings. The bottom plane of the figure depicts the genes that control the landscape. Adapted from Waddington (1957).

complex system of interaction underlying the epigenetic landscape, where interactions among genes control the landscape, metaphorically represented in the figure by strings (Fig. 7.2).

In the 1960s, Goodwin and Kauffman studied the cellular state as an attractor of protein expression dynamics (Goodwin, 1963; Kauffman, 1969). In these studies, the cellular state represented by the concentration of various protein (mRNA) species is considered, which mutually regulates their expression, as will be discussed in Section 7.4.

However, even if each valley is represented as an attractor of gene expression dynamical systems, the question remains: Since there are thousands of genes (or components) in a cell, the state of gene (protein) expressions should involve a large number of degrees of freedom and the dimension of the state space should be extremely high. In contrast, the valley of Waddington's landscape is represented by a single horizontal direction. How is a one-dimensional horizontal axis accounting for a macroscopic cellular state extracted? How is the landscape height represented, and how does the landscape change along the depth direction?

Looking at the entire developmental process, one might think it would be too simplistic to represent the process by such simple landscape. Nevertheless, it is often the case that the cell population of a given tissue or group can be represented by the branching of valleys in the landscape along with one or few directions. Indeed, Waddington's landscape is now taken seriously by experimental biologists. For example, the process of differentiation of cell populations in the nervous system, including neurons and glia, as well as the process of differentiation of blood cells, including leukocytes, platelets, and red blood cells, seems to be represented by branching to valleys along one direction. Then, the total developmental process with cell differentiation is represented in just few dimensions. Such dimensional reduction would not be possible if the entire description of the expression levels of all tens of thousands of protein species were needed by each to represent the cellular state. Thus, the landscape picture, a macroscopic description, implies that drastic dimensional reduction from large degrees of freedom of expression levels occurs in the developmental process.

Now, before going to the theoretical validation of the landscape picture, we first discuss if there are experimental data to support the existence of such macroscopic landscape parametrized by just one or few numbers. In fact, a series of experiments by Asashima et al. in constructing tissues from undifferentiated cells, suggests to support this picture, even though the extraction of the landscape may not be complete (Kaneko et al., 2008). In the next section, we describe these experiments and possible extraction of the landscape.

7.3 Tissue Construction Experiments from Undifferentiated Cells (A and E)

About two decades ago, Asashima et al. succeeded in constructing a variety of tissues by adding a population of undifferentiated cells of frog eggs into a solution of a protein molecule called activin for a certain time span. With this procedure, cell differentiation was induced, and tissues were derived (Ariizumi & Asashima, 2001). Specifically, cell populations taken from the animal cap at the animal pole of the egg were cultured in an activin solution (more precisely, activin A, which is a certain protein). Different tissues such as the myocardium, notochord, skeletal muscle, and heart were successfully generated depending on different concentrations of the activin solution under which cells were exposed. This study showed that multiple tissues were generated using a single control parameter, activin concentration. In addition, using retinoic acid (and in some cases a protein called ConA) and varying their concentrations, nearly all of the frog tissues were constructed. Using the axes of concentration of these components, different tissues (or cell types) can be ordered (see Fig. 7.3; ConA is not included here because it is needed only for just few exceptional tissues) (Kaneko et al., 2008). This means that tissues are represented on a two-dimensional phase diagram using two axes: activin and retinoic acid concentration.

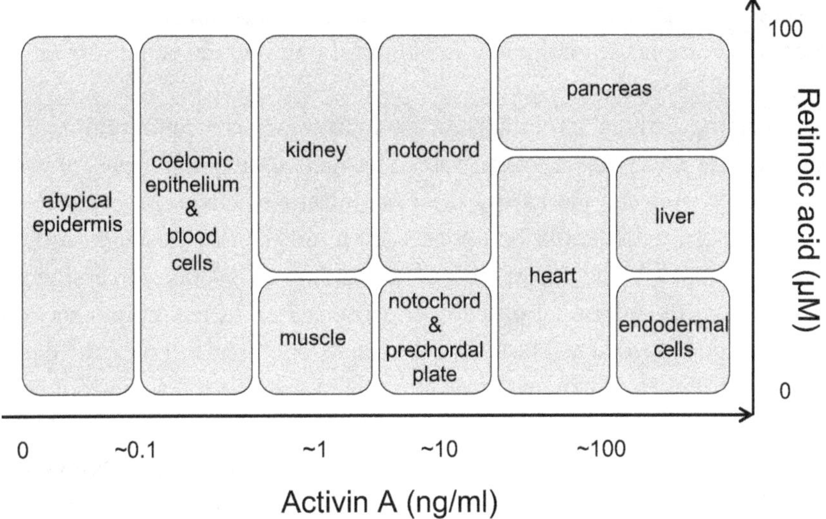

Figure 7.3 In vitro embryogenesis. A summary of the experimental results by Asashima's group (Asashima et al., 1989, 1990; Ariizumi & Asashima, 2001) is shown in the form of a phase diagram, in which different tissues are generated depending on the activin and retinoic acid concentrations. Xenopus presumptive ectodermal explants (animal caps) were isolated, and various concentrations of activin A (abscissa) were applied, followed by retinoic acid (ordinate), resulting in differentiation into various types of cells, tissues, and organs. Based on the study by Kaneko et al. (2008).

Interestingly, the order of the tissues on the axis of activin and axis of retinoic acid concentrations corresponded to the dorsal-ventral and anterior-posterior axes, respectively (although there was some slanting, i.e., with some angle): The order of the constructed tissues against the change in the concentrations corresponded to the spatial ordering of tissues during normal development. The control of the concentration of few chemical components works as parameters to allow for a stable arrangement of differentiated cells. What is even more interesting here is that although activin and retinoic acid play an important role in the normal developmental process, they do not take such high concentrations as were used in this experiment. The actual development is not controlled by these molecules alone, but by a more diverse set of components. However, as a result, the control of many components can be projected onto just the two concentration variables. It is important to note that the same tissue was formed as in the normal developmental process, even with these operations different from normal development. This suggests that tissues with a certain set of cell types exist as stable destinations and attractors in the dynamical system. Second, it is noted that the pathway to these stable states can be controlled by only a few parameters. This means that even though gene expression states have thousands of degrees of freedom and the developmental process is complex, biologically meaningful cellular states are regulated by a few degrees of freedom.

By immersing cells in a solution of activin at a high concentration, their states are destabilized, and paths to other states are opened, leading to the formation of differentiated tissue. By increasing the concentration of activin, paths to tissues that are harder to reach could be opened. It would then be possible to order the tissues according to the degree of their concentration. Now, one can assign a quantity to represent this ordering. Following the discussion in the previous section, this quantity may be regarded as plasticity (changeability). The initial undifferentiated state is highly plastic, whereas the differentiated states are less plastic, as in the bottom of the valleys. Between each valley, there is a hill whose height is larger if the valley is deeper. Now, by putting cells into the media with a higher concentration of activin, they can be kicked out to farther valleys intercepted by higher hills. In this way, the high and low plasticity of cellular states can be mapped to the concentrations of activins needed to generate the tissue of the states. Figure 7.4(a) is thus illustrated.

The landscape with its height as the plasticity in this figure corresponds to Waddington's epigenetic landscape, where the developmental time is represented by the depth axis in the figure, whereas the cell states were written on a horizontal axis. Previously, the one-dimensional coordinate was not specified. In the present figure, cell types are depicted in two-dimensional axes; the axes are given by the concentrations of activin and retinoic acid used in the experiment. These two-dimensional axes are correlated with the dorsal-ventral and anterior-posterior axes in the body.

Figure 7.4(a) shows the landscape after development, shaped by the development, whereas Fig. 7.4(b) shows the change of tissues along the direction of the activin axis (dorsoventral axis in the case of development), as it is difficult to see the change in the two-dimensional landscape. This figure shows the emergence and deepening of valleys in the landscape through development.

The obtained landscape is just a reinterpretation of the experimental results. Can we derive testable predictions based on this landscape picture of plasticity that can be verified by further experiments? Since plasticity is changeability, it is expected that when cells with higher plasticity are in contact with those with lower plasticity, the former change their states. We can check this prediction by taking a tissue at a deeper valley in the landscape and by contacting it in a shallower valley.

The tissue at a deeper valley is generated by placing cells in a solution of higher concentration of activin, whereas tissue at a shallower valley is generated through exposure to that at lower concentrations. These two tissues were physically contacted, to examine which are more subject to change. As a result of this experiment, the latter, that is the tissue at a shallower valley, was induced to change. Hence,

(a)

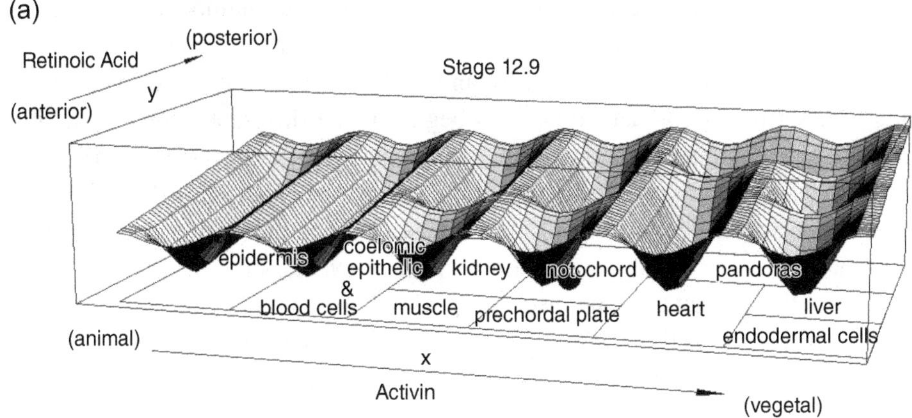

(b)

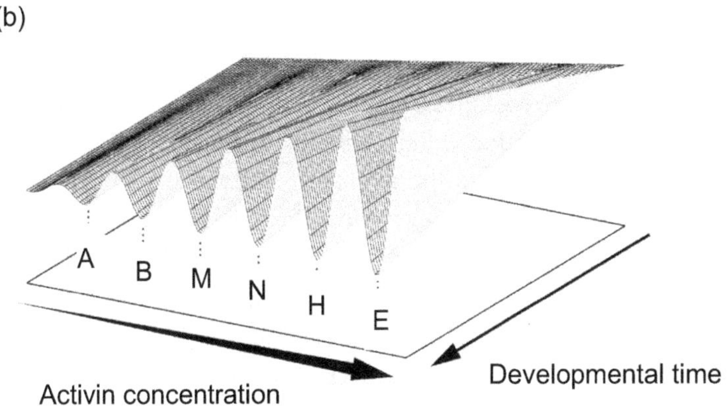

Figure 7.4 Developmental potential in relation to normal development and in vitro organogenesis caused by activin and retinoic acid (RA), with the corresponding fate map. The potential form at the end of the early developmental stage is plotted as a function of the two axes, which correspond to the concentrations of activin and RA. Based on Kaneko et al. (2008), (a) Two-dimensional landscape at the late stage of development, derived from Fig. 7.3. (b) Change in the developmental potential landscape. With developmental time, the deeper valleys were shaped. The concentration of RA was set to 0, and the epigenetic landscape was plotted. From the two-dimensional landscape, changes in the landscape only against the activin concentration are plotted. A, B, M, N, H, and E correspond to the tissues in Fig. 7.3 at the null retinoic acid.

the expectation from the theory that assumes the landscape height as plasticity, ordered by the activin concentration, is confirmed (Kaneko et al., 2008).

In normal development, it is observed that more plastic tissue is made to change under contact with less plastic cells. The most notable example of this is gastrulation: as the development progresses and the cell number increases, the aggregate of cells is folded, similarly to the buckling in mechanical processes. Then, cells that were located distantly and took different states are made to be contacted. With this contact, some cells begin to take different states. By taking the above landscape picture on plasticity, cells that have been contacted for a long period, that is, cells that have first started contact, are expected to be differentiated by going across higher hills in the landscape. Therefore, it is expected that the tissue order by activin concentration in the above experiment is in line with the order of the contact time in this developmental process. The ordering of tissues by gastrulation and activin-induced organogenesis agrees, as in the order of the notochord, kidney, and muscle. This suggests the validity of the idea of the order of plasticity.

Now, is there some quantity to characterize the plasticity deduced from the experiments? The plasticity corresponds to the degree to which the intracellular state is feasible to change in response to external signals; if the responsive change is larger, the plasticity will be larger. As a measure of such responses, the term competence is sometimes adopted in cell biology. This is the degree of change in gene expression by extracellular signaling molecules (Grainger & Gurdon, 1989). It is the rate of change in the degree of internal gene expression affected by external signaling molecules. In physics, this can be regarded as the response ratio, or susceptibility (e.g., the ratio of change in magnetization against the change in the external magnetic field). However, in the case of competence, the expression change is not necessarily given by the change in the concentration of a single specific molecule species, and so far, it has not been explicitly defined. Competence refers to the global change in all expression changes induced via a complicated signal transduction network through the receptor in the membrane. However, as will be discussed in the following chapters, we may expect the existence of few "macroscopic state variables" of the cell. We may expect the existence of a macroscopic quantity to characterize the plasticity or competence.

7.4 Coupled Dynamical Systems for Cell Differentiation (B, C, and E)

7.4.1 Waddington's Landscape and Dynamical Systems Model

As already mentioned in Section 7.2, Waddington suggested that this differentiation process can be understood in terms of the dynamical systems of gene expression in each cell (Fig. 7.1(b)). Following his insight, each valley is proposed to be interpreted as an attractor of an intracellular dynamical system for gene (protein) expression. The cellular state is represented by a set of gene (protein) expression levels that are reached by an attractor of such dynamical system. If the dynamical system has multiple attractors (several stable states), each of them

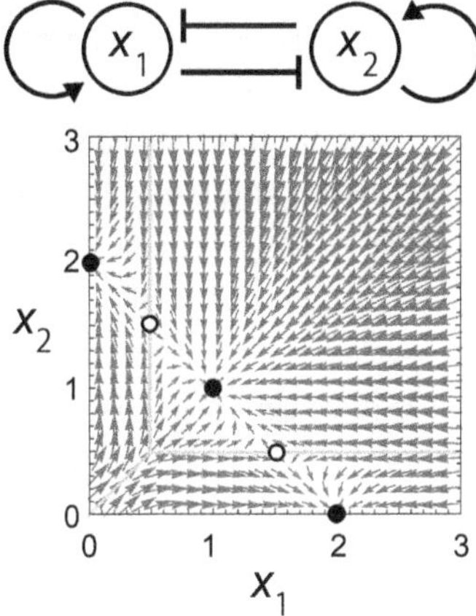

Figure 7.5 Example of a two-gene network with multiple attractors. X_1 activates its own expression and represses that of X_2, whereas X_2 activates its own expression and represses that of X_1. The dynamics of the concentrations of proteins X_1 and X_2 are represented by flows in a two-dimensional plane. There are three attractors, as shown by black circles in the figure. Based on Huang et al. (2009). The region intersected by the solid lines represents the basin for each attractor. (Courtesy of Jumpei Yamagishi.)

corresponds to a distinct cell type, and the process of falling into each valley is represented by the approach to each attractor, as already mentioned. Then, the cellular state remains in the vicinity of the attractor, even under internal noise or external perturbation.

As a simple model of such dynamical-systems with mutual activation and inhibition of protein expression, consider the network of two genes that regulate each other's expression, as shown in Fig. 7.5 (Huang et al., 2009). The state of the cell is represented by the dynamical system of the concentration of each protein, X_1 and X_2. The change in their concentrations is represented as a flow in a two-dimensional space. Depending on the initial conditions, the cellular state moves toward either of the three fixed points (attractors). Thus, it is possible to regard these three attractors as different cell types. If we adopt a more complex gene network, it is possible to create a system that has more attractors.

This view of each cell type as an attractor of a dynamical system is important and basic for considering the stability of cell types. However, this is not sufficient to fully understand cell differentiation. By this view alone, it is not enough to understand how the initial conditions for a cell to fall into each attractor are determined by the cell; therefore, the selection of the initial condition to provide each cell type is not included. In the attractor picture above, one possibility is to include noise in the gene expression dynamics. The cell state can then be switched to a different attractor by noise. The transition from one cell type to another is possible if each attractor has a different degree of stability against noise. Although this is one possible solution, further problems remain. First, the initial cellular state must be stable up to a certain stage, and then, as development progresses, differentiation needs to start. For this, the noise intensity should be appropriately tuned. It must be large enough for the stem cells to switch to different attractors. Still, since stem cells can self-renew, the noise level should not be very large. To achieve irreversible differentiation, the noise level would be somehow decreased, for retaining the cells in their states. Second, the population fraction of each cell type must be maintained within a certain range for the stability of the tissue level (cell ensemble level). Such stability crucially depends on the cell–cell interaction. To address this issue, study by single-cell gene expression dynamics will be insufficient.

Let us return to Waddington's landscape: The landscape with valleys along the horizontal axis (X-axis) is shaped along the depth axis (Y-axis) representing the developmental course. With this developmental time, the valleys are shaped successively and are deepened in a process known as canalization. Therefore, a fundamental question remains: given that the attraction to each valley along the X-axis is represented by gene expression dynamical systems, what does the Y-axis represent (slower) landscape change represent?

With the question in the depth axis in mind, there are three basic questions to resolve concerning the postulates of Waddington's landscape itself. First, there is the issue of hierarchical branching. Since the valleys are successively generated over developmental time, the shallower valleys are generated first and are then branched, and these branching processes are repeated. Second, Waddington argued that the developmental process itself, that is, the motion along with the shaping of valleys, is also robust to perturbation, and coined the term *homeorhesis* to represent such path stability. However, the mechanism contributing to the robustness of this shaping process, including successive branches, remains elusive. Third, the number fraction of each cell type is also robust to perturbations or initial conditions, as mentioned previously.

Considering these three postulates of the landscape, we need to understand how landscape change progresses along the depth (Y) axis. About seventy years after the proposal of the epigenetic landscape, we have now possible candidates that could cause such a slow landscape change. One candidate is cell–cell interactions. With the progression of development and the cell number increases, the influence of cell–cell interactions on the intracellular dynamics for each cell type becomes stronger. Slow modifications of intracellular expression dynamics can lead to novel attractors or an increase in their robustness.

Another potential source for slow landscape change is the change in chromatin structure with the epigenetic modification, one of the hottest topics in cell and developmental biology. Currently, epigenetics is mainly studied in terms of a variety of molecular mechanisms that affect the feasibility of the expression of each gene without the changes in DNA sequences, such as methylation or some other molecular modifications in DNA. These modifications change, for example, the openness of chromatin around the promoter of a gene, which affects the expression of genes, since, the efficiency of transcription binding promoter changes depending on the degree of openness. If the chromatin is more open around a gene, it is more feasible to express the actions of other genes on its promoter.

Below, we first discuss the issue of cell–cell interactions, and the second issue, epigenetic modification, will be discussed later in Section 7.5.

7.4.2 Coupled Dynamical Systems for Cell Differentiation (B)

We have investigated models with (i) intracellular reaction dynamics and (ii) cell–cell interactions for over 30 years (Kaneko & Yomo, 1994, 1997, 1999; Furusawa & Kaneko, 1998a, 2001, 2009, 2012a). When the concentration of chemical components in a cell oscillates over time, they can be differentiated into multiple types because of cell–cell interactions mediated by the exchange of chemicals between cells. Here, cells whose chemical concentrations oscillate in time can exhibit both replication (self-renew) and differentiation, that is, having pluripotency.

As the cell number increases, the increased cell–cell interactions lead to the differentiation of cellular states, in which temporal variations in chemical concentrations are reduced. The differentiated state of the cells is characterized by a decrease in the compositional diversity and a loss of concentration oscillation. From these models, we derived a dynamical-systems theory of pluripotency, irreversible differentiation, and regulation of the number ratio of each cell type.

There are a variety of choices in modeling each of the processes in (i) to (iii). We have a few methods that incorporate "internal reaction dynamics" and "cell–cell interactions" with an increase in the number of cells by division. To date, the cell differentiation process to lose pluripotency is commonly observed, if the internal reaction dynamics can exhibit concentration oscillation. Since we have already discussed the result that adopts catalytic reaction dynamics for (i) (Kaneko, 2006), we will discuss the results of models that adopt gene expression dynamics for (i).

7.4.3 Differentiation from Oscillatory States (B, C)

Here, we adopt the following setups: (i) Intracellular gene expression dynamics in which k genes mutually activate or repress the expression of some other genes and the concentration of protein species $x_i (= 1, 2..., k)$). (ii) Cells interact by exchanging diffusible components. The concentration of the corresponding component in each cell changes according to a diffusive interaction. (iii) The cell number increases by cell division, possibly altering the degree of cell–cell interactions.

We have studied a few models that incorporated these three points. Suzuki et al. (2011) examined the process (i), governed by a gene regulatory network (GRN) consisting of positive (activation) and negative (repression) paths (see Fig. 7.6(b) and also Appendix of Chapter 5). There are a huge number of possibilities for this GRN. For instance, by restringing the GRN with five genes and ten paths, there are possible 1.5×10^8 networks. By choosing all combinations of activation and inhibition paths. Suzuki et al. (2011) examined all possible networks. In this study, the number of cells was increased from one to 32, and then they examined if there were cells that took different states, which were defined as those showing different gene expression patterns; the criterion is given by the time average of x_i of each cell is sufficiently different. This criterion indicates whether the cellular states are in different regions in the state space, that is, they fall onto different attractors. Using this criterion, gene regulatory networks that showed differentiation were screened.

The results with differentiated cell types fall into two classes. In one class, when a cell is divided into two, the original attractor of the expression state that is stable in a single cell is destabilized by the cell–cell interaction. The homogeneous state of the single-cell attractor was no longer stable. This destabilization of the homogeneous state is explained based on the classical Turing instability (Turing, 1952).

In this case, however, the original cellular state is not reproduced. When there are multiple cells, the original single-cellular state turns to be always unstable. Hence, the cell could not reproduce the original state. Alternatively, the differentiated cell types derived from the original cell simply replicate themselves, and do not produce different cell types. This does not satisfy the requisite for pluripotent cells, which must have both replication and differentiation capacity. For pluripotent cells, choice of replication or differentiation depends on the situation surrounding them, for instance, on the cell–cell interactions. In the present class, there are no cells that fulfill both replication and differentiation, that is, the postulate of pluripotent stem cells. Hence, we do not discuss this case further.

In the other class, the original cell type, S, exhibits gene expression the oscillation. The change in components for a single isolated cell is given by a limit cycle attractor (stable oscillatory state). This oscillatory state remains even after the cell divides, up to a certain number of cells. However, as the cell number increases, expression oscillation of each different cell starts to take a different phase of the oscillation cycle due to cell–cell interactions. Here, the oscillations are not synchronized between cells, and their phases of oscillations are distributed (see Fig. 7.7).

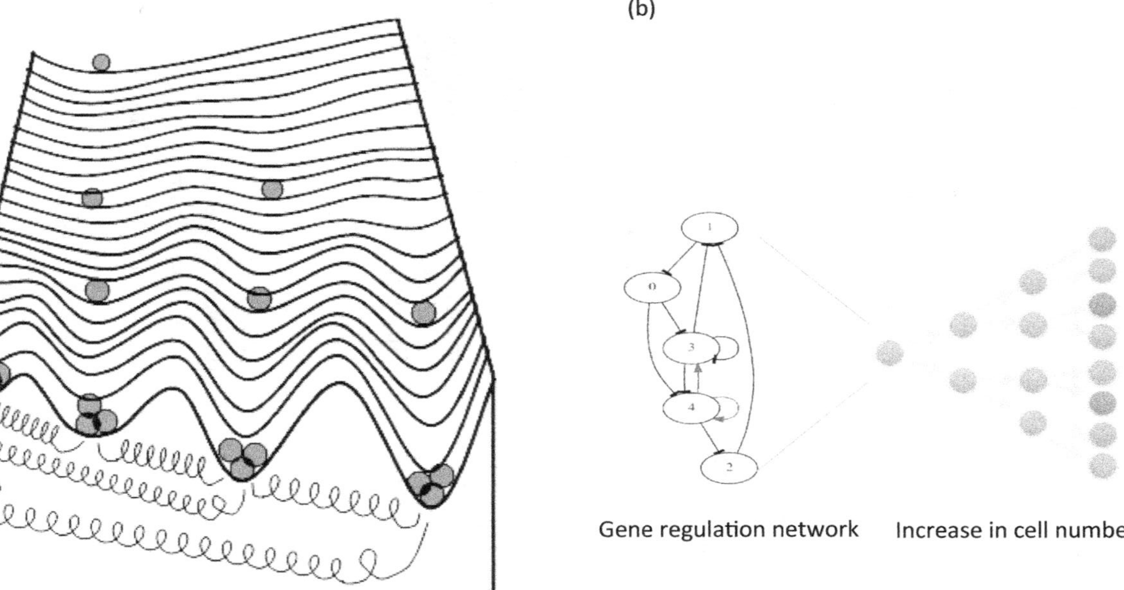

Figure 7.6 (a) Schematic representation of interaction-induced differentiation: The cell–cell interactions, represented by springs, induces a change in the landscape, to stabilize each cellular state. Based on (Kaneko & Yomo, 1997). (b) Schematic representation of a model with an intracellular gene expression network. As the cell number increases, the cellular states are differentiated. Based on Suzuki et al. (2011).

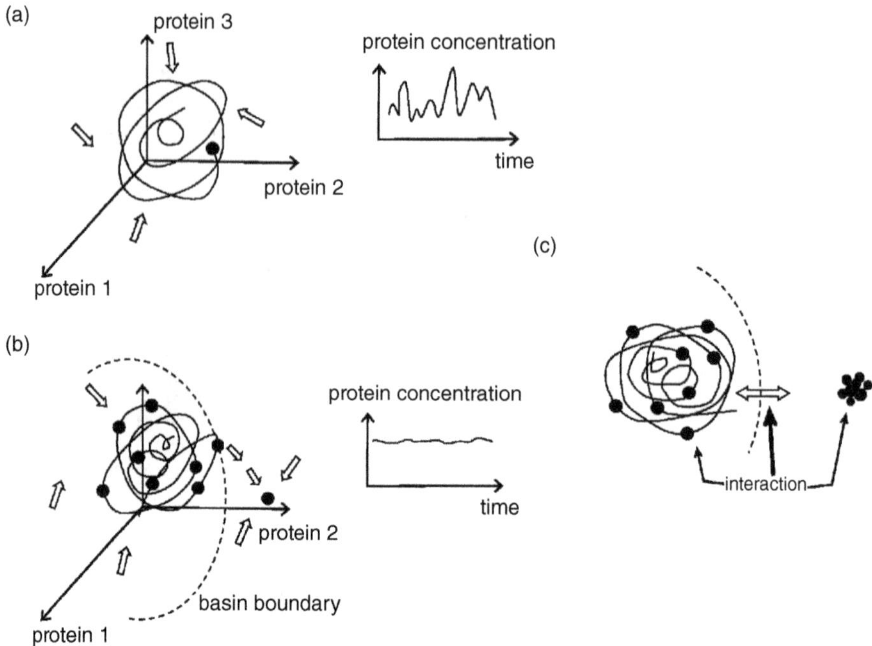

Figure 7.7 Schematic representation of differentiation from an oscillatory state in gene expression dynamics: (a) Phases of oscillations diversify by cells (b) with cell–cell interactions, a novel state in which oscillation ceases emerges (c) Two types of cells are mutually stabilized by cell–cell interactions. The original oscillatory state corresponds to the type S cell in the text, and the novel fixed state corresponds to the differentiated state A. Examples that support this schematic representation have been extensively observed in the model simulations of gene expression dynamics (Suzuki et al., 2011; Goto & Kaneko, 2013), For earlier examples see (Furusawa & Kaneko, 1998a, 2001; Kaneko & Yomo, 1994, 1997, 2000).

As the cell number increases further, cell–cell interactions lead some cells to switch to a distinctly different state. For such cells, the expression of some components is suppressed, whereas, for others, it increased (which we term "type A" cells). In these newly generated cell types, the oscillation of the component concentrations ceases, and they settle to nearly constant values. This differentiation is depicted in the state space of each component, as shown schematically in Fig. 7.7. Here, attraction to the original oscillatory state is lost due to the interaction term. Type A cells maintain the same expression profile when they divide thereafter. Contrastingly, the other type S maintains the concentration oscillation of some component concentrations.

With a further increase in the cell number, a certain percentage of S cells replicate to maintain the oscillatory state upon division. The remaining cells differentiate into type A cells, which self-renew. In summary, pluripotency is

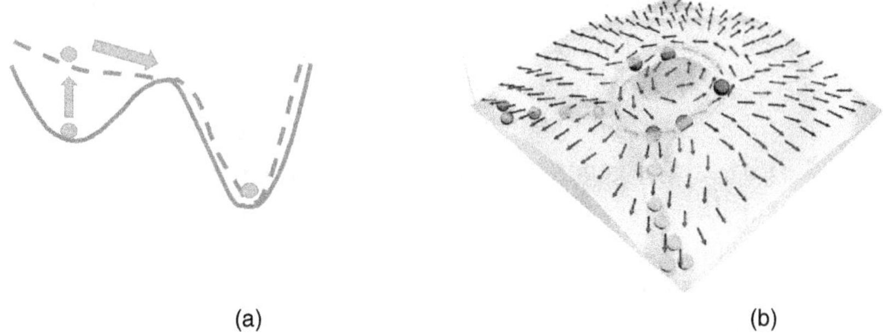

(a)   (b)

Figure 7.8 Schematic landscape representation of differentiation from (a) fixed points (b) an oscillatory state based on Furusawa & Kaneko (2012a). (Courtesy of Chikara Furusawa).

represented by the oscillatory expression, and its loss corresponds to commitment to differentiation.

In terms of dynamical systems, this differentiation is understood as a bifurcation from a limit cycle to a fixed-point attractor, induced by cell–cell interactions (Suzuki et al., 2011; Goto & Kaneko, 2013). Then, changes in the degree of cell–cell interactions alter the flows in the dynamical system in cellular state space. At some point, the flow of states of the two cells that take different phases of oscillation is moved apart. One moves beyond the basin of the original attractor, and the other moves into a new state in which oscillation is lost. Depending on the difference in the timing of the phases of oscillation, only one of them crosses the basin boundary and transitions to a new non-oscillatory state. The cell–cell interaction causes a bifurcation to a fixed-point state.

Contrastingly, the remaining cells stabilize their original oscillatory state by interacting with differentiated, non-oscillatory cells. Eventually, both cells settle into stable states under cell–cell interactions.[2]

An alternative picture of stem cell differentiation in terms of the fixed point is shown schematically in Fig. 7.8(a), where the valley becomes shallower and the ball rolls out. Here, if the first state is being pulled into a fixed point (the bottom of the valley), all the balls roll out of the original valley (state) if the landscape changes. The feedback must be introduced for the two cellular states to coexist: When a ball rolls out, the original valley has to be deepened again by the

---

[2] The simplest example is a case with only two genes, x and y, where x promotes the expression of x and y, whereas y suppresses the expression of x. Here, only the y component is diffused in/ out, leading to diffusive cell–cell interactions. In this case, if the parameters are taken appropriately, the concentrations of x and y oscillate in a single cell, and as the cell number increases, the phase of each oscillation starts to differ by cells. The expression levels of x and y also fall to almost fixed levels in some cells. This differentiation is understood as a bifurcation (*saddle-node bifurcation on an invariant curve* (SNIC)), resulting from the shift of the nullcline in each cell, caused by cell–cell interactions (Goto & Kaneko, 2013).

interaction to keep the other cell within it. This should occur before all cells flow out of the original valley. This postulates that the balls move over a broad area so that not all of them roll out all at once. This means that the landscape must change faster than the time scale with which the ball fluctuates around the valley, which would be quite difficult; in general, changes in the landscape due to cell–cell interactions are slower than those at intracellular dynamics, because the former is due to macroscopic changes in cell populations. Hence, a fixed-point picture would be difficult for a pluripotent state.

Contrastingly, if the internal state of cell changes owing to oscillations (or chaotic itinerancy (Kaneko, 1990; Tsuda, 1991; Kaneko & Tsuda, 2000, 2003; Furusawa & Kaneko, 2001) each cell takes a different state (composition and gene expression level) at each time. Cell–cell interactions only destabilizes only a certain range of states/compositions (Fig. 7.8(b)). As the internal states are broadly distributed, Not all cells are simultaneously destabilized. The oscillatory changes in the internal states of the cells are not much faster than the changes in the cell–cell interactions. Therefore, during the time scale (period) of oscillation, only some cell states are differentiated, whereas others remain in the original state. Then, the remaining cells can maintain their original states without differentiation.

Thus, gene expression oscillations will be relevant to stemness, that is, compatibility between potentiality for both proliferation and differentiation. The loss of stemness is characterized by the loss of oscillatory dynamics.

7.4.4 Indices for Pluripotency that Decrease through Hierarchical Differentiation (C)

In the above example, we are mainly concerned with a single stage of differentiation S→A, however, if we increase the number of genes and hierarchically organize the regulatory network, we can construct a system so that cells can be differentiated in several stages. How can the differentiation process be expressed in this case? Figure 7.9 shows a schematic diagram of this process. This schematic picture has been confirmed by simulations using specific gene regulation networks (Suzuki et al., 2011).

The chemical composition of the initial cell varies over a large range in the state space. As differentiation progresses, the chemical composition becomes specialized and fixed successively, as shown in the figure. The process of cell differentiation can be described as follows:

(1) From a state with a diverse composition of components to states biased to some components.
(2) From a state with higher to lower plasticity.

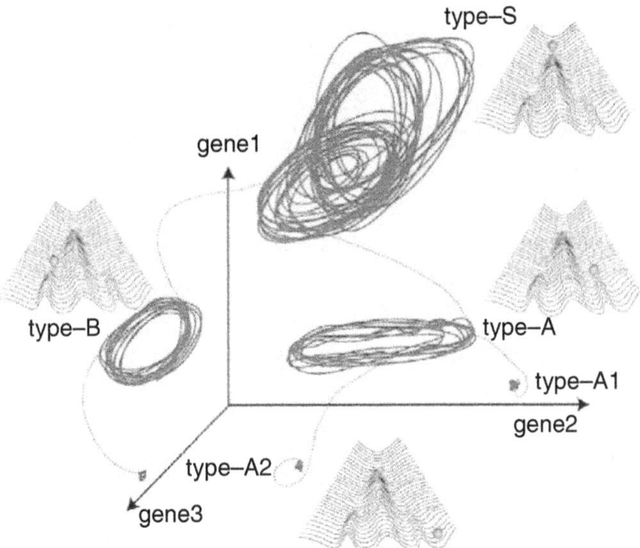

Figure 7.9 Schematic representation of hierarchical differentiation based on dynamical systems. As differentiation progresses, the region in which cellular states can vary is decreased. This is consistent with Waddington's landscape.

(3) From a state with a high temporal variability to a fixed state with low variability.

(4) From a state with larger heterogeneity over cells to that with fewer differences.

Corresponding to these four changes, the following indicators have been proposed to capture the irreversible differentiation process (Furusawa & Kaneko, 2001, 2009, 2012a).

(1) Decrease in compositional diversity in cells: Pluripotent cells express a wide variety of genes, whereas there are no specific genes that are predominantly expressed. In fact, as a measure of component diversity, the entropy can be introduced from the concentration of each component, say, as $-\sum_i p(i) \log p(i)$, where $p(i) = -x_i / \sum_j x_j$ is the fraction of expression of each gene i (Furusawa & Kaneko, 2001; Kaneko, 2006). Computing this "entropy" over different cells in the model shows that it decreases with the progress of differentiation. So far, we have confirmed this change in diversity entropy in the simulation. Still, it can be experimentally measured from the total gene expression level by transcriptome analysis. One can then examine whether the diversity of expression provides one measure of pluripotency.

(2) Decrease in plasticity: Plasticity concerns the degree of the changeability of a cellular state; in other words, the inverse of the environmental perturbation needed to switch a cell to a different type. In a theoretical model, it corresponds

to the minimum degree of perturbation in the interaction term required for a switch from one cell type to another.[3] Therefore, the reciprocal of this minimum perturbation may serve as an indicator of plasticity. In fact, in current cell models, as differentiation proceeds, the switch requires a larger perturbation indicating the decrease in plasticity.

(3) Decrease in temporal variation in component concentrations: Oscillations of protein concentrations decreased as cells were differentiated. The pluripotent cells in the model presented here demonstrated temporal oscillation in the concentration of some proteins. With this oscillation, the cellular state varies over a broad region in the state space. As differentiation progresses, the temporal variation decreases. The range of states cells can visit decreases with the developmental course (Fig. 7.5).

(4) Decrease in cell–cell variation: The timescale of temporal variation is rather long so that at each snapshot in time, the concentrations of some proteins differ distinctively across cells. By comparing cells at a given time, then, the cell–cell variation is large for pluripotent stem cells and decreases as differentiation progresses.

In the present volume, we put forward the view that plasticity and variation (fluctuation) are correlated. According to this view, variation in gene expression provides an index for plasticity. In pluripotent cells, gene expression levels not only fluctuate by noise but also vary spontaneously through oscillation. Thus, such cells are thought to have higher plasticity.

In summary, irreversible differentiation is described as a decrease of the diversity in intracellular components (gene expression), increase of the stability of an attractor, decrease of plasticity, as well as of temporal and cell-to-cell variations.

7.4.5 *Proportion Regulation; Macro-Micro Consistency (C)*

Notably, in the mechanism presented in the previous section, the timing and pathway of differentiation are robust to noise, implying the *homeorhesis* (Waddington, 1957). With cell–cell interaction, the differentiation frequency of a stem cell is autonomously regulated by the population of each cell type, resulting in robustness in the population ratio of each cell type against perturbations.

To investigate this point, we carried out simulations of the model in Section 7.4.2, by fixing the total cell number at N (instead of the simulation starting from a single cell) and setting the initial cell numbers of stem and differentiated cells to N_0 and $N - N_0$, respectively. By varying N_0, we examined whether each cellular

---

[3] A similar quantity is defined as an index of how much perturbation is needed for a given attractor to switch to another, as termed as "attractor strength" (Kaneko, 1998a).

state is stable in the presence of cell–cell interactions at each cell-type distribution. Then, it is found that there is a certain range of N_0 within which both types coexist stably (Suzuki et al., 2011).

For example, the states with $N_0 = 0$ and $N_0 = N$ are unstable, and the stem cells with oscillatory dynamics coexist with the differentiated cells only within the range $0 < N_l \leq N_0 \leq N_u < N$, where the upper and lower limits, N_u and N_l, are dependent on the network and parameters of the protein expression dynamics. This range $[N_l, N_u]$ is much smaller than $[0, N]$. This result implies that the cell number regulation works in the model. Furthermore, when starting from a single cell, the number ratio of each cell type falls in a much narrower range than the above range $[N_l, N_u]$, even if the system is simulated under a large magnitude of noise in gene expression dynamics and/or at each cell division event. The timing of differentiation from the initial cell type is also stable in each course, even though we simulate the developmental course with a large amount of noise. The developmental time course of differentiation, thus, is stable.

In other words, consistency between the stability of a cellular state and the population of cells implies that the cell-type ratio must be within a certain range, as the cell–cell interactions depend on the proportion of each type of cells present. Recall that differentiation started when the fraction of cells in the initial S state was 100 percent. Thus, if the ratio of S states is too high, the cell state is expected to be unstable. Alternatively, as the fraction of differentiated cell type A increases, the original S state is stabilized so that replication of S continues, increasing the ratio of S. Recall that state S is an attractor, as shown in Fig. 7.7, and the boundary of its basin (the range that the state is attracted to) moves with the degree of cell–cell interactions, according to the population ratio of each cell type. As the ratio of S cells increases, the cell–cell interactions bring the boundary closer to the attractor. When the ratio exceeds a certain value, the attractor and basin boundary touch. Then, the differentiation follows, and the population ratio of S-type cells can no longer increase. However, as the ratio of type-A increases, the boundary moves away from the attractor, and the S cell continues to replicate.

In this situation, as the cell number increases, changes in the interactions cause the state to collide with the basin boundary, leading to the regulation of the proportion of each cell type, as observed in the simulation. This is a universal property in a system in which cells are differentiated by expression dynamics and intracellular interactions. (Kaneko & Yomo, 1999; Furusawa & Kaneko, 2001; Nakajima & Kaneko, 2008; Pfeuty & Kaneko, 2014).

The more plastic a cellular state is, the more likely it is to change upon external perturbation, including the effect of cell–cell interactions. Hence, the regulation of the proportion of cells is more effective. In other words, the robustness of the cell population increases. This may be another example of the reciprocal relationship between plasticity and robustness, as described in Section 5.4. Here, it is important to note that the plasticity at the microscopic (single-cellular) level and the robustness at the macroscopic (cell population) level are reciprocal. This is a representation of macro-micro consistency.

Indeed, a stable developmental process in a multicellular organism requires robustness in the proportions of cell types. Even in cellular slime molds, a prototype multicellular organism, the number of spores and stalk cells remains within a certain range (Rafols et al., 2001). Moreover, although the above discussion does not take into the spatial arrangement of the cells account, it was noted that the differentiated cells also formed a stable spatial pattern when spatially local interaction was included (Furusawa & Kaneko, 2000b, 2002, 2003b).

7.4.6 Experimental Verification of the Theory (E)

Although the complete experimental confirmation of the presented theory is not yet available, there are some experimental results that may support it.

Recently, single-cell measurements of Hes1 protein concentrations were performed using imaging techniques. Kobayashi *et al.* (Kageyama's group) (2009) demonstrated the existence of oscillatory gene expression in embryonic stem cells over a few hours. Here, the phase of the Hes1 oscillation was found to control the cell fate choice toward neural and mesodermal differentiation. Furthermore, the observation of oscillations in Hes1-downstream genes suggests that the oscillations in these expression levels propagate through the gene network, leading to differentiation from ES cells. Interestingly, the oscillation found in ES cells disappeared in the differentiated cells, as is consistent with our theory. Furthermore, cell–cell communication via Notch-Delta signaling has been suggested to regulate the fate decision of neural progenitors under the control of the oscillatory expression dynamics of Hes1 and other genes.

On the other hand, Huang used time series transcriptome data and discovered that the expression dynamics exhibited transitions over quasi-stable states (Chang et al., 2008). The transitions are irregular, however, not purely random, and may involve components of irregular oscillatory dynamics.[4] Additionally, from the fluctuating expression level of the stem cell marker Sca1, they found slow-scale changes in cellular states, which are suggested to be regulated by cell–cell communication.

---

[4] The behavior here may suggest the chaotic itinerancy (Kaneko & Tsuda, 2003) commonly observed in high-dimensional dynamical systems, although the distinction between chaotic and noisy dynamics may be difficult here. As mentioned previously, the distinction between just noisy dynamics versus the dynamics with oscillatory components plus stochasticity can be made by the following procedure. The transition probabilities among three states A, B, and C are measured as $p_{A \to B}$ and $p_{B \to C} \cdots$. Then, compare

Furthermore, larger cell-to-cell variation in stem cells has been experimentally observed. Heterogeneity in expressions of Stella, Nanog, and Hes1 expressions was recently observed in embryonic stem cells (Chambers et al., 2007; Hayashi et al., 2008). This variation is consistent with the theory we discussed.

7.5 Epigenetic Fixation (D)

Consistency between processes with different time scales is also important in cell differentiation. Here, we discuss this issue concerning the epigenetic modification. As mentioned in section 6.4 DNA modifications change the openness of the chromatin structure, altering the feasibility of gene expression. Although these modifications do not alter the gene itself, they are sustained for a prolonged period; the timescale of epigenetic modification is much longer than that of protein expression and faster than that of genetic change. Accordingly, one can discuss whether the faster change in the expression may be embedded into the slower epigenetic change.

To date, the established mathematical model for epigenetic modification with gene expression dynamics is not available. Still, by the modification, the feasibility of the expression changes so that the threshold for the expression is altered. In contrast, epigenetic changes are influenced by protein expression and stabilize the expression state. It is natural to assume a feedback process such that if the expression level is higher (lower), the threshold for the expression is decreased (increased). Then, one can adopt the epigenetic feedback model in which the threshold $\theta_i(t)$ for the expression of gene i changes as

$$d\theta_i(t)/dt = -\gamma(\theta_i(t) - (\Theta_0 - x_i(t))), \tag{7.1}$$

as has been discussed in Chapter 6. Here, the timescale for the modification $1/\gamma$ is larger than the timescale of the change in x. With the above process, the on/off of the protein expression level $\{x_i(t)\}$ is solidified. Accordingly, the differentiated states are not easily perturbed by noise or other external disturbance (see Fig. 6.11 of Chapter 6).

We investigated a single-cell model that includes epigenetic feedback and gene expression dynamics controlled by a GRN and epigenetic modification (without cell–cell interactions) (Matsushita & Kaneko, 2020). As the epigenetic feedback regulation is stronger, more attractors with different expression patterns, that is, more cell types, are generated depending on the initial condition. Particularly, when the initial gene expression pattern shows temporal oscillation, several fixed

$p_{A \to B} p_{B \to C} p_{C \to A}$ with $p_{A \to C} p_{C \to B} p_{B \to A}$. If the two deviate significantly, then some oscillatory component with the circulation $A \to B \to C \to A$ exists.

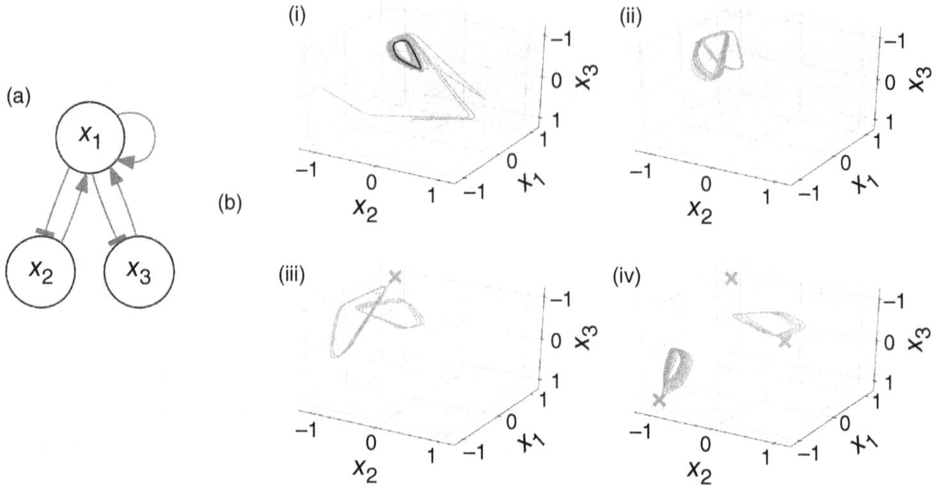

Figure 7.10 (a) Three-gene system, with mutual activation and inhibition ($J_{11} =$ 0.26, $J_{12} = J_{13} = -0.35$, $J_{21} = 0.4$, $J_{31} = -0.36$). (b) (i) Trajectories from different initial conditions, by distinct orbits. Initially, they approach the same limit cycle attractor. (ii) Trajectories are separated into two groups, depending on the initial conditions, for $40 < t < 50$. (iii) Further separation of the group of trajectories for $80 < t < 90$. (iv) Three trajectories reaching distinct fixed points, represented by ×, for $120 < t < 150$. With this change, θ_i is also differentiated. For each trajectory, θ_i approaches the fixed point x_i^*. Reproduced from Matsushita & Kaneko (2022).

expression patterns (fixed-point attractors) are generated by the epigenetic feedback (see Fig. 7.10). They are hierarchically organized as branching valleys in Waddington's landscape.

Then, we examined which attractors are reached by repeating the above dynamics under noise by starting a given initial pattern, to find that the fraction of each final fixed state was robust to the initial conditions and noise. The timing of hierarchical differentiation was also robust.

Using the change in expression dynamics under the slow epigenetic modification process, Waddington's epigenetic landscape is depicted explicitly, in which the landscape axis (horizontal axis in Fig. 7.1) is given by the principal component X of the expression pattern, the depth axis is given by the developmental time with slow epigenetic modification, and the height is given by $-log[P(X)]$ with $P(X)$ as the cell-number distribution of X. When the original attractor in the absence of epigenetic modification shows oscillatory dynamics (as a limit cycle attractor), the timing of branching to different cell types, the number of differentiated cell types, and the number fraction of each cell type are all robust to perturbations during development and to variations in initial conditions. Hence, the generated landscape satisfies the following three postulates implicitly assumed in

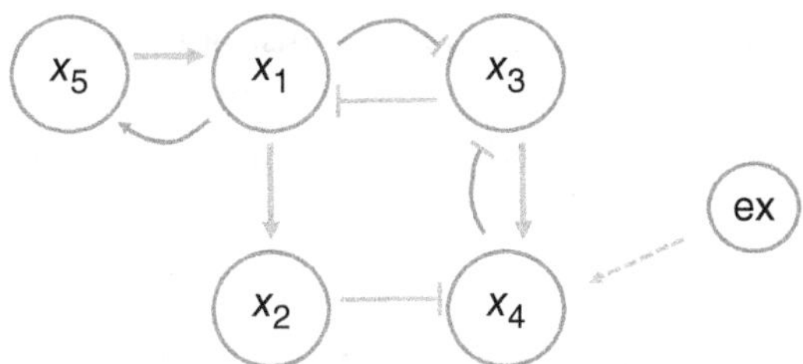

Figure 7.11 A gene-regulation network is simplified from that extracted from experimental data. X_1, X_2, X_3, X_4, X_5 correspond to the genes *Nanog, Oct 4, Gata6, Gata4*, and *Klf*, respectively. By appropriately choosing the reaction parameters, gene expression levels oscillate over time. With this model, cell differentiation progresses by cell–cell interactions, whereas through the epigenetic feedback process as given in the text, each cellular state is fixed.

Waddington's landscape: (i) hierarchical branching is supported by the hierarchical attractor generation from the limit cycle (see e.g., Fig. 7.10); (ii) homeorhesis is supported as this branching process is independent of initial conditions and robust to noise; and (iii) the cell-number robustness is demonstrated as $P(X)$ is also independent of initial conditions and robust to noise. As mentioned in Section 7.1, this robustness in the path and the cell-number distribution to perturbation is an essential requirement for the development of multicellular organisms.

Can we then verify the present dynamical-systems theory by specifically using the gene expression dynamics of the real cell? For this, detailed information on the actual gene expression network is required, which, however, is not sufficiently available. Still, the network is at least partially known (Dunn et al., 2014). Then one can simplify the known GRN to extract its core part to examine the gene expression dynamics (Miyamoto et al., 2015). Figure 7.11 is thus obtained. Unfortunately, the parameters in the reaction paths in the network are not yet available. However, with their appropriate choice, oscillatory gene expression is observed. Then, by introducing the cell–cell interactions as given in the model in Section 7.4, differentiation progresses through cell–cell interactions, as the cell number increases (Miyamoto et al., 2015). The differentiated cell type loses the oscillation, resulting in the biased expression for a few components.

Next, by introducing an epigenetic process, as shown in Eq. (7.1), the differentiation state is fixed, as θ_i is in(de-)creased. Thus, differentiation is fixed.[5] The differentiated state is then tightly fixed by epigenetic feedback.

In summary, the results in the present section, the processes with fast protein expression dynamics and slow epigenetic modification mutually stabilize

---

[5] In the paper, we adopt the rate equation of the Hill form, to correspond with experiments. By introducing the feedback process to change the threshold, as in Section 7.5, common epigenetic fixation is obtained.

each other. The cell types that are stable against noise or external perturbations are generated. Embedding changes at a faster time scale to slower processes shapes a robust memorized state and consistency between processes with different timescales.

7.6 Cellular Reprogramming: Operation to Reverse the Irreversibility (C–E)

The differentiated states are stabilized by the epigenetic feedback. This is desirable as the robustness of each cellular state and the ensemble of cells. Then, it may be difficult to reverse the differentiated state back to the original pluripotent state by a simple external operation, which, however, was carried out experimentally by Takahashi and Yamanaka (2006) as cellular reprogramming.

Indeed, they have succeeded in the reprogramming, by overexpressing few (4) genes for days, by using retroviruses, during which time it is suggested that chromatin is reconstructed. In this case, no direct manipulation of the epigenetic state was applied, even though the erasure of epigenetic memories is a key factor for cellular reprogramming. In other words, epigenetic memories are erased only by the perturbation of gene expression. Moreover, the number of overexpressed genes was just four, which are notably fewer than the total number of genes involved. By such simple manipulation, the cellular state successfully returns to the original pluripotent state, that is, the top of the Waddington's landscape, which could be an unstable point, as it is followed by differentiation. Thus, the following problem in cellular reprogramming needs to be solved: How can reprogramming robustly make cells head toward a pluripotent state via the overexpression of only a few genes?

Here, we discuss reprogramming in terms of dynamical systems theory. When the expression of some genes is lost in differentiated cells, their expression must be recovered for reprogramming. In the dynamical-systems picture, the state of pluripotent cells is located at an attractor involving the expression of a variety of genes, whereas for the state of the differentiated cells, the expressions of some genes are nearly zero. To regain the original pluripotency, the cellular state must be moved back toward the original attractor.

In considering whether it is possible to reverse the change, it is important to distinguish between variables that can and cannot be manipulated externally. If the cell state is described by the abundances of each component (protein expression) and epigenetic modification of the DNA, it is natural that the cellular state is reverted to the original by replacing all the abundances and epigenetic modifications by the original values in the pluripotent attractor. This, however, would be an unrealistic operation. In statistical thermodynamics, this corresponds to the following situation: after the irreversible increase of entropy (say after the temperature in space is homogenized), if all the velocities of molecules could be completely reversed, then time irreversibility would be reversed. The question we should address here is whether one can reverse the time's arrow under limitation of our capability in operations.

First, we consider a simple model in section 7.4, in which only the protein expression changes but not epigenetic modifications are taken into account. Now that for differentiated cells, the degree of freedom of the chemical components (gene expression) is reduced in the state space, it is natural that pluripotency will be recovered if the degree of freedom is restored through the introduction of several components to necessary concentrations. This can be achieved either by placing the component directly within the cell or by elevating several components; in the example of Figs. 7.7 and 7.9, a differentiated cell can be returned to its original S state if it is placed in an environment that supplies sufficiently the components that were reduced in the differentiated cells.

In fact, in a paper that proposed this dynamic system differentiation (Furusawa & Kaneko, 2001), it is predicted that pluripotency is regained from differentiated cells by compulsively expressing multiple genes. If there is no epigenetic process involved, the cell's pluripotency can be recovered by instantaneously adjusting the protein expression levels x_i to appropriate values that bring the system closer to its original state S. However, in the presence of epigenetic fixation, the process of reprogramming is more challenging because reversing each epigenetic modification by external means is difficult, akin to reversing all molecular motion in statistical physics. When epigenetic fixation is present, resetting gene expression patterns alone may not be sufficient to reprogram differentiated cells. Even if the gene expression pattern of a differentiated cell is reset to that of a pluripotent cell, the cellular state will quickly return to the differentiated state because the epigenetic modifications have not been altered.

In the theoretical model we discussed in the last section, epigenetic modification is represented by threshold variables θ_i. For example, when x_i is low, the threshold θ_i increases owing to epigenetic feedback. Then, even if x_i is increased instantaneously, x_i returns to a lower value, as θ_i is high. As the change in θ_i is slow, elevation of the expression of some protein levels over slow time scale of $1/\gamma$ is required to regain the pluripotency. In other words, it is not enough to change the state for a moment, but it is necessary to overexpress the gene for a sufficient time span. This would be difficult simply by changing the external environmental conditions instantaneously. From the point of view of molecular biology, it implies that short-term input from the outside will not alter the epigenetic modification, but sustained manipulation for the overexpression is required.

Hence, the overexpression of appropriate genes for a timespan beyond the time scale for epigenetic modification is postulated for reprogramming. Still, as epigenetic modification θ_i is not altered externally, this might not necessarily imply the regain of the pluripotency. Only if the feedback from x_i resets, θ_i to the original level, the state switches to the original pluripotent state, that is, the original valley in the landscape from which the differentiation starts.

The original pluripotent state turns to be unstable as it is followed by their progression toward various states. Hence, returning to the original state may sound difficult. In terms of dynamical systems, the pluripotent cell is expected to be represented by a saddle point, such that orbits are attracted from many directions but are departed along an unstable direction (manifold), which represent the cell differentiation process. To regain pluripotency by reprogramming, however, cellular states must be placed on the stable manifold of the saddle by common manipulations from different attractors (differentiated cell types). However, such manipulation would require finely tuned control. In contrast, reprogramming experiment by Takahashi and Yamanaka (2006) is mediated by the overexpression of a few common genes across various differentiated cell types.

Therefore, some dynamical systems concept overcoming a fine control to a saddle point is needed. Now, considering the significance of oscillatory dynamics, discussed so far, its possible role for cellular reprogramming was searched for, by examining if and how attraction to the unstable manifold is achieved by slow epigenetic fixation of the oscillation of fast gene expression. In fact, Matsushita et al. (2022) have recently shown that oscillatory gene expression dynamics with slow epigenetic modifications enable cellular reprogramming by the overexpression of only a few genes.

Usually, approach to a saddle point requires fine tuning, as some component along the direction of the unstable manifold, albeit small it be, would be amplified with time. In the standard dynamical systems of one-layer (θ_i in this case), this would be true. Here oscillation of fast gene expression variables (x_i) exists and the change in θ_i is driven by it. As the change in θ_i is slow, the influence of x_i is averaged over slow time scale, which will weaken the instability along the direction of unstable manifold. As the oscillation dynamics are extended beyond the stable manifold of a saddle point, global attraction to the vicinity of the saddle point may be facilitated by exploiting the interplay between fast gene expression and slow modification dynamics. Global attraction to the saddle point, that is, the top of the Waddington's landscape (or pluripotent state) is thus possible without tuning the perturbation, as long as the perturbation allows for regaining the oscillation. Here, the slow dynamics for epigenetic modification are driven by the temporal average of faster dynamics. By averaging out of faster oscillation, instability around the saddle point of slow dynamics is smeared out, to that the state comes closer

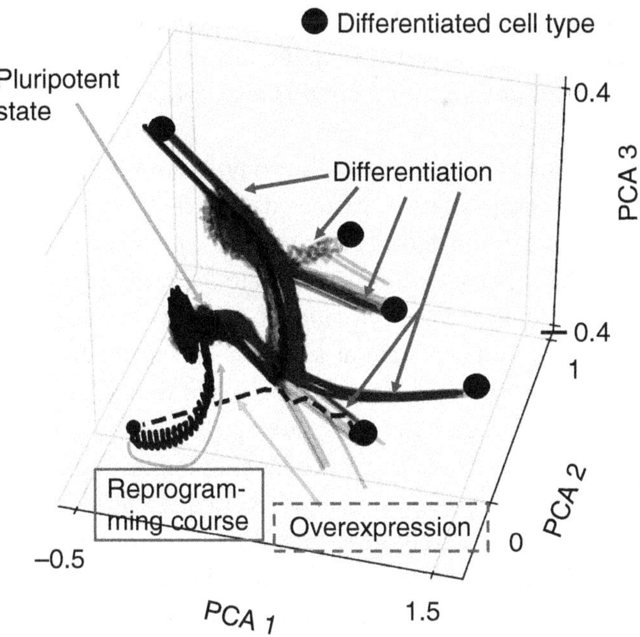

Figure 7.12 Time development of cell differentiation and reprogramming. Adopted is a r Random gene regulatory network model with N = 10, with randomly generated gene-regulatory matrix J_{ij}, where for $\theta_i = 0$, $x_i(t)$ oscillates in time. For plotting, we adopted a principal component analysis (PCA) of $\{\theta_i\}(i = 1, 2, \cdots 10)$ over 1000 trajectories starting from $\theta_i \sim 0$. The trajectory $\{\theta_i(t)\}$ was plotted by adopting the first, second, and third PCA modes. Trajectories starting from $\theta_i \sim 0$, five different fixed points are reached after transient oscillation dynamics, as shown in the trajectories to five differentiated states given by •. From one of the differentiated cell types, we added external input to three genes as reprogramming manipulation, with which θ_i changes as shown in the curve plotted by broken line. Then, the external input to x_i, that is, reprogramming manipulation is terminated. After this, the cell state was attracted to the initial pluripotent state, from which the differentiation starts again. Reproduced from Matsushita et al. (2022).

to the point and stays longer therein (for detailed account of this mechanism by dynamical-systems theory, please refer to [Matsushita et al., 2022]).

Now, we can explain the return to the top of the landscape by reprogramming, which would be seemingly unstable. Once the initial oscillatory state is regained, the memory of the cellular state before reprogramming manipulation is erased, because the oscillation state is attracted from different initial states. According to the study, regaining oscillation is the main requirement for reprogramming and elaborate manipulation to induce cells into a specific state is not necessary (Fig. 7.12). This explains the role of oscillations in the gene expression in pluripotent cells and epigenetic modification through the differentiation process, as well as how reprogramming is possible by overexpressing only a few genes among thousands. Timescale separation between the fast expression dynamics and slow epigenetic modification feedback required is also consistent with observations of previous studies.

The present dynamical-systems theory for reprogramming is also confirmed by using the gene expression dynamics extracted from the real cell in the last section (Fig. 7.11). After a differentiated cellular state is obtained as in the last section, we examined whether it was possible to return to the original pluripotent (oscillatory) state by maintaining the concentration x_i of some proteins X_i that were not expressed in this differentiated state at a high level ($x_i \sim 1$ in the model) over a time span of $1/\gamma$. We found that it was not sufficient to overexpress only a single component i, in order to regain the pluripotency. This is because the initial oscillation requires multiple components, and changing the level of expression of just a single gene is not sufficient. The simulation results of this model showed that by maintaining the expression of four genes at a high level for some time the cellular state returns to the initial pluripotent state (S state in Section 7.4).

In the model, the overexpression of the four genes corresponding to Oct4, Sox2, Klf4, and Myc was used to regain the pluripotency, as in the experiment. This, however, does not indicate that these four should be the minimum, nor does it rule out other possibilities. However, it would be universal that overexpression of multiple genes over the epigenetic timescale is required.

Theory about which of the many genes should be activated and the probability of reprogramming is not yet complete. However, if the multicomponent oscillatory state provides pluripotency, then one can say that pluripotency is restored by activating genes (groups) related to the feedback circuits responsible for the oscillatory gene expression.

Recall that thermodynamics gives a universal limit in reversing irreversible changes. The second law of thermodynamics (within the conditions for its application) has taught us that there is a universal limit for converting heat into work, as thermal motion cannot be controlled. Reprogramming, that is, reversing the biological irreversibility, may also have some limitations, as every intracellular state, especially epigenetic modifications, cannot be controlled. In the case of thermodynamics, some waste, that is, dissipation of heat, cannot be avoided to reduce the entropy within a system. Similarly, some waste may not be avoided to reverse all gene expression states. Even in the current reprogramming experiment, interferences between the activation of gene expression and epigenetic changes exist. Therefore, "waste" is expected to occur to reverse the committed differentiation. At this point, it may be interesting to note that the probability of successful reprogramming in the construction of Gurdon's somatic clones (Gurdon et al., 1975) or induced pluripotent cells following the experiment by Takahashi and Yamanaka

(2006) is rather low, and is achieved only for a small number of cells, even though it is not sure if such failure may be related with the above "waste." At any rate, understanding the possible limitations of reprogramming will be a major challenge to address in the future, in developmental biology.

7.7 Collapse of Consistency: Metamorphosis and Cancer (F)

Thus far, we have considered the process of cell differentiation and tissue development based on macro-micro consistency, between cell states and populations, and between gene expression dynamics and slower epigenetic modifications. Are there some cases in which this consistency is broken?

For example, some multicellular organisms exhibit metamorphosis. In larvae, the cell type and population first form a "consistent" state, where the system is nearly stable and is maintained during larval life. Only slow changes occur during that *metastable state*, and then it is destabilized at a certain time, resulting in a transition to an adult body composed of spatial patterns of different cell types. This means that the consistency between cells and tissue, once formed, is destabilized because of some slow change, such as the growth of an individual or the secretion of a hormone, followed by the emergence of a novel "consistent" state.

In this sense, metamorphosis can be regarded as the realization of a novel trait due to the collapse of consistency between cells and tissues. For example, Takagi & Kaneko (2005) examined a certain class of dynamical-systems models of cell differentiation discussed in this chapter, in which cell types and their number distributions first reached a quasi-stationary state as the cell number increased; then, at a certain stage with further development, they are destabilized, leading to a novel distribution of other stable cell types. Unfortunately, however, the understanding of metamorphosis has not progressed significantly. The viewpoint of consistency and its collapse in such hierarchical dynamical systems may shed new light on this problem.[6]

On the other hand, cancer cells illustrate another collapse of cell-tissue consistency. Indeed, cancer cells are thought to break the consistency with the organized tissue composed of normal differentiation process and grow in a "selfish" manner. Then, can we discuss cancer along the line of the interacting dynamical systems, as presented in the present chapter?

Indeed, Kauffman and his colleagues proposed a hypothesis that there are attractors that are not used in normal development and that they correspond to cancer cells (Kauffman, 1971; Huang & Ingber, 2006, and Huang et al., 2009). Even

---

[6] Often a teleological explanation is adopted for metamorphosis: the larval period is spent in water (underground), whereas adults live on land so that the organism has to change the tissues. However, it is first necessary to understand how the system can make such change that allows for metamorphosis.

if the idea is valid, some basic questions remain. First, cancer cells are often considered selfish; therefore, their relationship with other cells must be considered. Second, cancer cells usually accumulate genetic mutations. After the accumulation of genetic mutations, the cell exhibits slightly different gene expression dynamics. Then, they could be attractors of dynamical systems that are different from the original. How are these two features understood? Here, we discuss these issues along the lines of the present chapter, in terms of the collapse of consistency between cells and tissue.

Generally, gene regulation networks are complex and involve a large number of proteins that mutually influence the expression of others. Many attractors can often exist for such high-dimensional dynamical systems (as confirmed by simulations of random GRNs). The number of attractors might generally decrease as robustness increases through evolution. However, considering a large number of involved protein species, some attractors in gene expression dynamics, besides those visited during normal development, may exist. Indeed, in the earlier simulations by Furusawa and the author, such an attractor, which did not appear over normal development, was also identified. Such "aberrant" states are not generated in the normal developmental course and cell–cell interactions. They are not stabilized in the presence of other cell types and do not help other cell types. These additional attractors would then be selfish in the sense that they can grow without forming stable relationships with others. This is in strong contrast to normal cells in the study of Section 7.4, which were stabilized by cell–cell interactions and formed mutually stabilizing relationships.

Now, let us allow the use of the results from Chapter 8 here; evolutionary robustness: an adapted state throughout evolution increases robustness to noise and consequently to genetic mutations. As we shall see in Chapter 8, this is a property of attractors in a GRN that is achieved by natural selection over generations.

Indeed, most attractors in a randomly chosen GRN were not robust to mutations when the GRN was sufficiently complex, while robustness to noise and mutation appears as a result of evolution to produce the functional phenotype, as will be discussed in Chapter 8. The aberrant cell types here, however, are not functionally necessary for the survival of multicellular organisms, and thus are not under selection pressure to preserve their state. Hence, they have not had a chance to achieve high robustness against noise or mutation through evolution. Accordingly, the variability in gene expression levels of aberrant cells due to noise or genetic mutations will be much larger.

Through the division of such cells, their states would be more changeable by somatic mutations. Since the state is not yet optimized for growth speed or robustness, it can increase the growth rate and robustness can be enhanced by such mutations. Hence, some mutations can spread in the population: mutations that are beneficial for cancer cells will accumulate. With the accumulated mutations, the aberrant cell types gain robustness and a higher growth rate, resulting in the proliferation of "cancer" cells. This contrasts with the normal cell types that are expected to have already achieved robustness to noise and mutation through evolution, where somatic mutations cannot increase robustness or growth speed. Here, the cell type that first emerges as an aberrant attractor may correspond to a cancer stem cell.[7]

To summarize, the cell types discussed here are characterized as follows: (i) They do not appear throughout normal development but are generated as a single-cell attractor when perturbed sufficiently. (ii) They are selfish and do not form stable relationships with other cell types. (iii) Their phenotypes are vulnerable to noise or changes in cell–cell interactions. (iv) They are not robust against mutations. (v) They easily accumulate mutations, leading to increased robustness. (vi) They are in differentiated states compared with pluripotent cells, whereas they are different from terminally committed cell types.

Cells with these properties may fit the observations of cancer cells or cancer stem cells. In contrast to the hypothesis by Kauffman (1971) and Huang et al. (2009), the proposal here concerns intra-intercellular dynamics rather than single-cell gene expression dynamics and distinguishes cancer cells from normal cell types regarding robustness. This distinction led to the above six characteristic points of cancer cells.

According to point (i), a sufficiently large external perturbation can trigger the emergence of cancer (or cancer stem) cells. Point (ii) might explain how cancer cells are not useful for other normal cell types. Also, because the normal developmental course is a result of cell–cell interactions in the theory presented here, the appearance of cancer cells is expected to depend on cell–cell interactions (Rubin, 1990, 1994a,b). Relatively large variation in cancer cell phenotypes (Heppner, 1984; Reya et al., 2001) may reflect points (iii) and (iv). The phenotypes, even though they might be differentiated, could be changed more easily than normal committed cells. Furthermore, phenotypes change depending on the interaction with other cells and mutations. The most plastic cell-type corresponds to cancer stem cells (Jones & Baylin, 2002).

According to point (v), genetic instability in cancer cells is not a cause but a result of phenotypic instability in dynamics, due to weaker robustness in the phenotype. In this sense, genetic instability in cancer cells may be regarded as a genetic response to increasing reproducibility of the phenotype under a given environment, a kind of genetic assimilation in Waddington's sense

---

[7] The argument here does not include epigenetic modification. Probably it will be relevant to bridge between faster genetic dynamics and slower genetic mutation process.

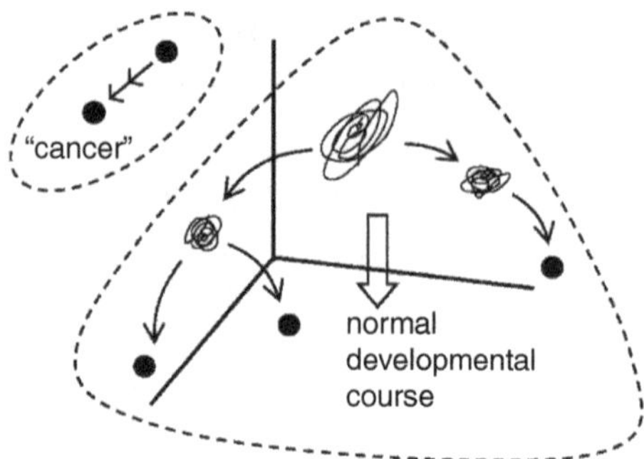

Figure 7.13 Schematic representation of normal differentiation process versus carcinogenesis: Besides attractors adopted in the normal developmental process, there are aberrant attractors with lower robustness. By accumulating mutations, such cells of aberrant attractors gain robustness and achieve recursive production.

(Waddington, 1957), as will be discussed in Chapter 8. Phenotypic changes resulting from gene expression dynamics would then induce epigenetic modifications (as discussed in Section 7.5) and later be fixed to genetic changes.

The hypothesis presented here has several consequences. According to the present picture, mutations would not be a direct cause of cancer cells, and the generation of cancer cells would be strongly dependent on cell–cell interactions and environmental variation. The density dependence of the frequency of tumor cells was reported by Rubin (Rubin, 1990, 1994a,b). Additionally, several experimental reports have shown that transplantation of cancer cells into other tissues, which modifies their interaction with other cells, normalizes the cancer cells (Mintz & Ilmensee, 1975; Kasemeier-Kulesa et al., 2008; Hochedlinger et al., 2004). From this perspective, it is expected that cancer cells do not accumulate mutations when they are first generated. At this stage, they revert to the original multipotent cells by appropriate operations. Such reprogramming has already been observed in nuclear transplantation and nuclear cloning (Blelloch et al., 2004; Li et al., 2003). If the change in cellular states can be observed by single-cell imaging, the present hypothesis can be confirmed with the help of microarray analysis. (see Fig. 7.11).

Thus far, the view on cancer cells presented here remains a hypothesis. There are neither specific theoretical models that confirm this hypothesis nor experimental demonstrations. Still, considering some related experimental results and the argument of evolutionary robustness in Chapter 8, this hypothesis may be promising (Kaneko, 2011) (Fig. 7.13).

7.8 Summary: Universal Properties Discussed in the Present Chapter

- During development, cells sequentially lose their ability to differentiate into other types of cells and are committed to different cellular states. This process can be described by a landscape picture in which the valleys are successively canalized. Here, the height of this landscape can be regarded as the variability of cells (competence or plasticity). On the other hand, there are only few spatial dimensions in which differentiation occurs in the landscape. These dimensions are possibly three or four. (A, E).

- As cells with oscillating gene expression proliferate and interact with each other, they differentiate into other expression states. Cells with an initial oscillation state have pluripotency, either self-renew or differentiate into other cellular states. Alternatively, cells that differentiate and lose their oscillations of expression simply replicate themselves, that is, they are committed. This differentiation can progress hierarchically. The pluripotent cells have a large diversity of expressed components, a large temporal variation in expression levels, and a large cell-to-cell heterogeneity; the irreversible progression of differentiation is characterized by the decrease in these three quantities (B, E).

- The relative proportion of each cell type is robust to the changes in initial conditions and perturbation by noise. This is because cell differentiation by intracellular expression dynamics and cell–cell interactions occurs to achieve consistency between cellular states and the cell ensemble. A reciprocal relationship exists between the plasticity of single-cell dynamics and the robustness of cell populations (C).

- Differentiation by protein expression dynamics is further stabilized by a feedback process from epigenetic modifications, such as DNA modification. With the epigenetic feedback regulation, oscillations in gene expression diminish, and are successively replaced by fixed cellular states (B, D).

- The irreversibly differentiated cell state can be reverted to a pluripotent state by recovering an oscillatory state, as a result of overexpression of multiple genes from the outside, as known as reprogramming in experiments (B, E).

- In a gene regulatory network consisting of many genes, it is expected that there will be attractors other than those that appear during normal development, which will be selfish cells that do not have a consistent relationship with other cells. They will be susceptible to genetic variation as they are not consistent with epigenetic or slow genetic changes. These cells may correspond to cancer cells (F).

8

Evolution of Phenotypic Plasticity and Robustness in Terms of Phenotypic Fluctuations

8.1 Introduction

When it comes to evolution, one might often focus on the genetic changes. In evolution, however, the object of selection is not the genes themselves, but the *phenotype*, a property that appears outside from each organism. The phenotype, for example, refers to the abundances of each protein in a cell, the growth rate of a cell, its mobility, its responses to external chemicals (e.g., bacterial resistance to antibiotics), and so forth. In multicellular organisms, the phenotype includes the developed shape and its function, motility, predator's capacity to get preys, and so forth. The survival rate and the offspring number of each organism are determined by such phenotypes.

Therefore, to discuss evolution, the primary question is not the change in genes per se, but rather the change in phenotype, even though the gene, of course, is involved in determining the phenotype. Hence, we focus on the phenotypic evolution in this chapter. At this time, it will be interesting to note that the feasibility of phenotypic evolution varies by each phenotypic trait from species to species. Some traits have been conserved throughout evolution, whereas some others are easier to evolve. Some species preserve their phenotype over long periods of time, known as "living fossil." Here this changeability of phenotype through evolution is not necessarily simply determined just by the rate of genetic change (substitution of DNA sequence for example).

Of course, the phenotype changes as the genes changes and in that sense the gene is an important factor in evolution that controls the phenotype. However, genes and phenotypes do not necessarily establish a simple one-to-one relationship. Phenotypes are generally formed as a result of the developmental dynamics of an organism's state. In unicellular organisms, the abundances of each component are a result of a variety of intracellular reactions, such as the dynamics of protein expressions. These shape the phenotype. In multicellular organisms, more complex developmental processes lead to the formation of cell types and their spatial patterns, which, in turn, lead to the formation of phenotypes. Depending on the nature of these dynamics that shape the phenotype, differences in genes may or may not result in a significant difference in phenotype.

To sum up, a phenotype is shaped through dynamics (developmental dynamics, in a broad sense), which provide a genotype–phenotype map. Hence, even if genes change at the same rate, the changeability of phenotypes by evolution is not identical, which depends on the nature of genotype–phenotype map.

8.1.1 Plasticity and Robustness to Characterize Evolution at a System Level

Evolution by natural selection refers to adaptive changes in both the genotype and phenotype over generations under a particular set of environmental conditions. Generally, it has the following structure:

(i) Genes control the dynamics to determines phenotypes (at least to some (a large) degree).

(ii) The selection process is governed by the fitness, which is determined by the phenotype and environment.

(iii) Phenotype itself is not inherited to the next generation (for most cases). Hence, the phenotype is transferred only by means of the genotype.[1]

If the mapping gene $\rightarrow$ phenotype in (i) is uniquely determined, then instead of considering the phenotype, one can discuss the evolution in terms only of the genes that determine the phenotype. In this case, one need only pay attention to the distribution of genes and their change: Recall that fitness is given as a function of the phenotype, as Fitness = F (phenotype). If the phenotype is given as a function of the genotype as Phenotype = G (genotype), then the fitness could be represented as Fitness = F (G (genotype)), that is, as a function of the genotype. Then, the problem of evolution could be simplified to the change in the population of genotypes.

However, as we saw in Chapter 3, phenotypes fluctuate even if individuals share the same genotype. The phenotype fluctuates from one individual to another around its mean value, and the variance is often quite large. This fluctuating phenotype is generated from the individuals sharing the same genotype, and its specific value as a result of fluctuation is not passed on to the next generation (as long as we do not take the epigenetic memory over generations into account). However, the magnitude of the fluctuations itself depends on the genotype, so that it can be

---

[1] Some phenotypes are transferred to the next generation; for instance, in a unicellular organism, the protein abundances are transferred to the next generation. There are also epigenetic changes that can be transferred. As discussed in Chapters 2 and 6, the memory of these changes is lost within a shorter time scale than that for the change in genes. (The time scale for the changes is ordered, as phenotype, epigenotype, and genotype from fast to slow).

inherited. This magnitude depends on the rigidity of the genotype–phenotype map. Indeed, this variability of the phenotype is related with the plasticity.

In fact, the genotype–phenotype mapping is the result of a developmental process. As the process involves stochasticity, the phenotype may not be uniquely determined by the genotype even under a given environment. This mapping may also change due to evolution. Hence, it is necessary to consider seriously the genotype–phenotype (G–P) mapping. Furthermore, this mapping is essential to discuss the following two concepts: plasticity and robustness, as already discussed in Chapter 2.

Plasticity (changeability or responsiveness): A phenotype generally changes depending upon external or internal perturbations. Plasticity generally refers to changeability against environmental variation. Through developmental dynamics, the influence of the environment is amplified or reduced depending on the biological system (Callahan et al., 1997; West-Eberhard, 2003; Kirschner & Gerhart 2005; Pigliucci et al., 2006). In other words, plasticity concerns how the G–P mapping is affected by environmental change.

Robustness: As already discussed in Chapter 2, Waddington proposed canalization, where a certain phenotype state is represented as a point at the bottom of a valley, and represented the developmental or evolutionary process used to shape a stable phenotype as the motion of a ball rolling along a valley. When the process is canalized, the stability of the reached phenotype is represented by the return of the ball to the bottom of the valley against perturbations applied to it (Waddington, 1942, 1957). In general, robustness is defined as the insensitivity of the phenotype against possible perturbations to the system (Wagner, 2000, 2005; Wagner, 1997; Barkai & Leibler, 1997; Alon et al., 2000; Siegal & Bergman, 2002; Ciliberti et al., 2007; Kaneko, 2007). Here again, these changes have two distinct origins: a nongenetic one, due to the stochasticity that can arise during the developmental process, and a genetic one, due to the genetic changes produced by mutations.

Besides the above question concerning the evolution of robustness, another issue to be addressed is the relationship between the two types of robustness concerning the changes on the developmental and evolutionary scales. Then, are developmental and evolutionary robustness (plasticity) correlated? Do the two types of robustness increase or decrease in correlation through evolution? Indeed, the search for a possible relationship between the two kinds of robustness is one of the main issues in Evo-Devo, a field that studies the relationship between evolution and development.

To answer such questions quantitatively, it is necessary to characterize the plasticity and robustness of a phenotype quantitatively. Here, recall that in general, a phenotype is not given by a single variable, but consists of several variables. For example, in a cell, phenotypes involve the concentrations of all the components, or expressions levels of a variety of protein species. The size, weight, or activity of an organism is also an example of phenotypes. The environmental conditions that influence phenotypes are also multidimensional, and there are many such factors, including concentrations of chemicals, temperature, and so forth. Genetic changes are also high-dimensional, and there is a huge combinatorial variety in DNA sequences.

Hence, to characterize plasticity and robustness, it is necessary to seriously take such multidimensional phenotypic changes into account. For instance, by taking a variety of environmental condition, and counting the number of such conditions in which the phenotype of an organism is preserved, one can define the robustness of the phenotype, as an important index for the evolution in the field. Similarly, the number of environmental conditions that lead to adaptive changes in phenotypes may give a measure of plasticity. In spite of the importance of such multidimensional characteristics, however, it is sometimes difficult to give a precise mathematical definition for such characteristics, because there are a huge number of possible changes in environmental conditions. Hence, to start a theoretical discussion, it is relevant to simplify the issue of plasticity and robustness by restricting our concern to the directed evolution under a given fitness function for survival, that is, under a fixed environmental condition. With this restriction, only the degree of change in one (or a few) variable(s) is studied, and then, possible relationships between the plasticity, evolvability, and robustness can be investigated quantitatively.

8.1.2 Variance of Phenotype as a Measure of Robustness

Following the above discussion, the phenotype distribution under a given environment and genetic background (which can also be distributed) needs to be investigated. This phenotype is a function of genotype and environment, and indeed is shaped by "developmental dynamics." Here, developmental dynamics refer generally to the process that shapes a phenotype. Not only does it refer to the standard developmental process in multicellular organisms, but also it can cover intracellular dynamics of unicellular organisms with cell growth and division. An example of such developmental dynamics is the protein expression dynamics leading to a certain composition of protein concentrations as a phenotype, mutually regulated by the expressed proteins, as described in Chapters 4, 5, and 7.

As already discussed in Chapters 2 and 3, such dynamics are generally complex and noisy. For instance, gene expression dynamics are noisy because of the molecular fluctuations in chemical reactions. Hence, the phenotype generated as a result of such dynamics is distributed among isogenic individuals (Oosawa, 1975;

Spudich & Koshland, 1976). As discussed in Chapter 3, variance of protein abundances over isogenic cells, as have been measured by using fluorescent proteins, are indeed rather large (McAdams & Arkin, 1997; Elowitz et al., 2002; Hasty et al., 2000; Furusawa et al., 2005).

Such stochasticity in the developmental dynamics might cause a loss of function or malfunction, if the fitness depends on the phenotype generated by the stochastic dynamics. The phenotype that is concerned with the fitness, then, is expected to maintain some robustness against such stochasticity in gene expression, that is, robustness to noise in its *developmental* dynamics. When the phenotype is robust to noise, this distribution of a (functional) phenotype would be sharper. Hence, the variance of an isogenic phenotypic distribution, denoted as V_{ip} here, will give an index for the robustness to noise in developmental dynamics (Kaneko & Furusawa, 2006; Kaneko, 2007).

Similarly, robustness against genetic change is measured in terms of fluctuations. Because of genetic mutations, the phenotype (fitness) is distributed. Since even the phenotype of isogenic individuals is distributed by noise in developmental dynamics, the variance of the phenotype distribution of a heterogenic population includes both the contribution from the phenotypic fluctuation in isogenic individuals and that caused by genetic variation. To distinguish the two, the average phenotype over isogenic individuals is first measured, and then the variance of this average phenotype over the heterogenic population is computed. This variance is a result of the genetic heterogeneity. It is called as genetic variance V_g (Futuyma 1986; Hartl & Clark 2007). Then, the robustness to mutation is estimated by this variance. If V_g is smaller, genetic change has little influence on the phenotype, implying a larger genetic (or mutational) robustness.

8.1.3 Response Ratio as a Measure of Plasticity

The phenotypic plasticity, as discussed in Chapter 2, represents the degree of change in the phenotype against variation in the environment, where we introduced the ratio of change in a phenotype against environmental change as a quantitative measure for it. For example, if the concentration (x) of a specific protein that depends upon the concentration (s) of an external signal molecule, the response ratio, R is given by $R = \Delta x/\Delta s$ (or $R = \Delta log(x)/\Delta log(s)$, if the response follows Weber's law as in Chapter 5).

By using this response ratio and the above variances to characterize the plasticity and robustness, respectively, the relationships between the two are explored here. With this characterization, the change in plasticity through the course of evolution is also studied (Visser et al., 2000; West-Eberhard, 2003; Weinig, 2000; Gibson & Wagner, 2000; Kirschner & Gerhart, 2005; Ancel & Fontana, 2002).

8.2 Evolutionary Fluctuation–Response Relationship (A–C and E)

8.2.1 Macroscopic "Derivation" from Stable Distribution (A)

Let us consider a distribution of a macroscopic phenotype variable, x (for example, this could be the growth speed of a cell, gene expression level, and so forth). As this variable is distributed, we introduce the distribution $P(x)$. Here the phenotype variable x is shaped as a result of a dynamical process under noise. This dynamical process depends on the genotype. If the genetic change is assumed to be represented by a change in a continuous parameter, a, then the evolutionary change in the phenotype of concern is expressed by the change in x against the change in parameter a.

As mentioned already, even among individuals sharing an identical genotype (i.e., parameter a), the variables, x, are generally distributed. Hence, it is necessary to study the distribution, $P(x; a)$. This is a conditional distribution of the phenotype variable, x, under a given parameter, a (e.g., genotype). Here the average and variance of x are defined with regards to the distribution, $P(x; a)$, for an isogenic population with a given genotype, a. The response adopted to define the plasticity is then given by the change in the average value of x against the change in a, while the (isogenic) phenotype variance, V_{ip}, is defined through this distribution, $P(x:a)$. Following the fluctuation-response relationship borrowed from statistical physics (see also Chapter 2), we discuss relationship between the response and V_{ip}.

To consider the relationship between phenotype change due to genetic variation and the phenotypic variance without genetic change, let us apply the formulation in Section 2.4.1, by assuming that a is a parameter that specifies the change in genotype (e.g., the number of substitutions in DNA sequence). Then Eq. (2.3) in Chapter 2, that is, the relationship

$$\frac{<x>_{a+\Delta a} - <x>_a}{\Delta a} \propto < (\delta x)^2 >, \tag{8.1}$$

implies that the change in the average phenotype due to the genetic change is proportional to the variance of the phenotypic distribution over clones V_{ip}. In other words, it suggests that the evolution speed of a phenotype (i.e., change in the average phenotype per generation) divided by the mutation rate is proportional to V_{ip}, the variance of isogenic phenotypic fluctuation.

Although the above formula is formally similar to that used in statistical physics, it is not necessarily grounded on any established theory. As described in Section 2.4.1, there are some assumptions in the above "derivation."

First, the distribution must be close to Gaussian. In Eq. (2.4), the form $P(x; a) \propto exp(-\frac{(x-X_0)^2}{2\alpha} + v(x, a))$ is assumed, where $v(x, a)$ represents a response term. Here, if the distribution of concern does not follow the Gaussian distribution, as long as the moment is finite, the distribution can be transformed to be nearly Gaussian, by suitable choice of (bounded) transformation of the variable. We can choose such transformed variable as a phenotype variable of concern. For example, if a measured variable z (say the concentration of some protein x) follows a log-normal distribution as often observed in biological systems, then, one can just adopt $x = log(z)$ as a phenotype variable in concern.[2]

Second, a bilinear coupling form in $v(x,a)$ is chosen, by assuming the deviation from the distribution at $a = a_0$ is small, so that $v(x,a)$ can be expanded as $v(x,a) = C(a - a_0)(x - X_0) + ...$, with C as a constant, higher order terms in $(a - a_0)$ and $(x - X_0)$ in $v(x,a)$ are neglected. Such neglect of the terms with $(x - X_0)^k(a - a_0)^m$ (with $k, m > 1$) may be justified if the deviations from a_0 and X_0 are small, but the validity of such an expansion in the distribution itself is, in some sense, a proposition here.

Furthermore, there is a strong underlying postulate here; the existence of a single scalar parameter a that controls the behavior of the phenotype x. In the case of environmental conditions, one can naturally adopt an environmental factor (say concentration of external resource or temperature) as a, but in the case of evolution, the validity of representing a genotype by a single parameter, a is not self-evident. Hence, one needs to check the above fluctuation–response relationship by microscopic numerical models and laboratory experiments.

8.2.2 *Microscopic Confirmation of the Relationship (B)*

To study the genotype–phenotype mapping at a *microscopic* level, the dynamical process for generating phenotypes needs to be explicitly considered. Here, this process is given by *developmental dynamics* in its general sense, as already mentioned. It involves a large number of variables, for instance, concentrations of many components. Through interactions among components at a microscopic level, a macroscopic phenotype or fitness is shaped, as a result of the developmental dynamics. For example, protein expression dynamics as mutual activation and inhibition over many genes will be considered as the *developmental dynamics*.

Phenotypes of each organism are determined by such dynamics, and accordingly the fitness (survival rate) is determined. The evolutionary process consists of selection according to the fitness and genetic change that causes the change in developmental dynamics. Genes control the equations of such developmental dynamics involving many degrees of freedom. Possible changes in each term in the equations or in parameter values correspond to variations in genes.

---

[2] In fact, Haldane (1949) suggested that most phenotypic variables should be defined after taking a logarithm.

Here, a population of organisms that may slightly differ in genes exists, the fitness of which may (slightly) differ by individuals. The offspring are generated according to the fitness value (i.e., those with a higher fitness value reproduce multiple offspring, and those with lower values do not produce any). In reproducing offspring, slight changes in genes are introduced, as changes in the terms or parameter values in the equation for the development dynamics. Below, we study the characteristics of evolution of two models with the above features, that is, phenotypic dynamics of many degrees of freedom and genetic algorithms for the population of organisms.

8.2.2.1 Catalytic Reaction Network

First, we use a model of the catalytic reaction network (see Section 5.3.1), and evolved it by selecting to have a higher abundance of a certain chemical component. The model is simulated by preparing $N(= 10^4)$ types of networks, measuring the amount of a certain predetermined component and selecting networks that generate higher values of the amount (e.g., networks of upper N/K ($K = 20$ if 5 percent)). From each network, K offspring are generated. When generating an offspring network from a mother's network, the network paths are rewired with a given, certain mutation rate. As a result, in the next generation, N cells (with slightly different networks) are generated which vary from each other with a small number of mutations, that is, changes in reaction paths. Among the mutants, those networks with a higher concentration of a given, specific chemical component are selected for the next generation. If this process is repeated, a cell with a larger amount of a predetermined component will evolve.[3]

Here, the amount of this component differs from cell to cell because of the stochastic catalytic reaction dynamics going on in each cell, even for cells that share the same network (gene). We computed the variance of the logarithm of concentration for identical networks, at each generation.[4] This gives V_{ip}. On the other hand, the evolution speed was computed by the increase in the logarithm of the concentration at each generation. In Fig. 8.1, we plotted the evolution speed versus V_{ip} over generations. One can see clearly the proportionality between the two (Kaneko & Furusawa, 2006).

8.2.2.2 Gene Regulation Network

Next, to examine the universality of the result, gene expression dynamics adopted in Chapters 4, 5, and 7 have been studied, which are governed by regulatory

---

[3] This is a method called genetic algorithm. Of these, it belongs to the simplest class.

[4] In this case the distribution of the abundance of X follows log-normal distribution. Hence *log(concentration)* was adopted as the phenotypic variable x, because the fluctuation–response theory is applied for the variable close to Gaussian distribution.

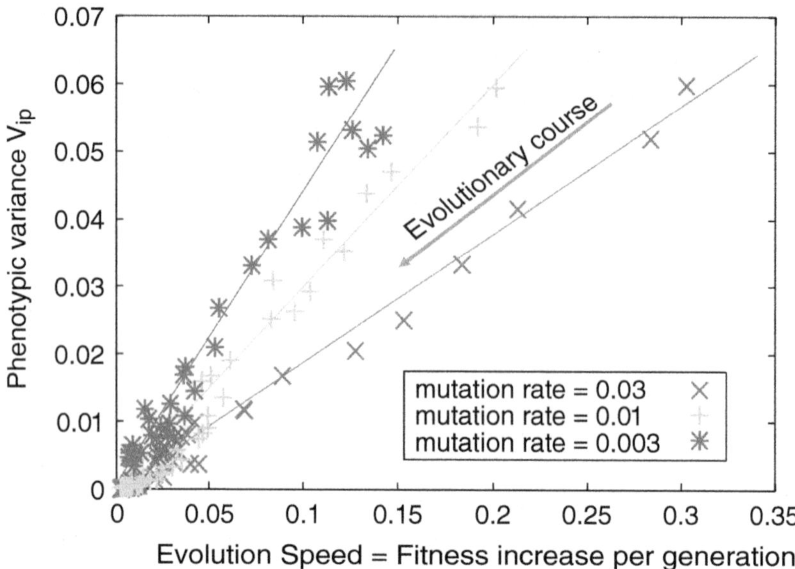

Figure 8.1 Evolution speed versus the isogenic variance of the fitness V_{ip}. The abscissa gives the evolution speed that is measured by the difference between the average phenotypes x of the two successive generations, where the average is computed from the selected 500 cells. x is given by the logarithm of the concentration of given component. The ordinate gives the fluctuation, that is measured by the variance of x of the clone of the selected cells, at each generation, computed by the distribution over 200 cells of the clone. Adapted from Kaneko & Furusawa (2006).

networks (Glass & Kauffman, 1973; Mjolsness et al., 1991; Salzar-Ciudad et al., 2000). To be specific, we adopted the following form of gene expression dynamics for each expression level x_i (see Section 4.10):

$$dx_i/dt = \gamma(f(\sum_{j}^{M} J_{ij}x_j) - x_i) + \sigma\eta_i(t), \tag{8.2}$$

where $J_{ij} = -1, 1, 0$, and $\eta_i(t)$ is Gaussian white noise given by $< \eta_i(t)\eta_j(t') >= \delta_{i,j}\delta(t - t')$. M is the total number of genes. The function f represents gene expression dynamics follows threshold-dynamics, say $1/(1 + exp(-\beta(x - \theta)))$ with $\beta \gg 1$ and θ as parameters (Kaneko, 2007, 2008) (see Sections 4.10 and 5.3.2). The value of σ represents the strength of noise encountered in gene expression dynamics, originating in molecular fluctuations in chemical reactions. The initial condition for this "developmental" dynamical system is set so that none of the genes are expressed. The phenotype is determined as a pattern of $\{x_i\}$ after the system reaches a stationary state (attractor).

The fitness F is determined as a function of x_i. Generally, not all the gene expressions contribute to the fitness. It is given as a function of the expression of only a set of "output" genes. For example, the fitness is assumed to be the number of expressed ("on") genes among the output genes $j = 1, 2, \cdots, k \leq M$, that is, x_j greater than a certain threshold (Kaneko, 2007). If the expression of x_j changes in time, we can take the temporal average of the number of expressed genes over time. (To be specific, the fitness F is given by taking the minus of the average number of output genes that are off; if all are on F is 0, which is the maximal fitness value, and $-M$ if all are off). Selection is applied after the introduction of mutations at each generation in the transcriptional regulation network, that is, the matrix J_{ij}.

Among the mutated networks, selected is a certain fraction of networks that has higher fitness values. Because the network is governed by J_{ij} which determines the 'rule' of the dynamics, it is natural to treat J_{ij} as a genotype. Individuals with different genotypes have different sets of J_{ij}. As the model contains a noise term, whose strength is given by σ, the fitness can fluctuate among individuals sharing the same genotype J_{ij}. Hence, the fitness F is distributed. This leads to the variance of the isogenic phenotypic fluctuation denoted by $V_{ip}(\{J\})$ for a given genotype (network) $\{J\}$, that is,

$$V_{ip}(\{J\}) = \int dF \hat{P}(F; \{J\})(F - \overline{F}(\{J\}))^2, \tag{8.3}$$

where $\hat{P}(F; \{J\})$ is the fitness distribution over isogenic individuals sharing the same network J_{ij}, and $\overline{F}(\{J\}) = \int F \hat{P}(F; \{J\}) dF$ is the average fitness.

This model was simulated under a sufficient level of noise σ (say 0.08). In this case, the fitness increases to the top level ($F = k$) with generations. The evolution speed at each generation is computed by the increase of the average fitness $\overline{F}$ at each generation. At each generation, the variance of the fitness is computed, which is defined above. In Fig. 8.2, this evolution speed versus variance is plotted. As is shown, they decrease while keeping proportionality (except the first few generations). Hence the proportionality between the two is confirmed.

8.2.3 Laboratory Experiment to Increase the Fluorescence(E)

Evolution experiments in the laboratory are now possible at a quantitative level (Cooper et al., 2003; Kishimoto et al., 2010). By using a population of (micro)organisms, under a given environment, it is possible to select those with a higher fitness. By repeating the selection over generations with genetic mutations, the changes of phenotypes (and genes) as well as of their distribution through evolution are measured.

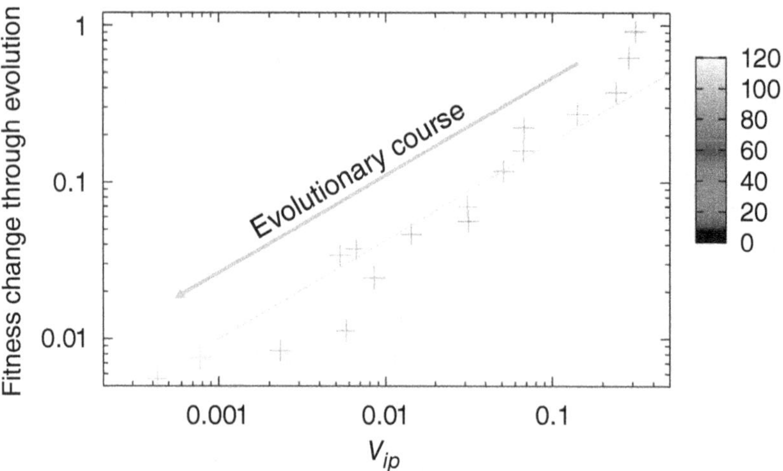

Figure 8.2 Evolution speed versus the isogenic variance of the fitness V_{ip}. The variance V_{ip} is computed at each generations 5,10,15,20, ... 100 and the increase in the fitness per 5 generation is computed as the evolution speed. The ordinate shows V_{ip} and the abscissa shows the increase in the fitness, at each generation. The top-right is the 5th generation, and with the generation both V_{ip} and the evolution speed decrease.

For example, by embedding a gene for a fluorescent protein in bacteria, a selection experiment to increase the fluorescence was carried out (Sato et al., 2003; Ito et al., 2009). Then, the increase in the fluorescence over generations was measured, which gave the evolutionary response. On the other hand, the variance of the fluorescence was measured by flow cytometry. From these two measurements, the relationship between the response and fluctuation was examined.

The relationship between evolution speed and isogenic phenotypic fluctuation was then studied. In this experiment, fluorescent protein gene was inserted into bacteria, and then selection experiment to increase the fluorescence was carried out. First, by attaching a random sequence to the N terminus of a wild type Green Fluorescent Protein (GFP) gene, a protein with (low) fluorescence was generated. The gene for this protein was introduced into *E. coli*, at the initial generation. Next, by applying random mutagenesis only to the attached fragment in the gene, a mutant pool with diverse cells was prepared. Then, cells with the highest fluorescence intensity were selected for the next generation. With this procedure, the (average) fluorescence level of selected cells increased over generations.

The evolution speed at each generation was then computed as the difference of the logarithm of the fluorescence intensity between the two generations. Here,

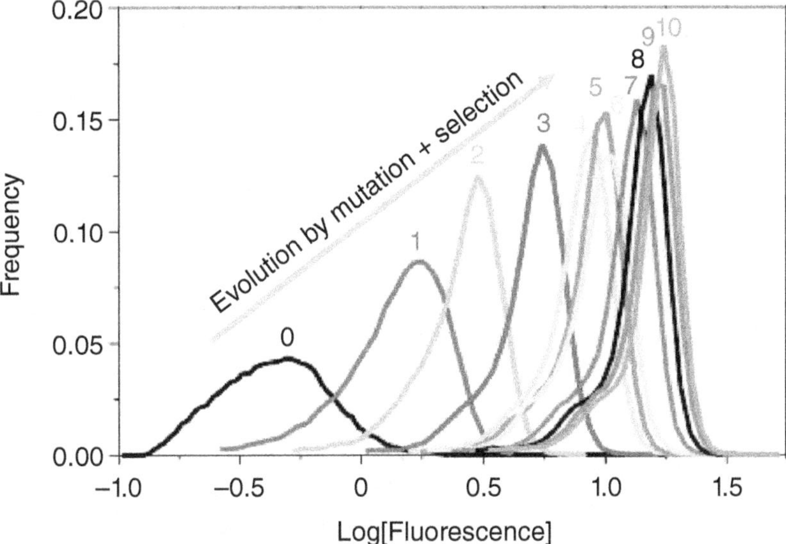

Figure 8.3 Histogram of the logarithm of the fluorescence intensity for each generation, for the experiment described in the text. The number above the peak of each distribution indicates its generation number. The fluorescence intensity of each *E. coli* in each generation was measured by flow cytometry. Adapted from Sato et al. (2003).

to observe the isogenic phenotypic fluctuation, a distribution of fluorescence of clone cells of the selected bacteria was measured. As the distribution is close to log-normal, the logarithm of the fluorescence intensity was adopted as the phenotype variable x, so that the distribution of $x =\log(\text{fluorescence})$ was nearly Gaussian. The distribution of x over the clones was measured with the help of flow cytometry. As shown in Fig. 8.3, the distribution was sharper with generations, whereas the evolution speed, that is, the increase of the log-fluorescence level x per each generation, also decreased. Then, the variance of fluorescence from this distribution was measured at each generation, from which the fluorescence variance versus the evolution speed per (synonymous) mutation rate was obtained.[5] In Fig. 8.4, the evolution speed (per generation) versus the variance multiplied by the synonymous mutation rate was plotted instead. The data support strong correlation between the two, suggesting the proportionality between the two (Sato et al., 2003).

---

[5] The applied mutation rate depends on the experimental procedure for mutagenesis. In this experiment two values of mutation rates are adopted depending on the generation. Hence, we need to measure the evolution speed per mutation rate.

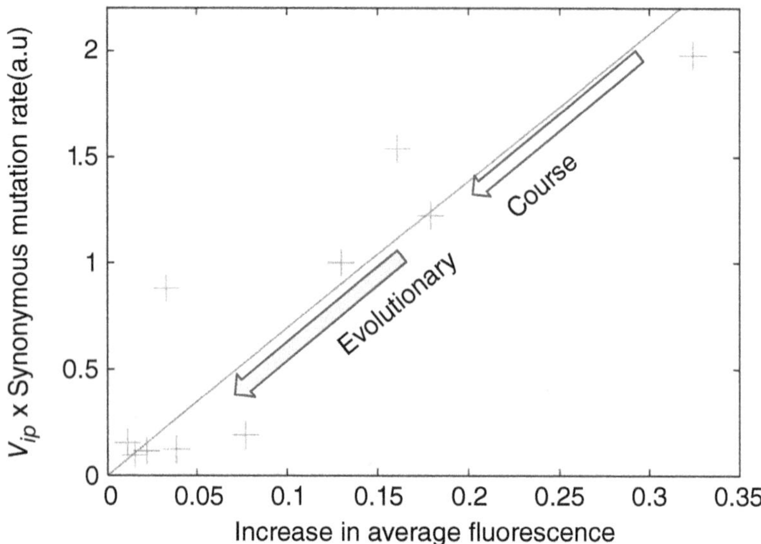

Figure 8.4 Evolution speed versus the isogenic variance of the fitness V_{ip}. The ordinate shows the variance V_{ip} multiplied by synonymous mutation rate at each generation. The variance is computed from the distribution of log(fluorescence) over isogenic bacteria population. The abscissa shows the increase in the fluorescence between the next and the present generations. Here the mutation rate was decreased at the 5th generation (see [Sato et al. 2003] for details). The right-top point is the data from the first generation, and with the generation, the evolution, as well as the variance speed decreases.

8.3 Relationship between Genetic and Developmental Robustness: V_g–V_{ip} Law (A–C and E)

8.3.1 Macroscopic Approach for V_g–V_{ip}-Evolution Speed (A)

8.3.1.1 Fisher's Theorem

The evolutionary fluctuation–response relationship mentioned above casts another question to be solved, because there is an established relationship between the evolutionary speed and phenotypic variance due to genetic variation. It is the so-called fundamental theorem of natural selection by Fisher (1930), which states that evolution speed is proportional to V_g, the variance of phenotypic fluctuation due to genetic variation.

According to the fundamental theorem of natural selection by Fisher (Fisher, 1930, 1958; Edwards, 2000), the rate of phenotypic change by genetic evolution is proportional to the phenotypic variance caused by genetic variation. Since genetic changes are passed on to the next generation, the larger the dispersion, the larger the "stride" of phenotypic change transmitted to the next generation through selection. Therefore, this relationship is easily expected and can in fact, be easily derived as follows.

Let $P_n(f)$ be the distribution of the fitness f at generation n. Then the average fitness is then given by

$$<f_n> = \int fP_n(f)df, \tag{8.4}$$

Where $< \dots >$ denotes the average over the distribution $P(f)$. Then the fitness distribution of the next generation is given by

$$P_{n+1}(f) = fP_n(f)/ \int fP_n(f)df = fP_n(f)/ <f_n> \tag{8.5}$$

Because the offspring number is proportional to the fitness. From this equation

$$<f_{n+1}> - <f_n> = \frac{\int f^2 P_n(f)df}{<f_n>} - <f_n>$$

$$= \frac{\left(\int f^2 P_n(f)df - \left(\int fP_n(f) \right)^2 \right)}{<f_n>} = \frac{<\delta f_n^2>}{<f_n>}, \tag{8.6}$$

where $<\delta f_n^2> = <(f_n - <f_n>)^2>$ is the variance of the fitness f_n. Hence, the rate of increase in the fitness per generation is proportional to its variance.

Similarly, the relation on any phenotypic variable g is obtained. Consider the average phenotype $<g_n> = \int gP_n(f)df$ at generation n. Then its change per generation is given by

$$<g_{n+1}> - <g_n> = \frac{\int fgP_n(f)df}{<f_n>} - <g_n>$$

$$= (<g_n f_n> - <g_n><f_n>) / <f_n> \tag{8.7}$$

This relation is called as Price equation (Price, 1970, 1972; Futuyma, 1986; Rice, 2004).

Note again that the above variance is that of *average* phenotype for each genotype over a **heterogenic** population. In contrast, the relationship in Section 8.2 states the proportionality between the evolution speed of phenotype and the variance over the clones, that is, *isogenic* individuals. In order for our relationship to be consistent with Fisher's theorem, the **phenotypic fluctuations among individuals with the same gene**, denoted by V_{ip}[6] should be proportional to **phenotypic**

---

[6] Phenotype fluctuations due to changes in the environmental variation are often denoted as V_e, and that due to random noise in the developmental process as V_{noise}, respectively. It can be said that V_{ip} here roughly corresponds to V_{noise}. (It is also sometimes referred to as fluctuating asymmetry). However, we do not know to what extent the variance among individuals with the same gene individual are due to noise during

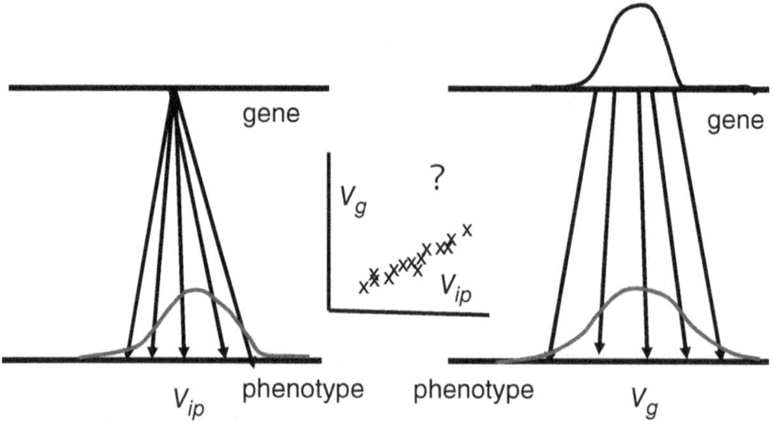

Figure 8.5 Schematic representation of V_{ip} and V_g. Both are the variance of phenotypes but the former is a result of noise across isogeneic individuals, whereas the latter is mapped from the genetic distribution. Here we seek for the proportionality between the two.

fluctuations due to genetic variation, V_g. Here, the former, V_{ip} concerns the variance of isogenic phenotypic fluctuation (see Fig. 8.5 for schematic representation). The latter, the phenotypic variance due to genetic mutations, is standardly denoted by V_g.[7]

In contrast, the evolutionary fluctuation–response relationship, proposed above, concerns phenotypic fluctuation of isogenic individuals, V_{ip}. Hence, the fluctuation–response relationship and Fisher's theorem are not identical. If both are correct, it is suggested that V_{ip} and V_g are proportional through an evolutionary course. Such proportionality, however, is not self-evident, as V_{ip} is related to the variation against the developmental noise and V_g against the genetic change (mutation). The relationship between the two, if it exists, postulates a constraint on genotype–phenotype mapping shaped through evolution.

Remark: Recall that the phenotype of the clones, that is, among isogenic individuals can vary (by noise), as already mentioned. Then, in order to estimate V_g precisely, we must carefully distinguish it from the variance by noise: If one simply compute the variance of phenotype over heterogeneous population, it includes both the variance by noise and noise.[8] To be precise, one first needs to obtain the mean phenotype over clones (i.e., over isogenic individuals), to eliminate the

generation. This is because small fluctuations in the environment can be amplified throughout the developmental process. In order to avoid any misunderstanding here, we use neither V_{noise} nor V_e.

[7] In conventional population genetics, the sexual reproduction is discussed, where the notation to separate the variance due to exchange of genes by sexual recombination from that by mutation is sometimes introduced. We do not go into details of such distinction here.

[8] If the variances of the two are originated independently, the variance of the simple heterogeneous population could be the addition of V_{ip} and V_g, but the independence is not necessarily justified.

influence by V_{ip}, and then, one obtains the variance of the mean over a heterogenic distribution, to get V_g.

8.3.1.2 Evolutionary Stability Theory for Genotype–Phenotype Mapping

To explore the possible relationship between V_{ip} and V_g, we formulate a phenomenological theory so that the evolutionary fluctuation–response relationship and Fisher's theorem are consistent. In Section 8.3.1.1 $P(x; a)$ was studied, the distribution of the phenotype x under given genotype a. Here the distribution of the genotype needs to be also taken into account, which itself is a result of selection process by the phenotype x. Considering a feedback process from phenotype to genotype, a two-variable distribution $P(x, a)$ both for the phenotype x and genotype a is introduced.

By using this distribution, V_{ip}, variance of x of the distribution for given a, can be written as $V_{ip}(a) = \int (x - \overline{x(a)})^2 P(x, a) dx$, where $\overline{x(a)}$ is the average phenotype of a clonal population sharing the genotype a, namely $\overline{x(a)} = \int P(x, a) x dx$.

V_g, in contrast, is defined as the variance of the average $\overline{x(a)}$, over genetically heterogeneous individuals and is given by $V_g = \int (\overline{x(a)} - <\overline{x}>)^2 p(a) da$, where $p(a)$ is the distribution of genotype a and $<\overline{x}> = \int \overline{x(a)} p(a) da$ is the average of $\overline{X(a)}$ over all genotypes.

Now by considering a robust and gradual evolutionary process, we postulate **evolutionary stability**: at each stage of the evolutionary course, the distribution has a single peak in (x, a) space. Indeed, when a certain region of phenotype, say $x > x_{thr}$, is selected, gradual evolution to increase x works if the distribution is concentrated around a peak. If the distribution is flat, or has multiple peaks, non-fitted mutants would remain, and the evolutionary process is hindered. This single-peaked-ness leads to a constraint on the above variances (see Fig. 8.6 for schematic representation). This stability condition is calculated by using the Hessian condition of the derivatives of $logP(x, a)$ in general (Kaneko, 2006). Here we derive the condition by using the Gaussian distribution again. Then, the distribution $P(x, a)$ written as:

$$P(x, a) = \widehat{N} \exp\left[-\frac{(x - X_0)^2}{2\alpha(a)} + C(x - X_0)(a - a_0) - \frac{1}{2\mu}(a - a_0)^2 \right], \quad (8.8)$$

with $\widehat{N}$ as a normalization constant. The Gaussian distribution $\exp(-\frac{1}{2\mu}(a - a_0)^2)$ represents the distribution of genotypes around $a = a_0$, whose variance is (in a suitable unit) the mutation rate μ. The coupling term $C(x - X_0)(a - a_0)$ represents the change in the phenotype x by the change in the genotype a. Recalling that the above distribution (8.8) can be rewritten as

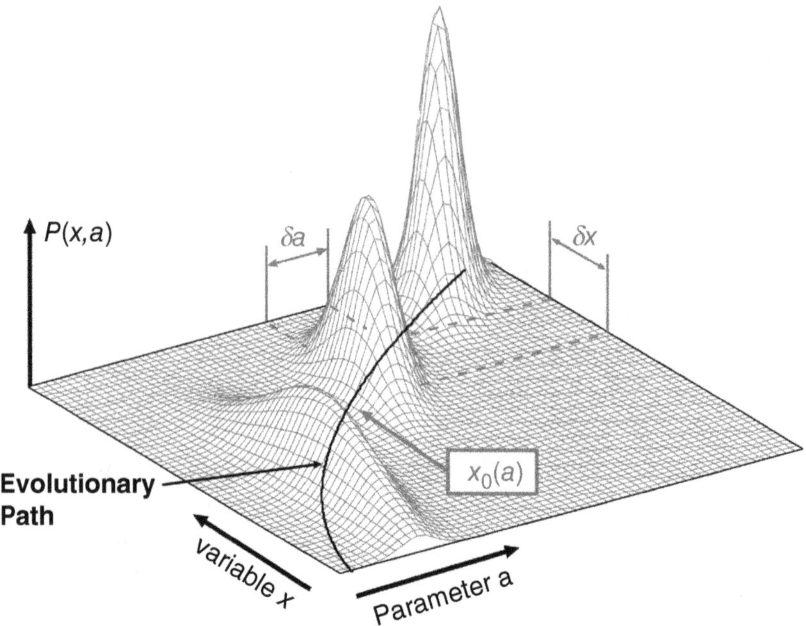

Figure 8.6 Schematic representation of the two-variable distribution $P(x,a)$ on phenotype x and genotype a. The thick curve shows $X_0(a)$, given by the evolutionary course. Three distributions $P(x,a)$ at three generations through this evolutionary course are depicted, as are seen in the change of peak positions. Adapted from Furusawa & Kaneko (2006).

$$P(x,a) = \widehat{N} \exp\left[-\frac{(x - X_0 - C(a - a_0)\alpha(a))^2}{2\alpha(a)} + \left(\frac{C^2\alpha(a)}{2} - \frac{1}{2\mu}\right)(a - a_0)^2 \right],$$

$$(8.9)$$

the average phenotype value for given genotype a satisfies

$$\bar{x}_a \equiv \int xP(x,a)dx = X_0 + C(a - a_0)\alpha(a). \qquad (8.10)$$

The evolutionary stability condition mentioned above postulates that the factor multiplied by $(a - a_0)^2$ be negative. Otherwise, the distribution against gene would be flattened and the distribution could not be concentrated around a functional phenotype X_0. This postulate leads to the condition $\frac{\alpha C^2}{2} - \frac{1}{2\mu} \leq 0$, that is,

$$\mu \leq \frac{1}{\alpha C^2} \equiv \mu_{max}. \qquad (8.11)$$

This means that the mutation rate has an upper bound μ_{max} beyond which the distribution does not keep a peak in the genotype–phenotype space. Beyond $\mu = \mu_{max}$, the distribution is no longer represented by the above Gaussian-type form centered at around (x_0, a_0), and has a long tail which extends to low-fitness values.

Hence, mutants with low fitness values appear. This is akin to the error catastrophe by Eigen (Eigen & Schuster, 1979). The catastrophe here, however, is due to errors in genotype–phenotype mapping in contrast to combinatorial problems in genetic space in the Eigen's case.[9]

By recalling $V_g = <(\overline{x(a)} - X_0)^2>$ and Eq. (8.8), V_g is given by $(C\alpha)^2 < (\delta a)^2 >$. Here, $(\delta a)^2$ is computed by the average $P(x,a)$, as $\mu/(1 - \mu C^2 \alpha)$. Then, it follows that

$$V_g = \frac{\mu C^2 \alpha^2}{1 - \mu C^2 \alpha} = \alpha \frac{\frac{\mu}{\mu_{max}}}{1 - \frac{\mu}{\mu_{max}}}. \tag{8.12}$$

If the mutation rate μ is small enough to satisfy $\mu \ll \mu_{max}$,

$$V_g \sim \frac{\mu}{\mu_{max}} V_{ip}, \tag{8.13}$$

is obtained by recalling that $V_{ip} = \alpha$. Thus the proportionality between V_{ip} and V_g is obtained.

Note that a few assumptions are required to obtain the above relationship. First, the genotype is assumed to be represented by a continuous parameter a. A candidate for such scalar parameter is Hamming distance from the fittest sequence, or a projection according to the corresponding phenotype (Sato & Kaneko 2007). Second, the existence and stability of a two-variable distributions in genotype and phenotype $P(x,a)$ (instead of $P(x;a)$) is assumed. Third, existence of error-catastrophe, that is, the threshold mutation rate μ_{max} is implicitly assumed, beyond which the stability condition is not satisfied. This means that the fitted state is rare in the genetic space.

Phenotypic variance of nongenetic origin is traditionally discussed as environmental variance V_e or fluctuating asymmetry. Here we consider the noise as a major source for V_{ip}, but one could regard this variance as a component of V_e(Futuyma, 1986; Hartl & Clark, 2007), which also measures the isogenic phenotypic fluctuation. Evolution speed under the existence of both V_g and V_e (or V_{ip} here) as well as the covariance of phenotypes are formulated in quantitative genetics (Price, 1970; Falconer, 1981). Here, in contrast, the proportionality between

---

[9] In the above formulation, the distribution is symmetric against the peak, as we assumed the Gaussian-type distribution. However, as the distribution is flatter, the deviation from Gaussian distribution will be amplified. Then, the assumption with symmetry will be lost. In fact, a fitted biological state is rarer than a non-fitted one. Then, as the distribution is flatter, non-fitted states are accumulated. In fact, according to the evolution simulation of the catalytic reaction network and gene-regulation network mentioned so far, most phenotypes obtained under large mutation rates take lower fitness values (Furusawa & Kaneko, 2006; Kaneko, 2007), and the evolution to increase the fitness cannot progress when the mutation rate is larger than some threshold. Hence, this loss of stability of fitted state under larger mutation rate is akin to the error catastrophe by Eigen (Eigen & Schuster, 1979).

the two variances is discussed, which evolves as the result of robust developmental process.

8.3.2 Microscopic Confirmation of V_{ip}–V_g Relationship (B,C)

So far, we discussed a macro-level phenomenology based on the experiments, whereas its microscopic origin is not so clear.

In the macroscopic theory in Section 8.3.1, we assumed a smooth genotype–phenotype correspondence behind the coupling term of a,x. In considering the concentration of the protein as a phenotype, when the abundances of protein concentration of the catalytic component change by noise, the rate of the associated reaction changes accordingly and the amount of protein synthesized as well. On the other hand, if the gene associated with the reaction is changed, the rate of the reaction, accordingly the amount of protein synthesized is changed. In this sense, changes by genetic alteration and due to noise can result in the similar consequence. This might justify the existence of a coupling term between x and a.[10]

Still, this argument is not so strong enough, whereas the microscopic support for such coupling term to give this correspondence has not yet been established at that moment. The experiment that gave us the impetus of the relationship of Section 8.3.1 may not be so accurate. Furthermore, the relationship between V_g and V_{ip} has not been directly confirmed therein. Hence, we explore the relationship by using the models discussed in Section 8.2, below.

8.3.2.1 Microscopic-Approach I:Catalytic Reaction Network

We first study the evolutionary process in the catalytic reaction network model discussed in Section 8.2, and examine the relationship between V_{ip} and V_g. In this study, we introduced 1,000 individuals with networks of slightly different reaction-networks (i.e., different genotypes) in each generation, and repeated a sufficient number of reaction-dynamics simulations (e.g., 1,000 times) for each of these 1,000 individuals to obtain the average fitness for a given genotype (network). Then we repeated the simulation over (1,000) different networks, to get the fitness distribution over the heterogenic population. From this distribution, the

---

[10] Still, one may find a bit strange how fast changes due to noise and much slower changes in genetic variation can cause similar changes at first glance. However, the premise of Einstein's Brownian motion theory is precisely the correspondence between fast changes due to microscopic noise and slow macroscopic changes due to external force. This is a consequence of stability of the equilibrium state. Similarly, the analysis in Section 8.3.1 is based on the evolutionary stability of the phenotype. Incidentally, it was Onsager's hypothesis that paved the way from Einstein's theory to linear response theory. According to Kubo (Kubo et al., 1985), "Fluctuations in thermal equilibrium should normally be only microscopic, and Onsager's hypothesis that these small fluctuations obey macroscopic laws may go far beyond empirical facts." In this sense, one might say that correspondence between noise-induced fluctuations and changes to genetic mutations follows the spirit of Onsager.

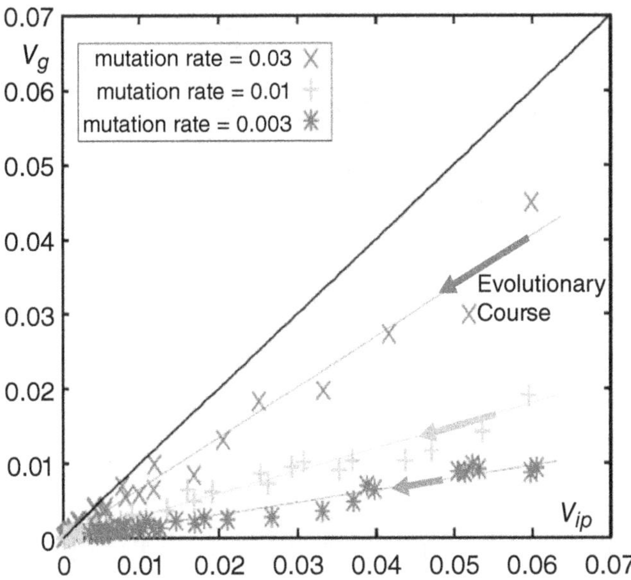

Figure 8.7 The relationship between V_{ip}, the variance of the phenotype x of the clone as measured in Fig. 8.3, and V_g, the variance of the phenotype x over 10^4 mutants from the selected cell. Plotted over generations in each course of evolution, given by a fixed mutation rate displayed in the figure. Adapted from Kaneko & Furusawa (2006).

variance, V_g, is obtained. On the other hand, the simulation results for the identical reaction network fluctuate around the mean value. This gives V_{ip}. (We used the variance of log(fitness) (i.e., log(concentration)) for both V_{ip} and V_g, as the fitness distribution is close to be log-normal).

The plot of V_{ip} and V_g for each generation is shown in Fig. 8.7. One can see the proportionality between the two clearly. As already mentioned, V_g increases approximately proportionally with the mutation rate μ. We have confirmed this increase, up to a certain mutation rate. Beyond this value of mutation rate, however, the distribution of the phenotype is flattened, and the evolution stops. Indeed, this occurs when V_g is reached approximately V_{ip}. Figure 8.8 shows how the distribution of this fitness is expanded as the mutation rate is increased. At this threshold mutation rate, the distribution almost flattens out, resulting in a phenotypic error catastrophe. This result agrees well with the macroscopic distribution theory in Section 8.3.1.

8.3.2.2 Microscopic Approach II: Gene Regulation Network Model

So far, we have discussed the significance of V_{ip} to the phenotypic evolution. This V_{ip} is originated in the noise in the intracellular process. Noise, however, seems to

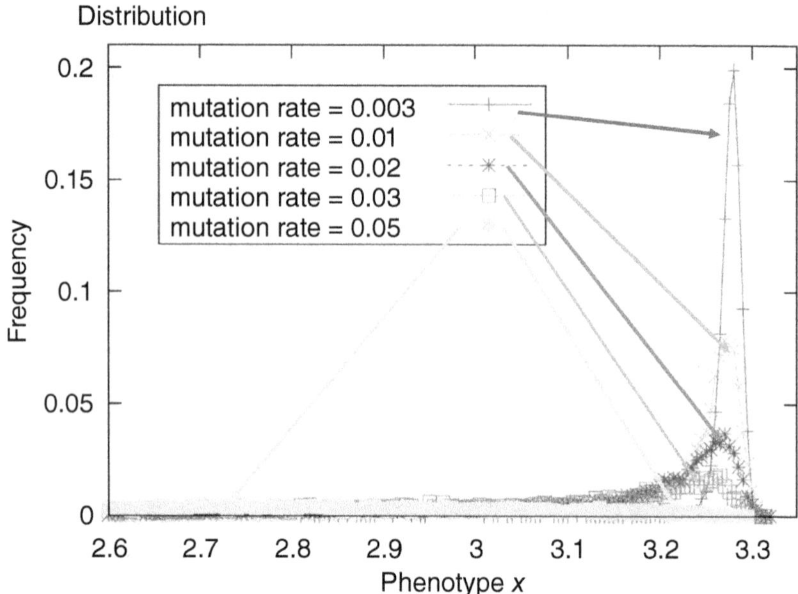

Figure 8.8 Distribution of the phenotype x over 10,000 mutants, generated with the mutation rates 0.003, 0.01, 0.02, 0.03, and 0.05. Around the mutation rate 0.03, the distribution is flattened, and the peak position starts to shift downward. Adapted from Kaneko & Furusawa (2006).

be an obstacle to maintaining an optimal state. On the other hand, the results of the Section 8.3.1 suggest a positive significance of noise to avoid the error catastrophe. Then, why is this so? Is the presence of noise relevant to the evolutionary stability of phenotype? In order to understand this point, it is useful to examine a model by changing the noise strength.

This problem could be investigated using the model in Section 8.3.2.1, but it is somewhat not easy to control the noise intensity as a parameter of the model.[11] In this subsection, we investigate the influence of noise on the evolution of noise by using the model of gene-regulation network in Section 8.2.2. We add a noise

[11] In general, the noise of chemical reactions comes from stochastic molecular collisions and state transitions. Even though it is stochastic for each reaction, the relative fluctuations decrease as the number of molecules increases, because the actual rate of the reaction approaches the average value with the increase in the number of molecules. For example, because the number of molecules N is proportional to the volume V and its standard deviation is usually proportional to $\sqrt{N}$ for a Gaussian distribution, the standard deviation of the concentration is proportional to $1/\sqrt{N}$, and the variance is reduced by $1/N$. Accordingly, by changing the number of molecules in a cell in the model of Section 8.3.2.1, it appears that the strength of noise could be controlled. However, both in this model and in a real cell, the number of molecules shows a log-normal-type distribution, as mentioned earlier. In this case, the standard deviation of the number of molecules is proportional to N. As a result, the concentration fluctuations are no longer dependent on N (or volume) (see Chapter 3). In the catalytic reaction network model, when the steady-state growth is reached, indeed, this is the case, so that the concentration fluctuations do not decrease with increasing cell size (total number of molecules N) in the model.

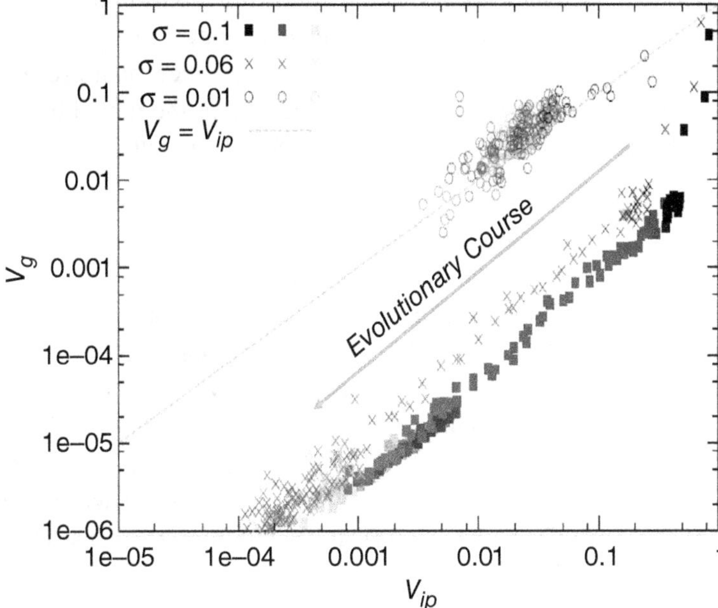

Figure 8.9 The relationship between V_g and V_{ip}. V_g is computed from the distribution of fitness over different $\{J\}$ at each generation, and V_{ip} by averaging the variance of isogenic phenotype fluctuations over all existing individuals. Computed by using the model in the text. Points are plotted over 200 generations. Gradual gray-scale change for each mark shows the increase in the generation, as shown in the sidebar. $\sigma = 0.01$ (∘), 0.06 (×), and 0.01 (■). For $\sigma > \sigma_c \approx 0.02$, both decrease with successive generations. See Kaneko (2007) for details.

term (Gaussian white noise $\eta(t)$ with $< \eta(t)\eta(t') > = 2\sigma\delta(t-t'))$ with the noise strength σ to Eq. (8.2), where the noise strength is a given parameter that can be controlled. We can then check the universality of the $V_g - V_{ip}$ proportionality in Section 8.2.

In the model, the genotype is represented by the matrix J, so that the phenotypic variance by genetic distribution is given by

$$V_g = \int d\{J\}p(\{J\})(\overline{F}(J) - <\overline{F}>)^2, \tag{8.14}$$

where $<\overline{F}> = \int p(\{J\})\overline{F}(\{J\})d\{J\}$ is the average of fitness over all populations.

In Fig. 8.9, V_g versus V_{ip} is plotted through the course of evolution (over generations). As shown in the figure, $V_g \propto V_{ip}$ holds, over generations, except those in the first few generations. Thus, the result of the GRN model is consistent with the phenomenological theory Section 8.2. According to the theory in Section 8.2, however, this proportionality appears only for $V_{ip} > V_g$, and indeed, noise of a sufficient level is needed to reduce the variance and increase the robustness. Hence the noise is relevant to avoid the error catastrophe that accumulates malfunction mutations in the phenomenological theory. Then, is such relevance of noise also true in the GRN model? To examine this issue, dependence of the model behavior upon the noise level was studied (Kaneko, 2007, 2008). The results are summarized as follows:

(i) There is a certain threshold noise level σ_c, beyond which the evolution of robustness progresses, and both V_g and V_{ip} decrease in proportion, where most of the individuals take the highest fitness value. In contrast, for a lower level of noise $\sigma < \sigma_c$, neither V_g nor V_{ip} decreases through the evolution (see Fig. 8.9). Mutants that have very low fitness values always remain. Some mutants from the fittest individuals take much lower fitness values.

(ii) At around the threshold noise level σ_c, V_g approaches V_{ip} and at $\sigma < \sigma_c$, $V_g \sim V_{ip}$ holds, whereas for $\sigma > \sigma_c$, $V_{ip} > V_g$ is satisfied where robust evolution progresses. Thus, the numerical results here are consistent with the phenomenological theory.

Then, how does the noise shape the evolution of robustness? Here, the temporal changes in gene expression are regarded as a process of starting from an initial state and reaching a final state (attractor) according to its dynamical system. As a result, the phenotype is shaped. In other words, the dynamical system provides a path from the initial state to the goal. Here, recall again that the expression dynamics (Eq. (8.2)) is complex, involving many genes. Hence, if the path is disturbed by noise, it will often be deviated from the original goal and settle on a different attractor, a goal point far from the original destination. If the original goal is an adapted, fittest phenotype, then the noise will be an obstacle against reaching it. Therefore, under a sufficient level of noise, the final attractor will be deviated to a different state as shown in a, b, c, d in Fig. 8.10(a), resulting in a low-fitness state. Then, through evolution, it is expected that a dynamical system with a flow that converges to the original trajectory will be shaped, in order to prevent such deviations, as shown in Fig. 8.10(b). In other words, under the evolution with a sufficient noise level, the stability against noise evolves through evolution, resulting in a lower V_{ip}.

On the other hand, if the noise is low enough, the trajectory will not be kicked away from the original orbit. Therefore, there is no necessity for evolution to a stable dynamical system that converges neighboring trajectories. Hence the flow in dynamical system remains as in Fig. 8.10(b).

Next, consider mutations to this gene network. Mutations can change the dynamical system of development. This corresponds to the addition of a new term or a change in parameter values in the dynamical system. As a result, the trajectory

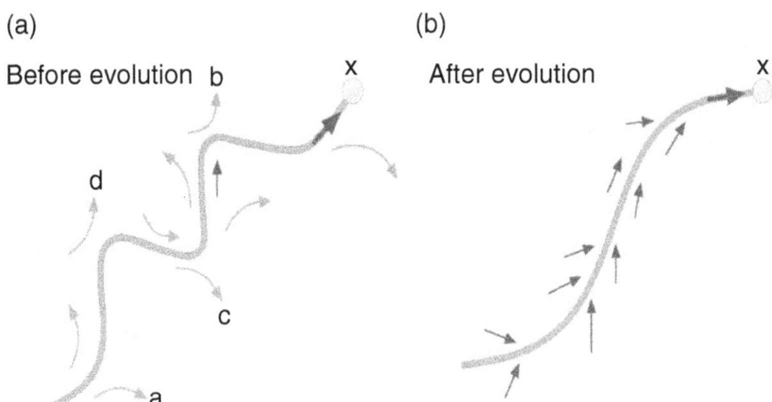

Figure 8.10 Schematic representation of the stability of an orbit reaching the final state. In general, developmental dynamics can have many attractors. Depending on the initial condition, each individual cell may reach a different phenotype. In the evolved system under a low noise level, the orbits are diverted to fall on different attractors by small perturbations (a). For an evolved system under a sufficient level of noise, global attraction to the target phenotype is shaped by developmental dynamics (b).

will be deviated. Then, the orbit in the dynamical systems evolved under a low-noise level (Fig. 8.10(a)) will be easily stepped out from the original orbit and may not reach the fitted goal. In contrast, in the orbit evolved under a sufficient level of noise, the convergence of trajectories is already strong enough as in Fig. 8.10(b) to avoid deviation from the original, so that even if the dynamical system is perturbed by mutation, the deviation from the original path will be little. In other words, when the robustness to noise has evolved and V_{ip} has been decreased, robustness to genetic mutation also evolves, resulting in the decrease in V_g. In summary, when the noise level is high, V_{ip} decreases and accordingly V_g also decreases. Hence the correlated decrease between V_{ip} and V_g is explained, even though a proportionality between the two cannot be derived precisely. To sum, this correlated decrease of the two variances can be understood as a process in which robustness against noise leads to robustness against genetic variation.

As mentioned, if the noise level is lower than the threshold σ_c above, robustness to noise has not evolved, and deviation from the original trajectory often occurs. Accordingly, perturbation to the orbit induced by mutation to the dynamical system will often cause deviation from the original path. This will result in mutants with lower fitness. Hence neither robustness to noise nor to genetic mutation evolves, so that both V_{ip} and V_g remain at a relatively high level.

The above argument is also supported numerically by computing the basin of attraction to attractors, as shown in Fig. 8.11. Note that an orbit of the network

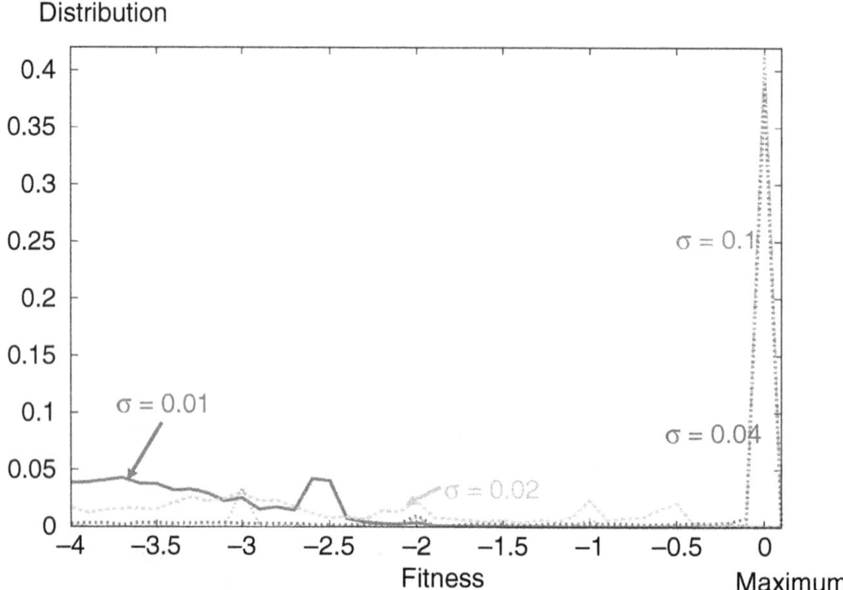

Figure 8.11 Distribution of the fitness value when the initial condition for x_j is distributed over $(-1, 1)$. We chose the evolved network, and for each network we took 10^4 initial conditions, and simulated the evolved gene expression dynamics without noise to measure the fitness value after the system reached an attractor. The histogram is plotted with a bin size 0.1. When the noise level is larger than σ_c, the distribution is sharply peaked at maximal F (In the model, there is symmetry against the sign change of x_j. Against such change in the initial condition, on/off of the target is reversed, leading to the fitness $-M - F$, where M is the number of output genes, so that the distribution for $F > -M/2 = -4$ is plotted here). By starting only with $x_j > 0$, all the points reach the highest fitness for $\sigma > \sigma_c \sim 0.04$.). On the other hand, for the networks evolved under a lower noise level $(< \sigma_c)$, the peak at the highest fitness is tiny. There exist many small peaks corresponding to attractors with lower fitness. Multiple attractors with lower fitness emerge. Reproduced from Kaneko (2007).

with the highest fitness, starting from the prescribed initial condition, is within the basin of attraction to an attractor corresponding to the fitted state. Hence, the basin of attraction to this target attractor is expected to be larger for the dynamics evolved under a high-level noise $(\sigma > \sigma_c)$. We have simulated the dynamics for the evolved fittest network, starting from a variety of initial conditions over the entire phase space, and measured the distribution of at each attractor. Distribution of the fitness for attractors over initial conditions is computed by simulating dynamical systems without noise, as is shown in Fig. 8.11.[12] For the network evolved under

---

[12] Due to the symmetry against the change of sign in x_i in the model, the distribution is symmetric between all-on and all-off states. In fact, by changing the initial condition from all-off to all-on, the orbit reaches an attractor in which none of target genes is expressed.

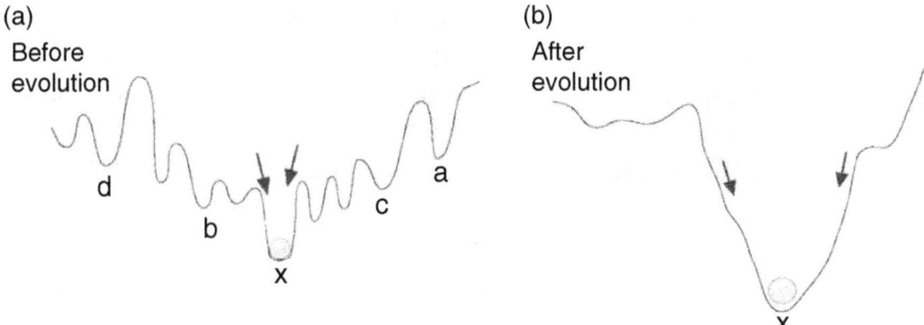

Figure 8.12 Schematic representation of the basin structure, represented as a process of climbing down a potential landscape. (a) Rugged developmental landscape, common in randomly chosen networks, remains under the evolution under low-level noise, in which case small perturbations can kick out the orbit to non-fitted attractors (a,b,c,d). (b) smooth developmental landscape is shaped by evolution under a high-level noise, in which the final fitted state is reached even under perturbations (b).

$\sigma > \sigma_c$, the distribution has a sharp peak at the fittest state. On the other hand, for the networks evolved under $\sigma < \sigma_c$, the peak height is very small, that is, the basin for the fitted attractor is tiny. There exist many small peaks corresponding to attractors having similar sizes of basin volumes.

The observation of the basin in Fig. 8.11 implies that for the networks evolved under $\sigma > \sigma_c$, attractors that give the highest fitness value are reached from a large portion of the initial conditions, whereas for those evolved for $\sigma < \sigma_c$, fitted attractors are reached only from a tiny fraction of initial conditions (i.e., in the vicinity of the given initial states). In the latter case, there coexist many attractors, besides the fittest one. Then, depending on the initial condition, the state is attracted to each of them.

This can be schematically represented as in Fig. 8.12. For $\sigma > \sigma_c$, the "developmental dynamics" gives a global, smooth attraction to the target (see Fig. 8.10(b)). In fact, such type of developmental dynamics with global attraction is known to be ubiquitous in protein folding dynamics (Abe & Go, 1980; Onuchic et al., 1995), gene expression dynamics (Li et al., 2004), and signal transduction networks (Wang et al., 2006).

On the other hand, according to the developmental landscape evolved at $\sigma < \sigma_c$, the final state (attractor) is diverted by small perturbations in the initial condition or during the process. Only from the vicinity of given initial conditions and under sufficiently low level of noise, the expression dynamics reach the fittest pattern, as is schematically shown in Fig. 8.12(a). A wide variety of different states are reached by changing the initial condition a little.

To sum, the smooth landscape of the developmental process (note this is not the fitness landscape) is shaped as the result of evolution, as shown in Fig. 8.12(b) whereas the landscape remains rugged under the evolution of smaller noise as shown in Fig. 8.12(a).

Note that from this landscape picture, we can repeat the earlier discussion on mutational robustness. Consider mutation to a network to alter slightly a few elements J_{ij} of the matrix J. This introduces slight perturbations in gene expression dynamics. For $\sigma > \sigma_c$, the dynamics with smooth, global attraction evolves, so that mutation causes only little change to the final expression pattern. In contrast, under the dynamics with orbits that are easily diverted, slight change in the dynamics can destroy the attraction to the original target, and the gene expression falls on a non-fitted pattern.

Hence, for a system evolved under a sufficient level of noise, the mutation to it does not change the fitness so much, that is, the fitness is nearly neutral against mutational changes. To confirm it numerically, the average fitness of mutants generated from individuals having the top fitness is computed, by taking a network with the top fitness evolved after generations. Here, m paths are changed randomly (i.e., change in a value of $J_{i,j}$ among $\pm 1,0$). The average fitness over such modified networks with m mutations is plotted as a function of m in Fig. 8.13. For $\sigma < \sigma_c$,

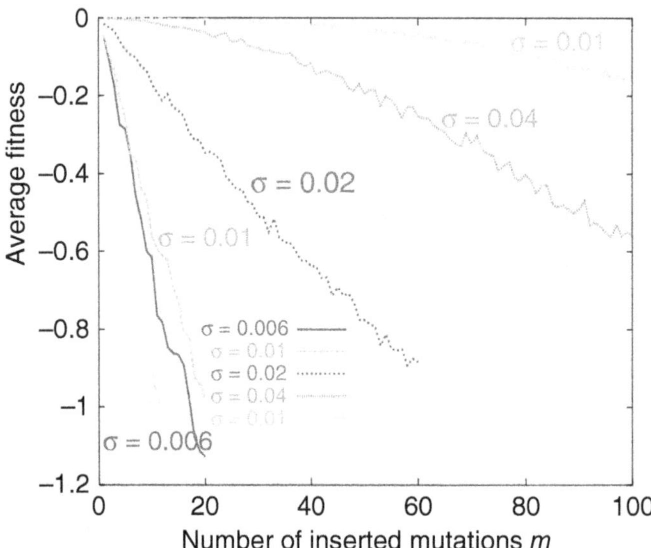

Figure 8.13 Decrease in the fitness plotted as a function of m, the number of added mutations, that is, of modified paths. The average fitness over 1,000 mutants generated from an evolved network is plotted. This evolved network is chosen from the top-fitness network evolved under the gene expression dynamics under a noise level σ. For $\sigma < \sigma_c$, the average fitness decreases linearly as $<\overline{F}>= -C(\sigma)m$, as a function of the number of mutated paths M. Here $C(\sigma)$ goes to zero as $\sigma \to \sigma_c$. Adapted from Kaneko (2008).

the average fitness $<\overline{F}>$ decreases linearly with the number of added mutations m, as $<\overline{F}>\sim -C(\sigma)m$ for small m. This slope $C(\sigma)$ decreases with the increase in noise intensity σ, and approaches 0 as σ approaches σ_c. For $\sigma > \sigma_c$, the linear component $C(\sigma)$ vanishes, and the fitness against mutation exhibits a plateau at $<\overline{F}>\sim 0$ for small m. In other words, the fitness landscape is 'nearly neutral' (Ohta, 2011).[13]

Remark on the decrease of the fluctuations

As we have seen, the degree of phenotypic fluctuations is reduced through the evolution at a given fitness under a given environment, both in experiments and simulations. Then the evolution speed of phenotype is also decreased. In terms of the proportional relationship between the fluctuation and the rate of evolution, the possibility of increase of the both is not ruled out, but in the simulations and experiments, this does not occur. With regards to this decrease, let us note the following points:

(i) Evolution of robustness: As the evolution leads to robustness to noise and mutation, V_{ip} and V_g decrease, as we have seen in this section.

(ii) Reduction of evolutionary capacity (decrease of V_g): if the state with higher fitness is rarer, generation of a phenotype with a higher fitness by mutation will be harder. Hence, the evolutionary capacity will be declined. On the other hand, as a result of the evolutionary stability, fraction of mutations that decrease the fitness will also be reduced. Accordingly, the fitness change by mutation will turn to be nearly neutral as posed by Ohta.

(iii) Negative feedback: The simplest mechanism to increase stability and reduce fluctuations will be a negative feedback in the gene regulation network. If there is such feedback, the increase in the concentration x of a component will affect some other component, which in turn decreases x, so that the fluctuation of x is reduced. In the case of gene control networks, negative loops such as $J_{ij}J_{ji} < 0$ or $J_{ij}J_{jm}J_{mi} < 0$ will provide such negative feedback. With such network, the perturbation to increase x_i will be compensated by such negative feedback.

Summing up the results of this section: *The reciprocal of V_{ip} gives an indication of the robustness of the phenotype against the noise, whereas the reciprocal of V_g gives an indication of its robustness against genetic variation. As both types of robustness increase in tandem throughout the evolution, the two variances decrease in correlation (in proportion).*

---

[13] Note also that such robustness to genetic change, that is, Waddington's canalization, was reported in simulations of Boolean gene expression dynamics, not by noise but by sexual recombination (Siegal & Bergman, 2002; Azevedo et al., 2006).

8.3.2.3 Universality: Replica-symmetry Theory

The results in the present section are expected to be general, as long as the fitness is determined through complex developmental dynamics, and dynamics to reach a fitted state is rare in the genetic space, and can also be perturbed by noise.

So far, we have confirmed this $V_g - V_{ip}$ law as a result of correlated increase in robustness to noise and to mutation in the gene regulation network. Another support is also provided by evolving-interacting-spin model which corresponds to a spin-glass model in statistical physics (mentioned in Section 6.1), which shares similarity with the gene regulation network model studied in this chapter. Here, spin variables S_i take up $(+1)$ or down (-1) values. The dynamics are governed by the interaction with other spins. If the two have ferro couplings $J_{ij} > 0$, the two spins tend to be aligned, which corresponds to the activating interaction, whereas if the two have anti-ferro couplings $J_{ij} < 0$, the two spins tend to be directed in the opposite, corresponding to the inhibitory interaction. By formulating the above interaction using a Hamiltonian ($H = -\sum_j J_{ij}S_iS_j$) for energy, with an inter-action matrix (J_{ij}), we can analyze the distribution of spin configurations under temperature (T), using $exp(-H/(kT))$. We then assign the fitness by a specific configuration of target spins and evolve the matrix J_{ij} through mutation and selection, to increase the fitness. It is then found that within a certain temperature range (Sakata et al., 2009; Sakata & Kaneko, 2020; Pham & Kaneko, 2023), the evolution of Hamiltonian produced the fitted spin configuration having robustness to mutation. This is akin to the observation in Section 8.3.2.2. In fact, the energy landscape in the evolved Hamiltonian at the intermediate temperature has a structure depicted in Fig. 8.12(b), whereas for lower temperature the rugged energy landscape as in Fig. 8.12(a) remains. In terms of statistical physics of interacting spins (Mezard et al., 1987), the former corresponds to the replica symmetric phase, whereas for the latter (i.e., that for lower temperature), the robustness is lost due to replica symmetry breaking resulting in the spin glass phase. Relatedly, evolution of protein structure has recently been investigated, which will be discussed in Section 9.8.

8.3.2.4 Robustness against Sexual Reproduction

So far, we have considered mutation as one source of genetic change. However, it is not the only source of genetic change. In sexual reproduction, recombination occurs between the genomes of two individuals, one male and one female. Furthermore, even in prokaryotes, exchange of genes between individuals often occurs without sexual recombination, through horizontal gene transfer (Takeuchi et al., 2014). Therefore, robustness to such genetic changes has to be included. In such a case, an argument similar to the present section can be applied, and phenotypic robustness to genetic exchange is expected to evolve. In this case, because the exchange occurs between individuals with different genomes, the stability depends on the gene distribution present in the population at the time.

The models presented so far are based on an asexual haploid population. On the other hand, the evolutionary stability theory on the distribution, $P(x,a)$, could also be applied to a population of individuals undergoing sexual reproduction. However, in sexual reproduction, recombination is the major source of genetic variation, rather than mutation. Thus, instead of the robustness to mutation, robustness to recombination (or genetic exchange) needs to be considered seriously.

As a schematic example, suppose the set of genes that can produce an adapted phenotype is in the region shown in Fig. 8.14. Then recombination can produce an offspring between the two individuals. If the set of adaptive genes is in the region shown in Fig. 8.14(a), the fitted phenotype is not robust to genetic exchange (recombination) because the genotype of some offspring from a pair of adapted parents is not always in the adaptation region. On the other hand, if the gene set in the adaptive state is a convex set, as shown in Fig. 8.14(b), the offspring generated by gene exchange between two adapted individuals will also be adaptive. Thus, the dynamical system of gene expression (development) is expected to evolve to shape such a convex set of fitted region, to ensure the robustness to gene exchange. In other words, a stronger constraint will be imposed when gene exchange (recombination) is taken into account.

Indeed, we have extended our GRN model to account for diploids with sexual recombination. Here, each individual had a pair of matrices, J_{ij}^1 and J_{ij}^2, and the gene

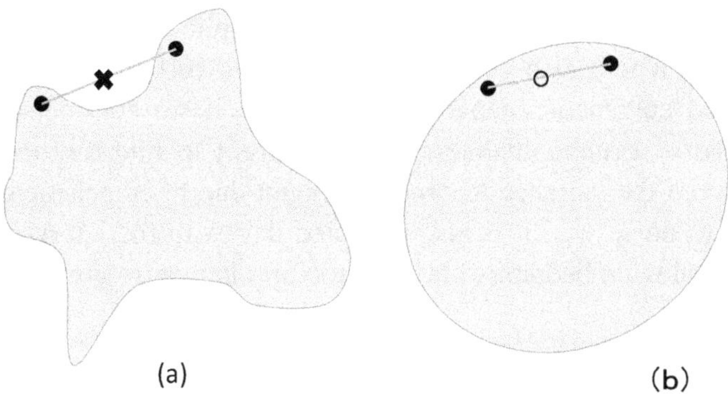

(a)   (b)

Figure 8.14 A set of genotypes that provide fitted states. (a) An example of a set in which fitted state is not sustained by recombination, that is, without robustness to recombination (b) An example of a set satisfying robustness to recombination (or horizontal gene transfer). The convex set. Under the evolution of recombination, such convex set of fitness landscape is expected to be shaped.

expression dynamics is given as a result of the average of the two matrices, instead of the summation $\sum_j J_{ij}x_j$ in Eq. (8.2). (see also [Siegal & Bergman, 2002; Azevedo et al., 2006] for the evolution simulations of Boolean gene expression dynamics under sexual recombination). By considering the recombination of two matrices from parents, GRNs were again numerically evolved to achieve a higher fitness. In this case, the proportionality between the two variances is again confirmed. Another noteworthy finding is that, in the case of heterozygotes, the robustness is further enhanced (the variances are further suppressed) (Okubo & Kaneko, 2022).

8.3.3 Consistency between Different Time Scales: Epigenetics (D)

So far, we have discussed evolutionary laws as a consequence of the genotype–phenotype consistency principle. Phenotypic change is a faster process than genetic change. Hence, this consistency can be regarded as a consequence of embedding of the fast process to the slow process, making the two consistent. In this regard, the theory in the present chapter is a quantitative version of the *genetic assimilation*, the term coined by Waddington. It is also inspired by Einstein's theory of Brownian motion based on the consistency between fast microscopic motion and slow macroscopic response (Kaneko, 2009).

Now, the different time scales in biological systems are not limited to phenotypes and genotypes. Epigenetics, which has received much attention in recent years, can be considered to belong to the intermediate timescale between the two (Fig. 2.4 in Chapter 2). They refer to the changes that last longer than those in protein expression levels, specifically, long-term changes in the modification in DNA, such as methylation or phosphorylation, that make a gene less or more likely to be expressed (see Chapters 6 and 7).

Since the theory in the present chapter can be applied to phenomena on different time scales, it will also be possible to apply it to the relationship between phenotypic and epigenetic changes. For example, if we consider an epigenetic change instead of genetic changes, one can expect to find a proportional relationship between the variance of protein amount due to epigenetic change, V_{epi}, and that due to noise, V_{ip}. It is also expected that with the intervention of such intermediate scales, embedding of fast phenotypic changes to slow genetic changes will be easier.

8.3.4 Laboratory Experiment (E)

So far there is no decisive laboratory experiment that directly proves the above V_g–V_{ip} relationship. This is because of the difficulty in the accurate measurement of V_{ip} and V_g through the *evolutionary course*. Instead, however, one can consider a similar relationship between isogenic and genetic variances across *different phenotypes (or traits)*, for an evolved individual (not over the evolutionary course). As will be seen in Chapter 9, the proportionality between the two variances across components is also derived theoretically, which are experimentally verified, to some degree.

Of course, if we assume Fisher's theorem, the proportionality between V_{ip} and evolution speed observed in experiments would implicitly suggest V_g-V_{ip} proportionality. Besides the experimental result in Section 8.2.1, there is a recent experimental verification of it in the evolution of RNA (Maeda et al., 2022).

8.4 Evolution to Sustain Plasticity and Fluctuations (C and F)

8.4.1 Maintaining Phenotypic Fluctuations and Restoring Plasticity

So far, we have shown that through the experiments and simulations of adaptive evolution under a fixed condition, fluctuations decrease through the course of evolution, and the evolvability is declined. Accordingly, robustness of the phenotype increases, and the plasticity decreases through evolution. Still, in nature, neither phenotypic fluctuations nor potentiality in evolution vanish. In the real biological systems, as we have already seen in Chapter 3, the fluctuations in the concentration of each protein remain to be at a substantially high level, and potentially of evolution is preserved. Then, why are the fluctuations not reduced, which, in principle, would be possible by evolving a negative feedback process appropriately? How are phenotypic fluctuations or plasticity sustained?

One possible factor for the preservation of plasticity or fluctuation is the environmental fluctuation. In the real evolution in nature, fitness and environment are not fixed, but variable. If the fluctuation is too much reduced, the plasticity to adapt to new conditions imposed by environmental changes will be lost. The plasticity of a biological system will be relevant to cope with the environmental change that alters the condition for the fitness. Hence, maintenance of fluctuations will be meaningful.

The second possibility, which is related with the first, will be the interaction among individuals. For each organism, most important factor for external environment will come from the interaction with other organisms, which can change in time. For instance, the species–species interaction changes effectively as the population of each species can change in time. This change in the effective interaction may lead to the maintenance of plasticity and evolvability.

The third point that might be overlooked in the theory is the possibility that the relevant degree of phenotypic variables would not be expressed as a single variable X and accordingly the corresponding genetic distribution would not by parametrized by a single scalar variable a: for a certain condition, such stable distribution of a pair of single variables breaks down somewhere, and the entanglement with other variables comes into play, enhancing the fluctuation.

In Section 8.4.2, we discuss the simulation results for the first possibility and briefly refer to the second case at Section 8.4.3. Then, at Section 8.4.4., we present the results of *E. coli* experiments concerning the third possibility.

8.4.2 Plasticity Increase by Environmental Variation (C and F)

We adopt the model in Section 8.2.2.2 (and 8.3.2.2), and introduce external environmental switches. As a simple example, we introduce the switch in the fitness condition. Specifically, after the evolution under the fitness to select all-on expressions for the target genes ($i = 1, 2, \cdots, M$) as adopted in Section 8.2.2.2, the fitness is switched to the condition that the first $M/2$ be on and the latter $M/2$ be off (i.e., the fittest gene expression pattern is $+ + + + - - - -$, instead of $+ + + + + + + +$). By this switch, the fitness of the network that evolved under the original fitness is drastically decreased, but after a few dozen generations, networks evolve to adapt to the new fitness condition.

Then, to see the evolution of phenotypic plasticity, the two variances of the fitness, V_{ip} and V_g, are computed (see Fig. 8.15). After switching the fitness condition, both of them first start to increase. During this increase, the proportionally between the two is roughly preserved (Kaneko, 2012b). The locus of (V_{ip}, V_g) over generations through the evolution follows in reverse, the course experienced through the decreases in V_{ip} and V_g. In later generations, both of them decrease again, following the proportionality. The proportionality law between the genetic and isogenic phenotypic variances is observed in the evolution both to increase and decrease the plasticity.

After this switch, the variances V_{ip} and V_g increase to restore the plasticity. Then, after several generations of rising fluctuations to recover the fitness, the previous process of decreasing V_{ip} and V_g proportionally takes place again. Here, if the noise intensity σ is large, more generations are needed to compensate the reduced fluctuations. On the other hand, if the noise intensity σ is smaller than σ_c in Section 8.3.2.2, the high fitness state with robustness cannot be reached sufficiently.

Now consider the case in which this environmental change by the fitness switch occurs repeatedly per some generation (say, 20 generations). Then, the decrease and increase of the variances, V_{ip} and V_g, are also repeated. Note that it takes more generations to adapt to a new fitness condition if the phenotypic variances are smaller. In our model, if the noise level in development is larger, the phenotypic variances have already been decreased during the adaptation under

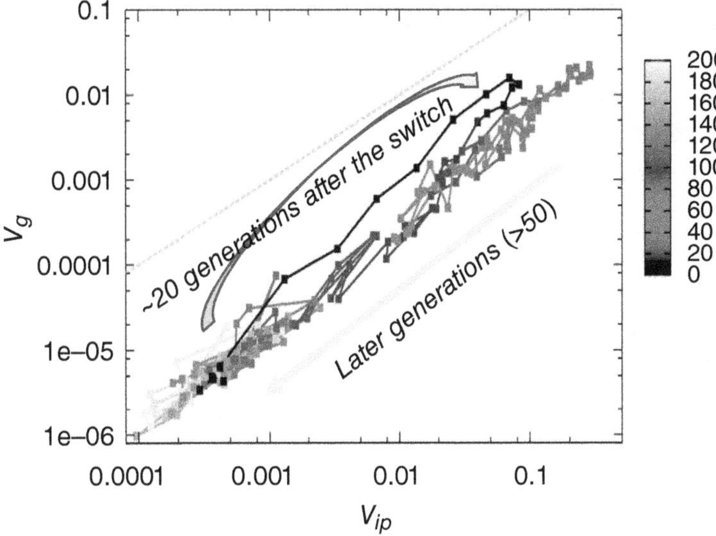

Figure 8.15 Change in variances after switching of fitness condition. After evolution under the fitness condition to favor $x_i > 0$ ("on") for all the target genes, $i = 1, 2, \cdots, M (= 8)$, as already studied, the fitness condition was switched at a certain generation to favor $x_i > 0$ for $i = 1, 2, .., M/2$ and $x_i < 0$ for $i = M/2 + 1, .., k$ (i.e., the fittest gene expression pattern was given by $+ + + + - - - -$, instead of $+ + + + + + + +$). The switching was applied after a sufficiently large number of generations when the fittest networks were evolved (i.e., with $x_i > 0$ for the target genes). The switch initially caused a decrease in the fitness, but after a few dozen generations, almost all of the networks evolved to adapt to the new fitness condition if $\sigma > \sigma_c$. The plot shows the variances of fitness V_g versus V_{ip} per generation after the switching of the fitness condition. The color represents the generation number from the switching. There is a correlation between the increases in both the variances after the switch. Then, there appears a proportional decrease as the adaptation to the new fitness condition progresses. Adapted from Kaneko (2012b).

given fitness condition. Hence, in this case, it takes more generations to adapt to a new fitness condition, and will not catch up the next switch. On the other hand, if the noise intensity σ is smaller than σ_c, robust evolution does not progress. Indeed, for continuous environmental changes, there will be an optimal noise level both to adapt sufficiently fast to a new environment and to evolve robustness of fitness for each environmental condition. In Fig. 8.16, the time course of the average fitness in a population is plotted when the fitness condition is switched every 20 generation. If the noise level is high, the system cannot follow the frequent environmental changes, and the fitness does not increase, as is expected. As the noise level is smaller, the fitness increases with the increase in the phenotypic variance. If it is smaller than σ_c, however, the fitness of some individuals remains low, because robustness to mutation is not

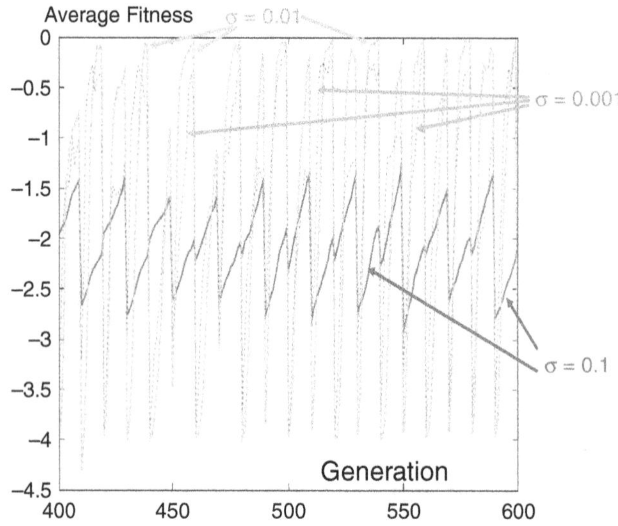

Figure 8.16 Average fitness is plotted per generation, where the fitness condition is switched every 20 generations between $+++++++++$ and $++++----$. The average of the mean fitness, $\overline{F}$, of each individual cell (over $L = 200$ runs) is computed over the total population ($N = 200$) at each generation. The noise level, σ, is 0.0001 (dotted), 0.01 ($\sim \sigma_c$; dashed) and 0.1 (solid). Adapted from Kaneko (2012b).

achieved therein. Indeed, if the noise level is near σ_c, the average fitness is optimal. The plasticity and robustness are compatible near the transition noise level, σ_c, where both V_g and V_{ip} are not reduced to lower values, and $V_g \sim V_{ip}$ (see Fig. 8.17).

One might think that the switching target conditions is a little bit different from the environmental variation. We therefore carried out evolution simulations, by modifying the gene expression dynamics of the model in Section 8.2.2 to include external inputs from the environment. Consider an external input to expression dynamics of each (some) gene. For it, an input term $I_i(t)$ is added to the time evolution of x_i. Then, by varying the input term $I_i(t)$ from generation to generation, the environmental change is considered (see [Kaneko, 2012b] for details.). In this case, there is also a critical noise intensity σ'_c at which the average fitness is maximized. As in Fig. 8.18, V_{ip}, around this critical value, V_{ip} and V_g go up and down, keeping roughly $V_{ip} \sim V_g$. On the other hand, if the noise is smaller than that, these variances remain to be large with $V_g > V_{ip}$, where robust fitted state is not achieved. If the noise level is larger than that, both the variances take smaller values, with $V_g < V_{ip}$, where adaptation to a novel environment (i.e., different input values) is not possible, so that the fitness remains low.

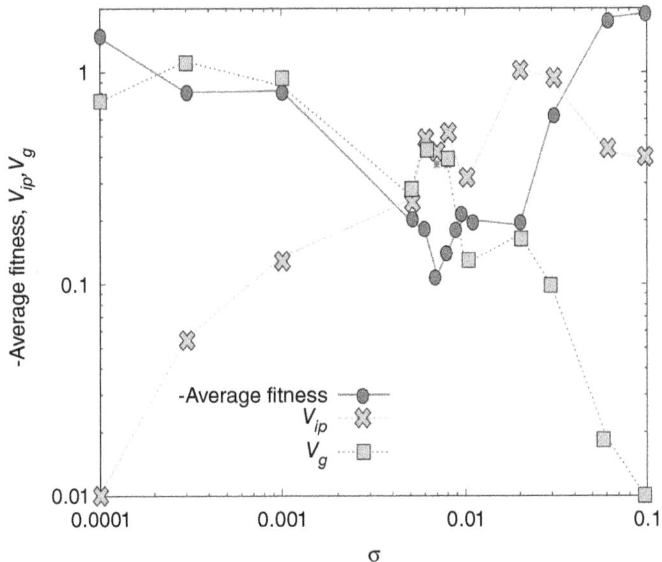

Figure 8.17 The temporal average of the average fitness ($\bullet$) and the variances V_{ip} (X) and V_g ($\blacksquare$), plotted as a function of noise level σ. The overall temporal averages over population and over generations are plotted. Instead of the fitness itself, its sign-inverted value, $- <F>$ is plotted by ($\bullet$). The temporal average is taken over 500–1000 generations, where the fitness condition is switched as in Fig. 8.18 (here for every 10 generations). At around $\sigma \sim 0.008$, the average fitness takes maximum, beyond which V_{ip} is larger than V_g. Adapted from Kaneko (2012b).

Thus, around the critical noise level, the average fitness is optimal, where moderate phenotypic variances are maintained to maintain adaptation to environmental changes.

8.4.3 Interaction Among Species

One of the most important *environmental* factors will be interspecies interactions. As a first step toward investigating phenotypic fluctuation in response to an interspecies interaction, a simple two-species system consisting of hosts and parasites has recently been studied (Nishiura & Kaneko, 2021). Hosts are expected to evolve to achieve phenotypes that optimize the fitness. As discussed in this chapter, a gene expression dynamics model for the host is investigated numerically to study the evolution of the genotype–phenotype map and phenotypic variances. Then, the robustness of the corresponding phenotype is increased by reducing the phenotypic variance. Next, parasites of different types are introduced to attack the hosts expressing the corresponding genes. Then, to avoid

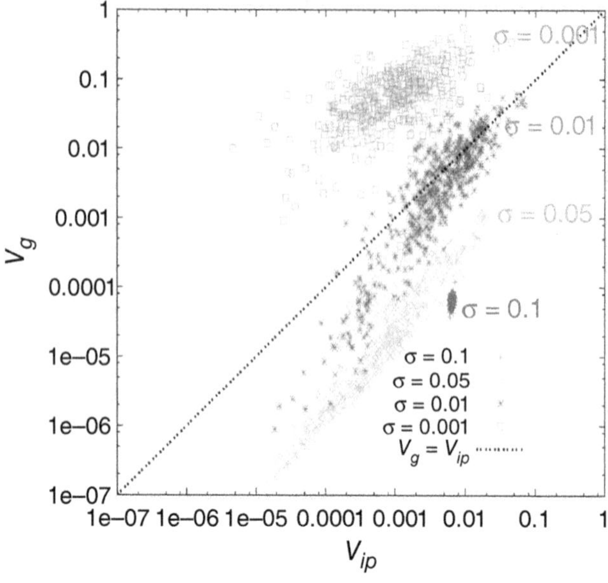

Figure 8.18 The variances of the fitness V_{ip} and V_g are plotted through the course of the evolution, under environmental variation. Each point is a result of one generation, and the plot is taken over 200–700 generations. The noise level σ is 0.1 (+), 0.05 ($\sim \sigma'_c$; ×), 0.01 (∗), and 0.001(□). In this case, input I_j is added to four genes, whose values are changed randomly within (0,0.8) per generation. Adapted from Kaneko (2012b).

the attack of parasites to a given phenotype, hosts need plasticity to change the phenotype.

A model combining host gene expression dynamics and host-parasite population dynamics is numerically investigated by combining the evolution of the host gene regulatory network as in Section 8.3.2.2. If the host-parasite interaction is weak, the fittest host phenotype will evolve to reduce phenotypic variances, as in Section 8.3.2.2. In contrast, if there is a sufficient degree of host-parasite interaction, the phenotypic variances of hosts will increase to escape the parasite attack. In the latter case, two strategies are observed: if the noise in stochastic gene expression is below a certain threshold, the phenotypic variance increases via genetic diversification, whereas above this threshold, the variance increases via noise-induced phenotypic fluctuations. In both cases, the phenotypic variance is increased by host-parasite interaction. Here, the increase in isogenic phenotypic variances V_{ip} due to host-parasite interaction reduces the growth rate at the single-host level; a trade-off between plasticity and growth rate is observed.

8.4.4 Interference among Degrees of Freedom (E and F)

As for the increase in phenotypic fluctuation through evolution, an interesting finding was made in the experiment to increase the fluorescence of protein in *E. coli* in the experiment in Fig. 8.3. By continuing the experiment to further increase the fluorescence, the evolutionary appearance of mutants with larger phenotypic fluctuation was observed (Ito et al., 2009). The experimental procedure is as follows: First, the experiment mentioned in Section 8.2.2 (Sato et al., 2003) yielded *E. coli* with genes that show high fluorescence. Using these *E. coli*, we further select *E. coli* with higher fluorescence, as we did before, by applying mutagenesis to the genes for the protein. However, there is one difference from the procedure before. In the previous experiment, we made each mutant to form a colony and the one with the highest fluorescence as a colony was selected. In other words, the selection was made by the average fluorescence in the population of clones. In contrast, this time, the selection was made at the single cell-by-cell level. Specifically, tens of thousands of cells were flowed into a cell-sorter and those cells with high fluorescence density (i.e., fluorescence/volume) were selected (the top 0.2 percent from approximately 2×10^4 cells are selected), as shown in Fig. 8.19(a). By repeating the process for four generations, the fluorescence density gradually increased. Then, the cells selected in this way at each generation were taken out and grown to measure the fluorescence density distribution over clone (isogenic) cells.

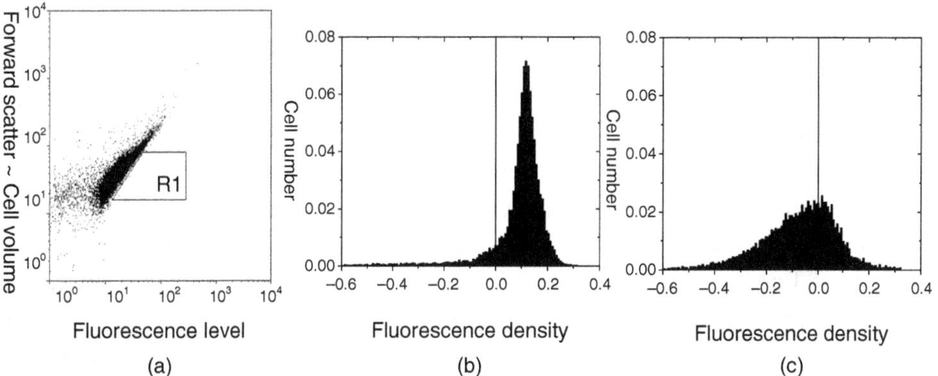

Figure 8.19 (a) the selection process of higher fluorescent cells by the cell sorter. The vertical axis represents forward scattering, which gives the cell volume approximately. The horizontal axis is the amount of fluorescence. To increase the fluorescence density, cells with a higher value of the horizontal axis divided by the vertical axis (*R*1 region in the figure) are selected. (b and c) The two examples of fluorescence distribution of given isogenic bacteria are plotted. Some have a high mean and low variance, and others a lower mean and higher variance. Reproduced from Ito et al. (2009).

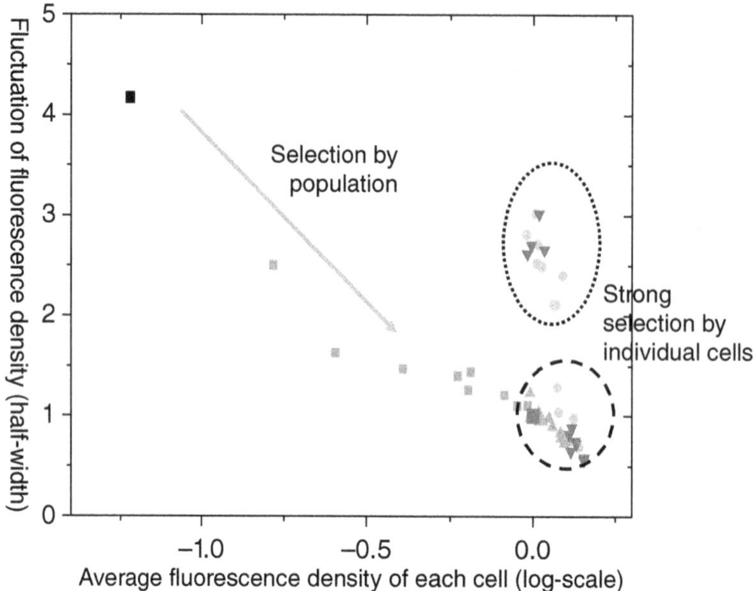

Figure 8.20 Mean fluorescence and its fluctuations (half-width computed as the width of half-maximum), plotted through the course of evolution. Increase in the fluorescence in the previous adaptive evolution (selection at the mean as in Fig. 8.3) is plotted up to 10 generations (■, and then the procedure is changed to the individual selection, as mentioned in the text. The variance (half-width of the distribution) and average of fluorescence of mutants are plotted after 2 more generations (▲), after 3 (•) and after 4 generations (▼). Both an increased group of fluctuations and a decreasing group of fluctuations appear. Adapted from Ito et al. (2009).

With this evolution experiment, a new type of mutants was uncovered, which has lower average of fluorescence and increased fluctuations (Fig. 8.19(c)), besides the previous type with high fluorescence average and decreased fluctuations (Fig. 8.19(b)). (See Fig. 8.19(b,c) for the comparison of fluorescence distribution of the isogeneic cells of the two types). In Fig. 8.20, we plotted the means and fluctuations of fluorescence over clones of the selected mutants, at each generation (as a measure of fluctuations, half-width of the distribution was used). Hence, in addition to mutants that reduced the fluctuation as in the previous experiment, we observed the emergence of mutants with enhanced fluctuations. This mutant with broad distribution (which we term *broad mutant*) is a distinctly different type from that in the previous evolutionary series. Furthermore, genetic analysis shows that the broad mutants appear independently from different mutations. In this sense, the emergence of broad mutant is not a rare event.

From the point of view of selection, it makes sense that mutants with increased fluctuations are selected at a certain rate if they arise. Even if the average

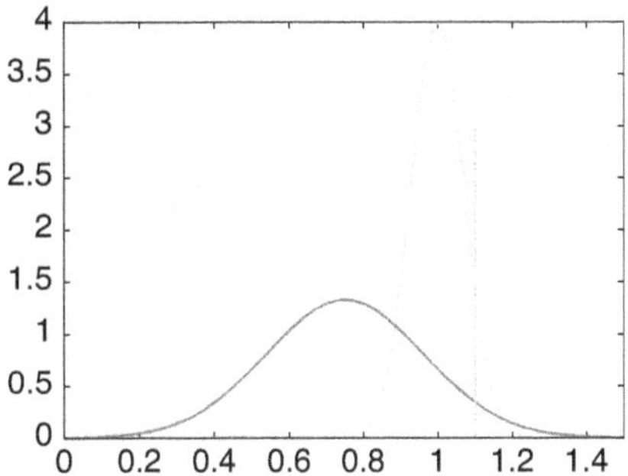

Figure 8.21 Schematic diagram showing that individual-level selection can select mutants with high variability. If the fitness is sufficiently high on an individual basis, it may also be selected from types with low mean values but high fluctuations. If we select individuals with a degree of fitness greater than or equal to 1.1 (dotted line) in the figure, the types with low mean values and large fluctuations (solid line) are also selected for the next generation to the same extent as those with high mean values and small fluctuations.

fluorescence density is lower, there is a certain fraction of individuals whose fluorescence values are sufficiently higher than their average, by taking advantage of large variance. Some of them may have higher fluorescence than mutants with high mean values and small fluctuations (Fig. 8.21). Therefore, strong selection at the individual level (i.e., by selecting only a small fraction from many individuals) can give rise to mutants with elevated fluctuations.[14]

Now the next question is how the fluctuation of the fluorescence density increased. There are many factors that determine the fluorescence density of the protein in the cell: (1) the number of plasmids, because the gene for fluorescent protein is embedded in a plasmid in this experiment (2) the concentration of the protein expressed from the gene, (3) the rate at which the protein is folded without aggregation in the cell, to function, and (4) the degree of fluorescence emitted by the folded protein. Upon examination, it was demonstrated that the factor (3) is most dominant to the rise in fluorescence in the previous evolutionary experiments. In other words, genes that yield the proteins that fold well without aggregation have been selected. However, the origin of the large fluctuation turned out not to be (3). After detailed examination, it was found that the process (2) is most influential to the rise in the fluctuation. In fact, it was confirmed that the mRNA concentration of the fluorescent protein gene fluctuates more. This may seem surprising, because at first glance it seems to violate the flow of the central

---

[14] In other words, in order to evolve mutants with a high fitness state with stability, it is more effective to select at an ensemble level rather than at an individual level, or to adopt weak selection pressure just by eliminating few with low fitness rather than strong selection pressure with choosing those with top fitness. Indeed, this has been known since Darwin's time (Darwin, 1859).

dogma of gene →mRNA→ protein: the fluctuations in the concentration of mRNA increased even though the selection was based on the protein. Now, how is this seemingly reverse flow possible? Upon closer examination, it was found that cell growth (volume increase) fluctuated more, resulting in the enhancement of fluctuations in the mRNA concentrations (amount/volume) (Ito et al., 2009). In fact, it has been found that the growth curves of many mutants exhibit large fluctuations depending on each individual cell.

Indeed, if the fluorescent proteins are expressed too much in the cell to increase the fluorescence, it may cause a negative effect on the cell, in particular, on its volume growth. Therefore, it is likely that the steady growth is perturbed more by internal or external noise, leading to larger fluctuations in the cell volume growth in highly fluorescent cells. The results suggest that the broad distribution of growth rates of isogenic cells result in increased fluctuations in the mRNA concentration. In other words, if we optimize the expression of fluorescence in the protein, the fluctuations will be decreased, but if we optimize it too much, the stability of the cellular growth process may decrease resulting in the increase of the fluctuations, because the balance among processes is disturbed.

Generalizing this result, one can pose the following hypothesis (Fig. 8.22): With the unidirectional evolution to increase a certain property, one phenotypic variable that has major contribution to this property is shaped. By projecting the

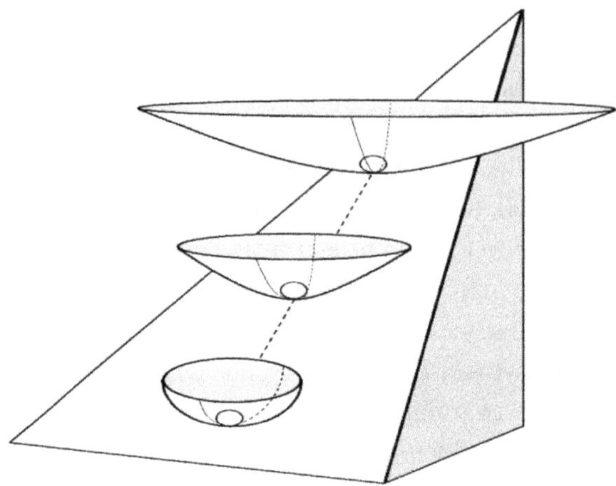

Figure 8.22 A schematic picture of how, under strong selection, decrease in phenotypic fluctuations in one variable leads to decrease the stability of other processes, leading to the increase fluctuations. As the fitness is increased, the valley along the given direction is sharper, to decrease the fluctuations along the direction, but those in a different direction (lateral in the figure) increase at a certain point.

phenotypic change by evolution onto this axis of this variable, the theory of the distribution function of $P(x,a)$ so far discussed will be validated, and the fluctuation of the variable decreases with the adaptive evolution. Accordingly, the phenotypic variable has a sharper distribution against the noise and mutation. Now, this variable is related with some other variables, since intracellular processes involve many variables mutually related with each other. Too much stabilization in the direction of one variable would make it difficult to maintain the balance with respect to other variables. This is because if we impose a severe condition upon a one-dimensional direction in a dynamical system with many variables, the fraction of such dynamical systems that maintain stability with respect to other degrees of freedom will be decreased. As a result, the stability in other directions will be reduced, resulting in an increase in fluctuations.

This increase in fluctuation may not be good in terms of stability, but it may also introduce a novel evolutionary potential by increasing the plasticity of other processes. Although the above experiment ended with the emergence of mutants with larger fluctuations, it will be interesting to pursue a further evolution from those.

8.4.5 Collapse of Consistency and Speciation (F)

Interaction-induced plasticity: As mentioned in Section 8.4.3, an important source to sustain the plasticity is the interaction among individuals, which may introduce a change in the *environmental conditions* faced by each individual organism. Because the interaction depends on each phenotype, the degree of interaction may be enhanced by the phenotype variation, while this variation in the interaction term can enhance the phenotypic variances. Then, the diversity in interaction and phenotypic plasticity may be mutually reinforcing, leading to both genetic and phenotypic diversity. This will also be important to understand the origins of the phenotypic and genetic diversity in nature.

Speciation: The most important source of diversity in ecological systems is introduced by speciation. Although in this chapter we have focused on gradual evolution maintaining a single-peaked distribution of phenotypes, in speciation, the distribution no longer maintains a single peak. Rather, the distribution of phenotypes changes from a single peak to a double peak. Indeed, we saw in the study of cell differentiation in Chapter 7 that the phenotypes of isogenic cells can be bifurcated into different types. Let us first consider such interaction-induced bifurcation of phenotypes for reproducing individuals. This differentiation is not speciation, because the individuals with different phenotypes still share the same genes. After reproduction, they can switch to the alternative phenotypes.

At this point, consider the genetic change after the phenotype bifurcation. Will genetic separation follow? In fact, we have previously shown that robust sympatric speciation is possible as a result of genetic separation following interaction-induced phenotypic differentiation, where genetic assimilation of the phenotype bifurcation leads to the two groups being both genetically and phenotypically separated (Kaneko & Yomo, 2000, 2002b; Kaneko, 2002).

In this theory of speciation, the genotype–phenotype consistency first collapses, resulting in two different states arising from the same genotype.[15] At this stage, the one-to-one correspondence between genotype and phenotype is broken. Later, however, as genetic evolution continues, the genotypes of the two phenotypes will be differentiated because each has a different preferred (increased degree of adaptation) genotype for the two phenotypes. As a result, the one-to-one genotype–phenotype correspondence is restored, and the two species are fixed with different genotype–phenotype pairs (see Fig. 8.23). Such a speciation process has also been numerically demonstrated (Kaneko & Yomo, 2000; Kaneko, 2002), while its experimental verification is still awaited. Furthermore, it will be important to reformulate the problem by extending the distribution theory in this chapter.

On the other hand, the appearance of mutants with high plasticity in Section 8.4, which can be interpreted as a result of over-optimization, will also weaken the tight correspondence between genotype and phenotype. In this case, the phenotypes are divided into two types, one with large fluctuations and the other with small fluctuations. The former has a greater evolutionary potential and can establish a new geno/phenotype. Thus, a new species can be generated as a result of the increase in plasticity with the breaking of genotype–phenotype consistency.

8.5 Summary: Universal Properties and Laws Discussed in This Chapter

- Evolutionary fluctuation–response relationship: Consider the evolutionary process under fixed environmental conditions, where genetic change leads to phenotypic change, and fitness is given as a function of phenotype. In this case, the variance V_{ip}, the fluctuation of the phenotype due to the noise is proportional to the rate of evolution of the phenotype. (A).
- Law of V_g-V_{ip} proportionality: The above relationship implies that the phenotypic variance due to nongenetic origin (noise), V_{ip}, and that due to genetic variation, V_g, are proportional. During evolution under fixed environmental conditions and fitness, both V_g and V_{ip} decrease in proportion (A, C).

---

[15] In a dynamical system, this is a bifurcation phenomenon, a process of splitting from a stable state (valley) into two valleys.

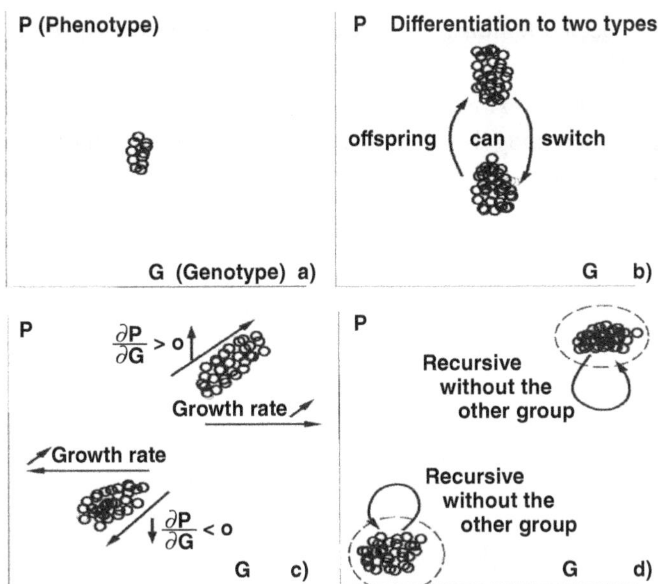

Figure 8.23 Schematic representation of the speciation scenario obtained from our simulation and theory. A pair (phenotype, genotype) is plotted successively with time: (a) the stage of interaction-induced phenotypic differentiation, (b) the stage of genotype–phenotype feedback amplification, (c) the stage of genetic fixation and (d) formation of two groups (*species*) distinct both in phenotype and genotype.

- Correlation between robustness to noise and to mutation: The above relationship is understood by noting that the acquisition of phenotypic robustness to noise through evolution also leads to robustness to genetic variation. Evolution under a fixed condition expands the basin for an attractor that gives adapted phenotypes, while mutation is nearly neutral after evolution (B, C).

- Phenotypic error catastrophe: In general, it is not easy to design a fit phenotype that is adapted to the external environment because developmental processes (or gene expression dynamics) are generally complex. Thus, as the mutation rate increases (or the noise level in the developmental dynamics decreases), an error catastrophe occurs where it is no longer possible to maintain the fit phenotype. This catastrophe occurs around the point where V_g exceeds V_{ip} (C, F).

- Larger variation maintained under environmental variation: While phenotypic variances and evolvability decrease under fixed environmental and fitness conditions, they rise and fall repeatedly when environmental conditions are varied over generations. In this case, both fluctuate while maintaining $V_g \sim V_{ip}$ proportionality. Phenotypic plasticity and evolvability are thus maintained under environmental variation (C, F).

- Interference between multiple processes: Strong selection under fixed evolutionary conditions can lead to the appearance of mutants with increased phenotypic variance. This may be due to over-optimization to obtain the fit phenotype, which may break consistency with other processes and reduce robustness (E, F).

9

Direction and Constraint in Phenotypic Evolution

*Dimensional Reduction and Global Proportionality in Phenotype
Fluctuation and Responses*

9.1 Introduction

In Chapter 8, we discussed the evolutionary fluctuation–response relationship,
which states that if phenotypic variance due to noise is greater, then the phenotypic evolution progresses faster. This suggests a correlation between short-term
phenotypic dynamics and long-term evolutionary responses. This, in some sense,
is a quantitative expression of Waddington's genetic assimilation (Waddington,
1957), under the spirit of Einstein and Onsager (Einstein, 1906, Kubo et al., 1985):
As in the Brownian motion, response upon (selection) force is proportional to the
phenotypic variance prior to the application of force (evolution).

Can we push this viewpoint forward to determine the direction of phenotypic
evolution in a high-dimensional phenotypic space (i.e., with a large degree of freedom)? Can one predict which traits are likely to evolve among many components,
before evolution progresses?

To answer the question, we first recall the characteristics of responses of phenotypes with a large degree of freedom, as presented in Chapter 2. There, a general
relation in phenotypic responses of cells over many components was presented,
demonstrating that global proportionality exists among all logarithmic changes in
concentrations against adaptation to different strength of stresses of given environmental condition. The relation therein was discussed as a consequence of the
steady exponential growth of cells. It was shown that when a cell grows and divides
while maintaining its composition, the abundances of each component increase at
the same rate; this constraint supported the global proportional relationship.

In the present chapter, we first present the experimental findings in which global
proportional changes across all components are confirmed even across many **different** types of environmental conditions. This result cannot be explained by
the constraint of steady growth alone. Some other constraint beyond the steady
growth is needed. Here we will see that this constraint is a consequence of robustness in a phenotypic state that is shaped throughout evolution. The phenotypes
of cells (or biological systems) can change mostly along only one or in a few

dimensions, although the original phenotypic space is high dimensional as a consequence of the diversity in the constituents. We demonstrate this evolutionary dimensional-reduction both through theory and simulations, whereas its consequence is consistent with experimental observations.

This constraint in environmentally induced phenotypic changes is extended to those induced by evolution. We demonstrate that long-term phenotypic changes via evolution and those via short-term adaptation are highly correlated. Global proportionality in the phenotypic changes by environmentally induced adaptation and those by genetically induced evolution is confirmed across all components, both in simulations and laboratory evolution experiments.

As is also observed in Chapter 8, response and fluctuation are *two sides of the same coin*, as has been demonstrated by statistical mechanics. Hence, a similar correlation in concentration fluctuations is expected across all components. Then, one may expect a proportional relationship between fluctuations of concentrations by genetic mutation and those by noise across all components. If this is the case, we can expect the global proportional relationship between response and fluctuations and between evolution and environmental adaptation, across all components, as summarized in Fig. 9.1. Demonstration of this relationship by simulation and experiments, as well as theoretical formulation for it is the aim of this chapter.

Recall that the variances in each trait (phenotype) due to genetic variation are proportional to the evolution rate of the trait according to the fundamental theorem of natural selection by Fisher (1930) (see Section 8.3). Hence, the evolution rate of each trait is correlated with its variance by noise. As the variance is predetermined

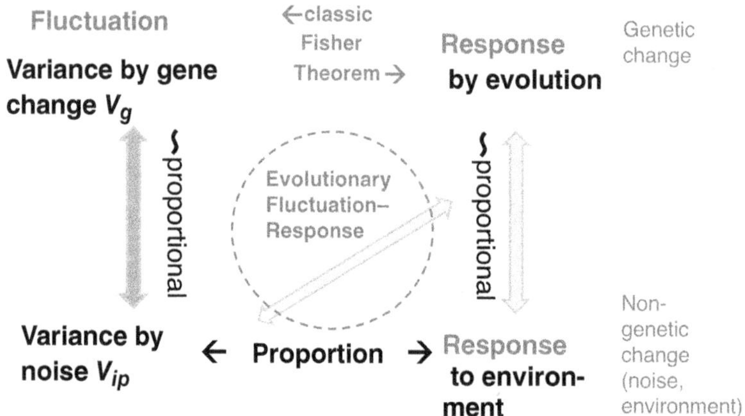

Figure 9.1 Schematic representation showing that 2-by-2 proportionality relationships among genetic/nongenetic fluctuations and responses hold over all components. The arrows indicate proportional relationships.

before mutation and selection, it is then suggested that the evolutionary potential of each trait is determined by the phenotype changeability due to environmental variation or noise, before genetic changes occur. This enables the prediction of phenotypic evolution. Among the high-dimensional phenotypic space, evolution progresses along the direction in which the variation by the noise or environmental response is larger, which is predetermined before genetic mutation. Although genetic variation itself is random and undirected, phenotype evolution tends to show directionality.

Furthermore, if the results obtained from laboratory evolution and evolutionary simulations are universal, it opens up the possibility that changes in cellular state during environmental change and evolution can be described by a limited number of "macroscopic variables." Put simply, despite the many components within a cell, its overall state can be effectively described by a few variables, similar to the concept in thermodynamics. This is a first step toward a macroscopic, universal understanding of the biological state, as expected in Chapters 1 and 2. Toward the end of this chapter, we explore the idea of a phenomenology in which the growth rate (and so on) acts as a macro potential, similar to free energy, while the environment and evolution serve as external variables.

The developed phenomenology of phenotypic evolution is not merely for theoretical satisfaction, but has significant implications for biology. If phenotypic evolution can be expressed in terms of a small number of state variables, it becomes possible to analyze the direction of adaptation and evolution. That is, we can discuss the course of adaptation and the overall trajectory of evolution. Once such a theory is established, it can describe the likelihood of different evolutionary outcomes and even predict the direction of evolution.

9.2 Global Proportional Changes in Gene Expression beyond the Simple Theory (C and E)

In Chapter 2, we compared responses to a given type of environmental condition with different strengths of stress, and found that the coefficient of proportionality equals the ratio of growth-rate changes. Let us now consider whether this relationship might hold under different environmental conditions. However, when examining the state space of each component's concentration, changes in different environments follow different directions, as shown in Fig. 9.2. Attempting to linearize and compare these trajectories would be problematic because they appear to diverge in different directions. Therefore, the use of theoretical derivations similar to those presented in Chapter 2 seems impractical in this context.

In fact, this can also be seen by using the equations of state variables X_j (say log-concentration of the component j) under an environmental vector $\mathbf{E}$. Recall

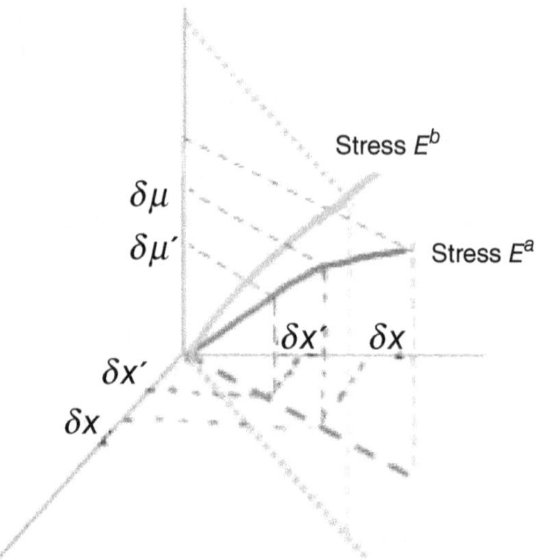

Figure 9.2 Schematic representation of the theoretical analysis: Changes in gene expression in a high-dimensional state space under different types of environmental stresses E^a and E^b are presented. Upon each environmental change, the phenotypic changes in component concentrations follow a curve satisfying the constraint as mentioned in Chapter 2. For a different environmental vector, the locus in the state space follows a different iso-μ line, and the theoretical analysis cannot be applied.

the derivation of Eqs. (2.6–2.10): According to the study in Chapter 2, the relation between δX_j and δE^a is written as

$$\sum_j J_{ij}\delta X_j + \gamma_i^a \delta E^a = \delta\mu(\delta E^a), \tag{9.1}$$

by using the Jacobi matrix $J_{ij} = \partial F_i/\partial X_j$ at E_0 and the susceptibility $\gamma_i^a = \partial F_i/\partial E^a$. Here γ_i^a depends on the direction of the environmental change a, that is, the type of environmental (stress) condition a for E^a. (Note also that linearization on the deviation from the original state X and the original environmental condition is assumed).

Then, comparing two types of environmental conditions E^a and E^b, we get

$$\frac{\delta X_j(\mathbf{E}^a)}{\delta X_j(\mathbf{E}^b)} = \frac{\delta\mu(\mathbf{E}^a)}{\delta\mu(\mathbf{E}^b)} \cdot \frac{\sum_i L_{ji}(1 - \gamma_i^a/\alpha^a)}{\sum_i L_{ji}(1 - \gamma_i^b/\alpha^b)}, \tag{9.2}$$

instead of Eq. (2.11), where $\delta\mu(E^a) = \alpha^a\delta E^a$ and $\delta\mu(E^b) = \alpha^b\delta E^b$. Now, since $\gamma_i^a \neq \gamma_i^b$, the proportionality as in Eq. (2.11) cannot be derived in general. Therefore, it is likely that the above theory cannot suggest the common proportionality.

Nevertheless, it is possible to plot the experimental results as in Chapter 2. We plot the change in log expression $\delta X_i(E)$ and $\delta X_i(E')$ of each mRNA under different environmental conditions over different stresses, where E represents one type of environmental stress (e.g., starvation), E' represents another type of stress (e.g., high temperature) and $\delta X_i(E)$ represents the log change in each expression level x_i from the original unstressed environment O to a given stress E (i.e., $logx_i(E) - logx_i(O)$). The plot based on the experiment of Matsumoto et al. (2013) is shown in Fig. 9.3(a–c), where two examples for $\delta X_i(E)$ versus $\delta X_i(E')$ are shown with the types of stress as osmotic pressure vs. temperature and starvation vs. osmotic pressure (Kaneko et al., 2015). The correlation between the two is clearly discernible, although the variation is greater than for the same type of environmental change compare with Fig. 2.6.

To further support the proportionality, we plotted $\delta X_j(\mathbf{E})$ for the changes in concentration of thousands of proteins (rather than mRNA) under different conditions using proteome analysis (based on data from Heinemann's group (Schmidt et al., 2016)) in Fig. 9.3(d,e). Interestingly, in both cases, a strong correlation between $\delta X_j(\mathbf{E}^a)$ and $\delta X_j(\mathbf{E}^b)$ was still observed for all components j, even under different environmental conditions. Although points for some genes deviated from the common proportionality, lowering the correlation coefficients as compared with those for the same type of stress, the global proportionality still held for most genes (note that Fig. 9.3 shows more than 1,000 points, so that most of these points fit on a single line). Thus, the global proportionality is still valid. (Additionally, other data suggested such a correlation in phenotypic changes under different environmental conditions (Keren et al., 2013; Stern et al., 2007)).

Is this slope related to the change in growth rate $\delta\mu(E)/\delta\mu(E')$ as in Chapter 2? First, in Fig. 9.3(a–e), the slope predicted by the change in growth rate is plotted as a straight line, which implies that the proportionality coefficient is almost identical to the ratio of the growth rate changes. Next, since the experiment on which Fig. 9.3(d,e) is based determined protein changes for 20 different environments, plots corresponding to Fig. 9.3(d,e) can be made across all pairs of different environments, to determine the slope. Then the relationship between the slope and the growth-rate ratio is obtained, as shown in Fig. 9.3(f). One can see that the slope of the logarithmic change in the concentration of many of the components agrees (at least approximately) with the change in growth rate. In other words, it is approximately written as

$$\delta X_i(E^a)/\delta X_i(E^b) = \delta\mu(E^a)/\delta\mu(E^b), \tag{9.3}$$

across all components i or in other words,

$$\delta X_i(E) = c_i\delta\mu(E), \tag{9.4}$$

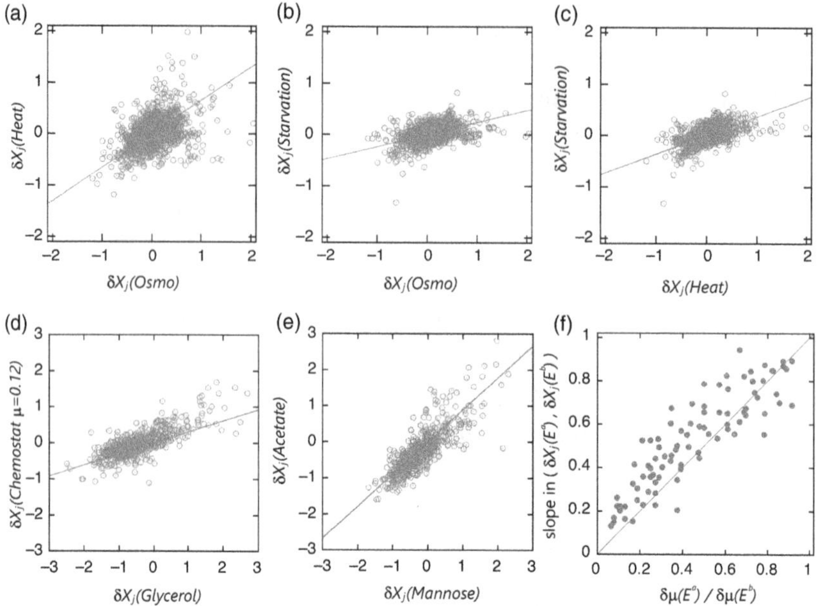

Figure 9.3 Common proportionality in *E. coli* mRNA or protein expression changes. Panels (a–c) show examples of the relationship between mRNA concentration changes $\delta X_j(\mathbf{E}^a)$ and $\delta X_j(\mathbf{E}^b)$. $\delta X_j(\mathbf{E})$ represents the log-transformed expression changes of the jth protein between the environment $\mathbf{E}$ and the original (unstressed) condition O. (a) changes due to osmotic versus heat stress, (b) changes due to osmotic versus starvation stress, and (c) changes due to heat versus starvation stress. Panels (d) and (e) show examples of the relationship between protein concentration changes $\delta X_j(\mathbf{E}^a)$ and $\delta X_j(\mathbf{E}^b)$. $\delta X_j(\mathbf{E})$ represents the log-transformed expression changes of the jth protein between the environment $\mathbf{E}$ and the standard condition O. In (d) and (e), the M9 minimal medium with glucose (as carbon source) is chosen as the original condition. (d) The relationship between the glycerol carbon source condition and the chemostat culture with $\mu = 0.12\mathrm{h}^{-1}$, and (e) the relationship between the mannose carbon source condition and the acetate carbon source condition. For (a–e), the straight line represents the slope calculated from the ratio of the observed growth rate changes. (f) Relationship between the slope of the change in protein expression and the change in growth rate. The abscissa represents $\delta\mu(\mathbf{E}^a)/\delta\mu(\mathbf{E}^b)$, while the ordinate is the slope in $\delta X_j(\mathbf{E}^a)/\delta X_j(\mathbf{E}^b)$. The slope was obtained by fitting the protein expression data. Reproduced from (Furusawa & Kaneko, 2018). (a–c) are based on the data by Matsumoto et al. (2013), and (d–e) are on the data by Schmidt et al. (2016).

where c_i depends on the component i, but not so much on the type of environment E.

Of course, some genes show strong environment-specific changes for each component, and in fact there are some components that are out of proportion in Fig. 9.3. In other words, there are some (but few) components that move more than expected by the growth rate change in the environment E', that is, c_i is uniquely large in that environment. Nevertheless, the majority of components changes in proportion to the growth rate, and c_i is not much dependent on the type of environment. In other words, only a small number of components show environment-specific changes, while majority of components changes with a single constraint: the growth rate and each component change in balance. The number of components in a cell is large. However, the number of dimensions in which the cell state can change is reduced to a much smaller number, even though the dynamical system of a cell involves a very large number of degrees of freedom.

Of course, in these experiments, cells are set at a steady growth condition. However, the steady growth condition alone cannot explain the relationship (9.3) observed in the experiment. Because gene expression dynamics are very high dimensional, this correlation suggests that a strong constraint exists in adaptive changes in expression dynamics that cannot be explained by the simple theory assuming only steady growth. The global proportionality is beyond the scope of the simple theory presented in Chapter 2.

Furthermore, as already mentioned, the results in Chapter 2 hold even though the cell-growth rate is reduced to about 20 percent of the original. The linear regime is quite large. To sum up, the question remains: how can one understand a broad range of linearity regimes as well as proportionality under different environmental conditions? To answer these questions, some factor other than steady growth must be evaluated.

Of course, cells are not only constrained by steady growth but also are a product of evolution. Through evolution, cells can efficiently and robustly reproduce themselves under external conditions. Therefore, the above two features that may be termed *deep linearity* may result from evolution. In Section 9.3, we examine the validity of the hypothesis that evolutionary robustness constrains intracellular dynamics to exhibit global proportionality in the adaptive changes of many components.

9.3 Emergence of Global Proportionality through Evolution: Formation of a Dominant Mode (C)

9.3.1 Models for Numerical Evolution

The above hypothesis that "deep linearity" is a result of evolution is difficult to examine experimentally, as the experimentally available data are only from organisms that currently exist as a result of evolution, which cannot be compared with the data before evolution. Hence, we used numerical evolution for some models in which the phenotypes emerge from high-dimensional dynamics.

To this end, we again utilize simple cell models that are adopted in Chapters 5 and 8. They consist of a large number of components. We numerically evolve them under a given fitness condition to determine how the phenotypes of many components evolve.

In the model, genes govern the network structure and parameters for the reactions that establish the rules for such dynamical systems. The phenotype of each organism, as well as the growth rate (fitness) of a cell, is determined by such reaction dynamics, whereas the evolutionary process consists of selection according to the associated fitness and genetic change in the reaction network (i.e., rewiring of the pathway).

As already discussed in Chapter 8 the cellular state is represented by the numbers of k chemical species, that is, $(N_1, N_2, \cdots, N_k)$, whereas their concentrations are given by $x_i = N_i/V$ with the volume of the cell V (Furusawa & Kaneko, 2003, 2012). There are $m(<k)$ resource chemicals $S_1, S_2, \cdots, S_m$ whose concentrations in the environment and within a cell are given by $s_1, \cdots, s_m$ and $x_1, \cdots, x_m$, respectively. Each reaction leading from one chemical i to another chemical j is assumed to be catalyzed by a third chemical ℓ, that is, $i + \ell \to j + \ell$. The resource chemicals are transported into the cell with the aid of other chemical components named as "transporters." Here, we assumed that the uptake flux of nutrient i from the environment is proportional to $Ds_i x_{t_i}$, where chemical t_i acts as the transporter of nutrient i, and D is a transport constant. For each nutrient, there is one corresponding transporter, represented by $t_i = m + i$. The other $k - 2m$ chemical species are catalysts synthesized from other components via the catalytic reactions mentioned above. The catalytic reactions transform nutrients into cell-component chemicals. With the uptake of nutrient chemicals from the environment, the total number of chemicals $N = \sum_i N_i$ in a cell increases. A cell, then, is divided into two cells when the total number of molecules exceeds a given threshold. Hence, the cellular growth rate depends on the catalytic network, which is determined by genes.

The evolutionary procedure is applied to increase the fitness, that is the growth rate of a cell. We prepared n parent cells with slightly different reaction networks, which were randomly generated with a given connection rate. We applied stochastic reaction simulation of the above model and selected n/L cells with high growth rates. From each of the n/L parent cells, L mutant cells were generated by replacing a certain fraction of reaction paths, whose rate is determined by the mutation rate. Next, we obtained n cells of the next generation, which contained slightly different reaction paths. For these cells, we repeated the same procedure to obtain the population of the second generation, and so forth.

The simulation of evolutionary dynamics was performed under a given set of nutrients $s_1^o, \cdots, s_m^o$ whose concentrations were set at $s_1^o = s_2^o = \cdots = s_m^o = 1/m$, at which evolution progresses so that the cell growth rate, that is, inverse of the average division time, is increased.

9.3.2 Emergent Global Proportionality through Evolution

Using the model described in Section 9.3.1, we analyze the response of the component concentrations to the environmental change from the original condition. Here, the environmental condition is given by the external concentration $\{s_1, \cdots, s_m\}$. We then changed the condition to $s_j^{(\varepsilon, \mathbf{E})} = (1 - \varepsilon)s_j^o + \varepsilon s_j^{\mathbf{E}}$, where ε is the intensity of the stress and $\mathbf{E} = \{s_1^{\mathbf{E}}, \cdots, s_m^{\mathbf{E}}\}$ denotes the vector of the new, stressed environment, in which the values of component $s_1^{\mathbf{E}}, \cdots, s_m^{\mathbf{E}}$ are determined randomly to satisfy $\sum_j s_j^{\mathbf{E}} = 1$.

For each environment, we compute the reaction dynamics of the cell to obtain the concentration $x_j^{(\varepsilon, \mathbf{E})}$ in the steady growth state, from which the logarithmic change in the concentration $\delta X_j^{(\varepsilon, \mathbf{E})} = \log(x_j^{(\varepsilon, \mathbf{E})}/x_j^0)$ is obtained; the change in growth rate μ, designated as $\delta\mu^{(\varepsilon, \mathbf{E})}$, is also computed, where x_j^o is the concentration for the original condition s_j^o.

We next examine whether changes in $\delta X_j^{(\varepsilon, \mathbf{E})}$ satisfy the common proportionality trend across all components under a variety of environmental changes for all possible ranges in $0 \leq \varepsilon \leq 1$. We examine the degree of proportionality both for the un-evolved (random) networks and evolved networks, under given environmental conditions. We also test whether the proportion coefficient is consistent with $\delta\mu^{(\varepsilon, \mathbf{E})}$.

First, we compute the response of expression to the same type of stress, that is, the same vector $\mathbf{E}$ with different intensities ε. In Fig. 9.4(a), we plot the correlation between changes in component concentrations $\delta X_j^{(\varepsilon_1, \mathbf{E})}$ and $\delta X_j^{(\varepsilon_2, \mathbf{E})}$ caused by different magnitudes of environmental change ($\varepsilon_2 = \varepsilon_1 + \varepsilon$, with $\varepsilon > 0$). For a small environmental change ($\varepsilon = 0.02$), the correlation is sufficiently large both for the random and evolved networks, whereas for a larger environmental change ($\varepsilon = 0.08$), the correlation coefficients are significantly smaller for the random networks. Next, the ratio of the growth rate changes $\delta\mu^{(\varepsilon_1, \mathbf{E})}/\delta\mu^{(\varepsilon_2, \mathbf{E})}$ is computed, to examine if it fits with the slope in $(\delta X_j^{(\varepsilon_1, \mathbf{E})}, \delta X_j^{(\varepsilon_2, \mathbf{E})})$ across all components. Figure 9.4(b) shows the ratio of the slope to $\delta\mu^{(\varepsilon_1, \mathbf{E})}/\delta\mu^{(\varepsilon_2, \mathbf{E})}$ (which turns to be unity when Eq. (9.3) is satisfied) as a function of the magnitude of environmental change ε_1. These results demonstrate that for the evolved network, Eq. (9.3) is maintained under large environmental changes, whereas it holds only against small changes for random networks. These results confirmed the expansion of the linear regime by evolution.

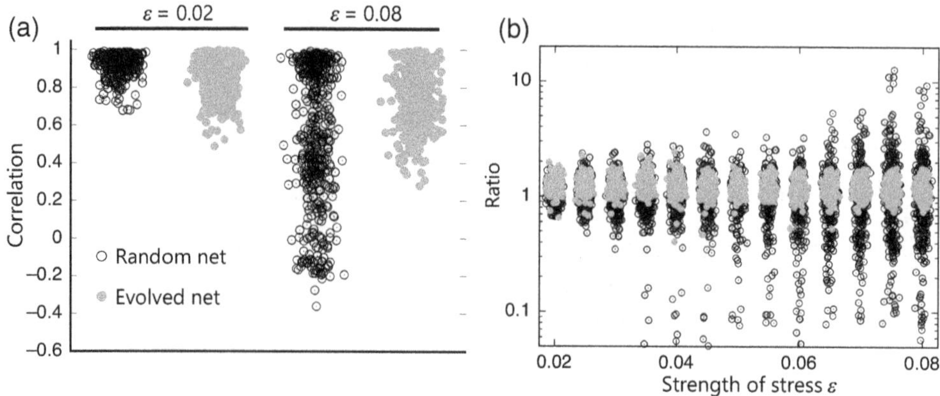

Figure 9.4 Common proportionality in concentration changes in response to the same type of stress in the catalytic-reaction network model. (a) Coefficient of correlation between the changes in component concentrations $\delta X_j^{(\varepsilon_1,\mathbf{E})}$ and $\delta X_j^{(\varepsilon_2,\mathbf{E})}$ with $\varepsilon_2 = \varepsilon_1 + \varepsilon$. For the random and evolved networks, the correlation coefficients with a small ($\varepsilon = 0.02$) and large ($\varepsilon = 0.08$) environmental change are plotted, which were obtained using 100 randomly chosen environmental vectors $\mathbf{E}$. (b) Ratio of the slope in the concentration changes to the growth rate change as a function of the intensity of stress ε. The ratio in the ordinate becomes unity when Eq. (9.3) is satisfied. For both (a) and (b), the gray and unfilled circles are the results for evolved and random networks, respectively. Courtesy of Chikara Furusawa. Adapted from (Furusawa & Kaneko, 2018).

Next, we examine the correlation of concentration changes under different types of environmental stresses. Figure 9.5(a) shows examples of $(\delta X_j^{(\varepsilon,\mathbf{E^a})}, \delta X_j^{(\varepsilon,\mathbf{E^b})})$ obtained by three networks from different generations. For the initial random networks, there is no correlation, whereas a modest correlation emerges in the 10th generation. Later, after evolution, common proportionality is observed; for instance, in the 150th generation, the proportionality hold over two digits. To demonstrate the generality of the proportionality over a variety of environmental variations, we compute the coefficients of correlation between $\delta X_j^{(\varepsilon,\mathbf{E^a})}$ and $\delta X_j^{(\varepsilon,\mathbf{E^b})}$ for a random choice of different vectors $\mathbf{E^a}$ and $\mathbf{E^b}$. Figure 9.5(b) shows the distributions of the correlation coefficients obtained by the random network and evolved network (150th generation). Remarkably, global proportionality is observed even under different environmental conditions that had not been experienced through the course of evolution.

The relationship between $\delta\mu^{(\varepsilon_1,\mathbf{E^a})}/\delta\mu^{(\varepsilon_2,\mathbf{E^b})}$ and the slope in $(\delta X_j^{(\varepsilon_1,\mathbf{E^a})}, \delta X_j^{(\varepsilon_2,\mathbf{E^b})})$ is presented in Fig. 9.5(c), whereas the ratio of the slope in $(\delta X_j^{(\varepsilon_1,\mathbf{E^a})}, \delta X_j^{(\varepsilon_2,\mathbf{E^b})})$ to $\delta\mu^{(\varepsilon_1,\mathbf{E^a})}/\delta\mu^{(\varepsilon_2,\mathbf{E^b})}$ is shown in Fig. 9.5(d) as a function of ε_2. The slope of δX agrees rather well with the growth rate change as given by Eq. (9.3) for the evolved networks as compared to the random networks.

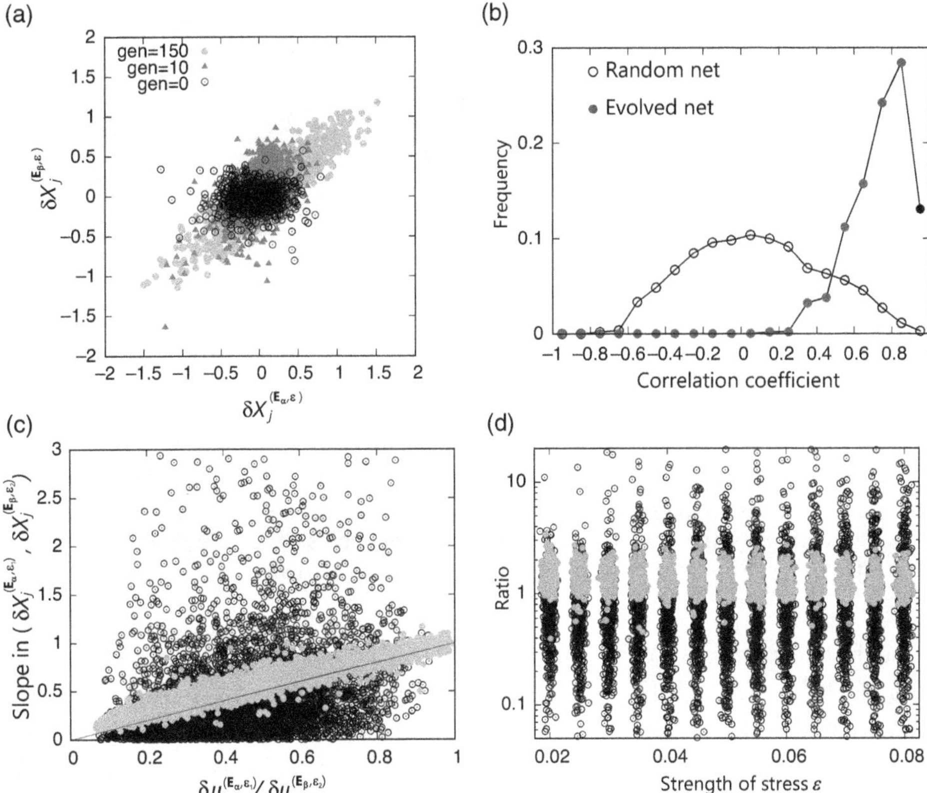

Figure 9.5 Common proportionality in concentration changes in response to different types of stress in the catalytic-reaction network model. (a) Concentration changes across different types of environmental stresses. For three different networks from different generations, $(\delta X_j(\mathbf{E}^a), \delta X_j(\mathbf{E}^b))$ are plotted by means of different randomly chosen vectors $\mathbf{E}^\mathbf{a}$ and $\mathbf{E}^\mathbf{b}$. (b) Distributions of coefficients of correlation between the changes in component concentrations $\delta X_j^{(\varepsilon,\mathbf{E}^a)}$ and $\delta X_j^{(\varepsilon,\mathbf{E}^b)}$, for a random network and evolved network (150th generation) obtained from 1,000 pairs of $\mathbf{E}^a$ and $\mathbf{E}^b$. The magnitude of environmental change ε is fixed at 0.8. (c) Relation between the ratio of growth rate change $\delta\mu^{(\varepsilon_1,\mathbf{E}^a)}/\delta\mu^{(\varepsilon_2,\mathbf{E}^b)}$ and the slope in $(\delta X_j^{(\varepsilon_1,\mathbf{E}^a)}, \delta X_j^{(\varepsilon_2,\mathbf{E}^b)})$. (d) Ratio of the slope in the relation between concentration changes and growth rate change as a function of the intensity of stress ε in the case of different types of stress. For (c) and (d), gray and unfilled circles are the results for evolved and random networks, respectively. Courtesy of Chikara Furusawa. Adapted from (Furusawa & Kaneko, 2018).

The global proportionality over all components across various environmental conditions suggests that $\delta X_j^{(\varepsilon,\mathbf{E})}$ across different environmental conditions $\mathbf{E}$ are constrained mainly along a one-dimensional manifold after evolution has progressed (even) under a single environmental condition. To verify the existence

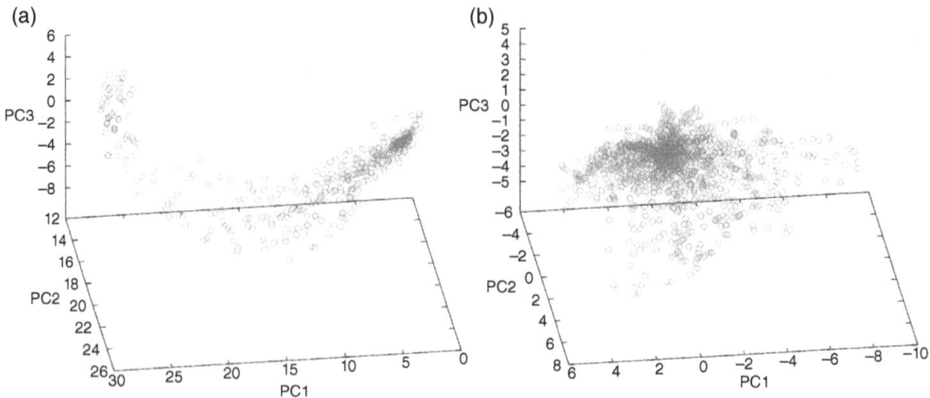

Figure 9.6 The change in $X_j^{(\varepsilon,\mathbf{E})}$ with environmental changes in principal-component space. Component concentrations $X_j^{(\varepsilon,\mathbf{E})}$ at randomly chosen various $\mathbf{E}$ and ε values are presented for (a) evolved and (b) random networks. In (a), the contributions of the first, second, and third components were 74%, 8%, and 5%, respectively. Reproduced from Furusawa & Kaneko (2018).

of such constraints, we carry out the principal component (PC) analysis of the data of $\delta X_j^{(\varepsilon,\mathbf{E})}$ across different environmental changes $\mathbf{E}$ and ε. In Fig. 9.6(a), the data are plotted in the space with the first three PC axes. In the evolved network, high-dimensional data from $X_j^{(\varepsilon,\mathbf{E})}$ are located along a one-dimensional curve.[1] In contrast, the data from the random network are scattered, and no clear structure is observed, as shown in Fig. 9.6(b). Furthermore, for the evolved network, the value of the first PC correlates well with the growth rate (Furusawa & Kaneko, 2018).

We then examine the evolutionary course of the phenotype projected on the same principal-component space, as depicted in Fig. 9.6(a). As shown in Fig. 9.7(a), the points from $\{X_j\}$ generated by random mutations in the reaction network are again located along the same one-dimensional curve. Furthermore, those obtained by environmental variation or noise in the reaction dynamics also lie on this one-dimensional curve, as shown in Fig. 9.7(b). Thus, the phenotypic changes are highly restricted, both genetically and non-genetically, within an identical one-dimensional curve.

As shown in Figs. 9.6 and 9.7, variation in the concentration due to perturbations is much larger along the first PC than along the other components. This suggests that relaxation is much slower in the direction of the first component than in the other directions.

---

[1] In this example, the contribution of the first PC reached 74 percent in the data.

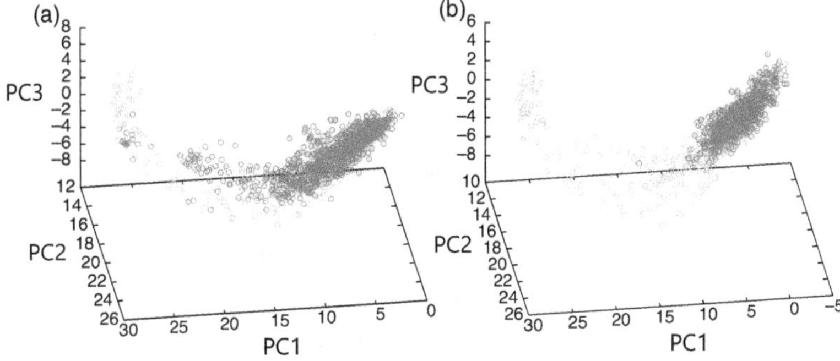

Figure 9.7 Change in component concentration X_j because of (a) mutations and (b) noise in reaction dynamics. In (a), mutations were added to the evolved reaction network by randomly replacing 0.5 percent of the reaction paths. The dark dots show the concentrations of components after mutations, which are projected onto the same principal-component space, as depicted in Fig. 9.6(a). The light gray dots represent the concentration changes caused by environmental changes for the reference, which are identical to those shown in Fig. 9.6(a). The dark dots in (b) represent the concentration changes observed at each cell division, which are caused by the stochastic nature of reaction dynamics. Fig. 9.6(a). Reproduced from Furusawa & Kaneko (2018).

In summary, we have observed emergent global proportionality which is far beyond the trivial linearity in response to tiny perturbations. After evolution, the linearity region expands to a level with an order-of-magnitude change in the growth rate. Additionally, the proportionality over different components across different environmental conditions is enhanced through evolution. In this global proportionality, evolutionary dimension reduction in phenotypic dynamics underlies phenotypic changes across a variety of environmental conditions, genetic variations, and noise. The changes are confined to a common one-dimensional manifold, given by the first principal-component mode which is highly correlated with the growth rate.

9.4 Evolutionary Dimensional Reduction Hypothesis (C)

Based on the observation of global proportionality and dimensional reduction from the high-dimensional phenotypic space in Section 9.3, the following hypothesis is proposed (Furusawa & Kaneko, 2018):

Phenotypic dynamics involve a large number of variables, and their state space is generally high-dimensional. However, phenotypic changes induced by environmental perturbations are constrained mainly along a low-dimensional (often one-dimensional) manifold (major axis). Relaxation dynamics of the phenotypes along this manifold are much slower than those across the manifold. Further, phenotypic changes induced by evolutionary changes (i.e., due to genetic changes that govern the phenotypic dynamics) also progress mainly along this manifold. The fitness gradually changes along the manifold, whereas it rapidly decreases across the manifold.

Indeed, the results described in Section 9.3 support the hypothesis, where the dominance of the first principal mode in a phenotypic change emerges after evolution, and the phenotypic changes as a result of mutation are constrained along the principal-mode axis (see Fig. 9.7). Moreover, expression data from bacterial evolution studies support this hypothesis, as described later.

This hypothesis is plausible considering the evolutionary robustness of phenotypes: First, in most cases, phenotypes (e.g., concentrations of chemicals) are shaped as a result of complex dynamics involving a large degree of freedom. For example, these dynamics can be determined by catalytic-reaction or gene-regulatory networks, which are determined by genes. In general, such networks that have higher fitness are rarer, and thus a mutation-selection process is needed to achieve higher fitness. Because of the complexity of the dynamics, stochastic perturbations may influence the final phenotype and will be an obstacle to keep the high fitness, unless the evolved networks reduce the influence of perturbations (Kaneko, 2012a).

As evolution progresses and a higher-fitness phenotypic state is reached, the state will become robust to perturbations. Otherwise, due to the inevitable noise in the dynamics, a rare fitter state will not be maintained. Increased robustness to perturbations is expected to result from evolution (see also (Ciliberti et al., 2006; Kaneko, 2007)). Accordingly, in the state space, the dynamics after evolution must provide a flow of strong attraction to the selected (fitter) phenotype against perturbations of most directions, as shown schematically in Fig. 9.8. Such a strong contraction toward the attracted state is shaped by evolution. However, there is (at least) one exceptional direction that does not have such a strong attraction, namely the direction along which evolution has progressed and will continue to increase the fitness. Along this direction, phenotypic states can be changed rather easily by perturbations. Otherwise, it would be difficult for evolution to progress. Thus, as shown schematically in Fig. 9.8, relaxation is slow along the direction of evolutionary change, while change is much faster along other orthogonal directions.

Now we consider the relaxation dynamics to the original state (attractor) at a given generation. For instance, the reaction dynamics are represented as $dX_i/dt = F_i(\{X_j\}) - \mu$, as given in Eq. (2.7) (Chapter 2). The linearized relaxation dynamics near the steady state is given by

$$d\delta X_i/dt = \sum_j J_{ij}\delta X_j - \delta\mu,$$

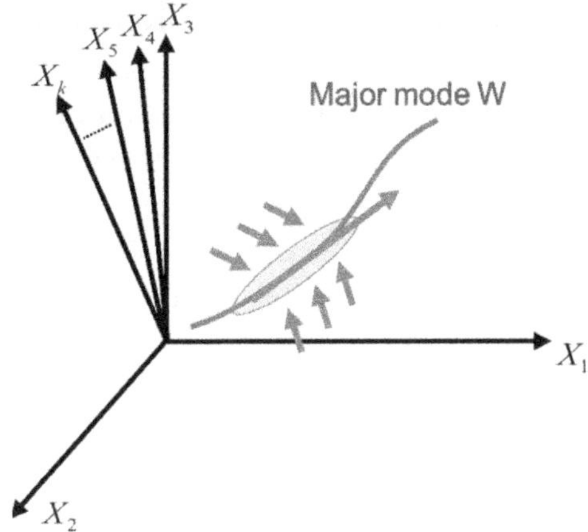

Figure 9.8 Schematic representation of the dimension-reduction hypothesis. In the state space of **X**, dominant changes are constrained along the mode **W** following the major axis and its connected manifold, whereas the attraction to this manifold is much faster.

The relaxation dynamics is represented by a combination of eigenmodes with negative eigenvalues $\{\lambda_i\}$'s. The magnitude of the (negative) eigenvalues will be large, except for one (or a few), while that along the evolution direction will be much closer to zero; that is, relaxation along the evolution direction is slower. Therefore, the variance along the first principal component will be dominant, as shown numerically in Fig. 9.6.

The present hypothesis indicates that only one mode dominates in the dynamics. Although the original dynamics are high dimensional, most changes occur along the one-dimensional manifold **W**, corresponding to the eigenvector for the eigenvalue closest to zero (or its nonlinear extension). Let us denote this direction as $\mathbf{w_0}$. The major change in the dynamics is projected onto this axis $\mathbf{w_0}$, as in the collective motion for a macroscopic variable in statistical physics (Mori, 1965). From this dominance of the single dominant mode W, the global linear relationship in Eq. (9.3) is naturally derived across different conditions **E**. Because δX_j's changes in all components X_j, are constrained along the dominant mode W, they are given by the projection of the change in W onto each X_j axis.

In fact, by employing the formulation of Chapter 2 and Section 9.2, the relationship (9.3) is derived according to the above argument: Let the eigenmode of change (the right eigenvector for the eigenvalue λ_k of the Jacobi matrix J) be $\mathbf{w^k}$ ($k = 0, \cdots, K - 1$). Now one (or a few) eigenvalues are near zero. Let $\mathbf{w^0}$ be the right eigenvector and $\mathbf{v^0}$ be the left eigenvector. Here, the manifold W is an extension of its tangential counterpart $\mathbf{w^0}$. The change along the (right) eigenvector corresponding to this eigenvalue close to zero is larger than the change in other directions which decay quickly. Since high-dimensional phenotypic changes due to various environmental changes occur mainly along $\mathbf{w^0}$, each component change $\delta X_j(\mathbf{E})$ is proportional to w_j^0, which, is the projection of the change along each axis X_j. Since this projection itself does not depend on the type of environment $\mathbf{E}$, $\delta X_j(\mathbf{E})/\delta X_j(\mathbf{E'})$ is independent of each component j.

Note that the growth rate $\delta\mu$ also changes along $\mathbf{W}$ (along w^0), and has one-to-one correspondence with it. Then along this eigenmode w^0, $\delta W \propto \delta\mu(\mathbf{E})$. Hence, we get

$$\frac{\delta X_j(\mathbf{E})}{\delta X_j(\mathbf{E'})} = \frac{\delta\mu(\mathbf{E})}{\delta\mu(\mathbf{E'})}. \tag{9.5}$$

For detailed derivation, see the footnote below[2] or the original paper (Furusawa & Kaneko, 2018; Sato & Kaneko, 2020).

To confirm the separation of a single or few eigenvalues, we have numerically computed the Jacobian matrix of the reaction dynamics the catalytic-reaction network employed in the model of Section 9.3.2, at each generation. (Here we used ordinary-differential, rate equations corresponding to the stochastic reaction process [Sato & Kaneko, 2020].) The inverse of all eigenvalues are plotted in Fig. 9.9, which shows that one eigenvalue is closer to zero, after evolution. The separation of one eigenvalue is evolved. In fact, the changes of high-dimensional concentrations upon environmental or genetic changes are constrained along the eigenvector of this separated eigenvalue.[3]

In summary, we can explain two basic features observed in experiments and simulations using the above theoretical formulation:

---

[2] The sketch of the derivation, by using linear algebra, is as follows. By using $\mathbf{L} = \mathbf{J}^{-1}$, $\delta\mathbf{X} = \mathbf{L}(\delta\mu\mathbf{I} - \gamma\delta E)$, follows, where $\mathbf{I}$ is an unit vector $(1, 1, 1, ..1)^T$. The relaxation dynamics are represented by a combination of eigenmodes with negative eigenvalues. By denoting the eigenvalues of matrix $\mathbf{L}$ as λ_k, with the corresponding right (left) eigenvalues $\mathbf{w}_k$ ($\mathbf{v}_k$), respectively. Then matrix L is represented by $\sum_k \lambda_k \mathbf{w_k v_k^T}$. This gives $\delta\mathbf{X} = \sum_k \lambda^k \mathbf{w_k}(\delta\mu(\mathbf{v_k^T} \cdot \mathbf{I}) - (\mathbf{v_k^T} \cdot \gamma)\delta E)$. According to the present hypothesis, the magnitude of the smallest eigenvalue of $\mathbf{J}$ (denoted as $k = 0$) is much smaller than that of the others. As the change in $\delta\mathbf{X}$ is nearly confined to the major axis corresponding to its eigenvector, the major response to environmental changes, is represented only by the largest eigenvalue of L (i.e., smallest eigenvalue of $\mathbf{J}$) with the corresponding eigenvector $\mathbf{w_0}$. Using the reduction to this mode $\mathbf{w_0}$, we obtain

$$\delta\mathbf{X} = \lambda^0 \mathbf{w_0}(\delta\mu(\mathbf{v_0^T} \cdot \mathbf{I}) - (\mathbf{v_0^T} \cdot \gamma)\delta E). \tag{9.6}$$

Because the dynamics are robust to perturbations from various directions except for the major axis $\mathbf{w_0}$, they should decay rapidly except for this direction. It is expected that γ, the sensitivity of $\mathbf{F}$ to an external change, has a larger value for directions other than $\mathbf{w_0}$, that is, the γ vector consists mostly of components of $\mathbf{w_j}$ at $j \neq 0$. Recalling that $\mathbf{v_0} \cdot \mathbf{w_j} = 0$ for $j \neq 0$ (according to linear algebra), thus the last term in the above equation will be negligible, giving:

$$\delta\mathbf{X} = \lambda^0 \mathbf{w_0}(\delta\mu(\mathbf{v_0} \cdot \mathbf{I})). \tag{9.7}$$

Accordingly, the global proportionality relationship Eq. (9.5) is reproduced.

[3] The inner product between the eigenvector and the direction of the vector for the first principal component is close to unity.

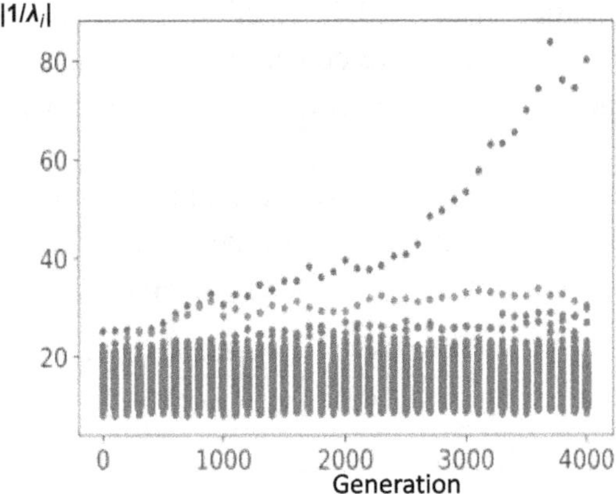

Figure 9.9 Evolutionary changes of the eigenvalues of the Jacobian matrix. The inverse of each eigenvalue is plotted per generation, for the case of a system with 100 components. For more detailed analysis, see Sato & Kaneko (2020).

(1) **Overall proportionality is observed in expression level changes across most components and across various environmental conditions**. This is because high-dimensional changes are constrained to changes along the major axis, that is, eigenvector $\mathbf{w_0}$.

(2) **There is an extended region of global proportionality**. Because the range in variation along $\mathbf{w_0}$ is large, the change in the phenotype is constrained to points near this eigenvector, causing the proportionality range of phenotypic change to extend via evolution. Furthermore, as long as the changes are nearly confined to the manifold along the major axis, global proportionality reaches the regime nonlinear to $\delta\epsilon$.[4]

9.5 Global Proportionality between Responses by Environmental and Evolutionary Adaptations (C–E)

9.5.1 Theory

According to the hypothesis in Section 9.4, the change due to genetic variation would also be constrained along with this major axis $\mathbf{w_0}$, as the most changeable direction $\mathbf{w_0}$ is the direction in which evolution has progressed and will progress.

---

[4] This expansion of linearity reminds statistical physicists of the foundation of linear response theory (Kubo, 1957). After this establishment, van Kampen criticised that the validity of linear regime would be quite small from the view of microscopic dynamics (van Kampen, 1971). In contrast, by considering macroscopic robustness of equilibrium state, and macro-micro consistency a la Onsager, the validity of linear regime is expanded to a macroscopic level, to assure the validity of linear response theory. As discussed in this chapter, the linear regime just by steady growth as described in Chapter 2 is quite small (see Fig. 9.4). Once macroscopic robustness of cellular state is achieved through evolution, the linear regime is expanded, so that the global proportionality works.

Indeed, in the simulation described in Section 9.3.2, the phenotypic changes caused by the mutational change are constrained along the manifold spanned by the first principal mode in the environmentally induced phenotypic changes (see Fig. 9.7).

Accordingly, we expect to observe global proportionality between the concentration changes induced by a given environmental condition (stress) ($\delta X_j(Env)$) and those due to genetic evolution ($\delta X_j(Gen)$), as given by

$$\frac{\delta X_j(Gen)}{\delta X_j(Env)} = \frac{\delta\mu(Gen)}{\delta\mu(Env)}. \tag{9.8}$$

For example, when cells are subjected to an environmental stress *Env*, the growth rate is reduced so that $\delta\mu(Env) < 0$, whereas the expression levels change with $\delta X_j(Env)$ accordingly. Next, the cells evolve under this given stress over several generations along with genetic changes. After n generations under genetic evolution, the growth rate recovers to some degree so that the growth rate shows a difference of $\delta\mu(Gen)$ from the original (non-stressed) state, satisfying $0 \geq \delta\mu(Gen) \geq \delta\mu(Env)$. The accompanied expression change, denoted by $\delta X_j(Gen)$, is then expected to satisfy

$$\frac{\delta X_j(Gen)}{\delta X_j(Env)} = \frac{\delta\mu(Gen)}{\delta\mu(Env)} \leq 1, \tag{9.9}$$

across most components j in a similar manner as in Eq. (9.5). Because $|\delta\mu(Gen)|$ is reduced with the progression of evolution, changes in the components introduced by the environmental change are reduced. Thus, there is an evolutionary tendency that the original expression pattern is recovered. This is reminiscent of the Le Chatelier's principle in thermodynamics.[5]

We next examine if the above relationship would hold in numerical simulation and bacterial evolution experiments.

9.5.2 Verification by the Reaction-network Model

We again employ the catalytic reaction network model in Section 9.3. After evolving the cells as described in the section under the given environmental condition, we switch the nutrient condition at a given generation (denoted by the arrow in Fig. 9.10). This causes the growth rate to decrease, which is later recovered through genetic evolution over generations. We compute the phenotypic changes induced by the environmental and evolutionary changes to examine the validity of the above relationship.

---

[5] For Le Chatelier principle, see Chapter 1.

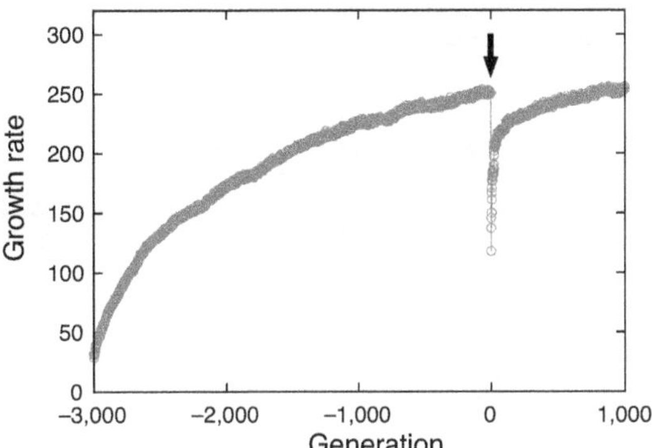

Figure 9.10 Growth rate (fitness) increase over generations from generation −3,000 to 0 under the environmental condition $\{s_1, s_2\} = \{0.5, 0.5\}$. At generation 0, the nutrient concentrations in the environment were changed to $\{s_1, s_2\} = \{0.9, 0.1\}$. This drastically decreased the growth rate at generation 0, which was later recovered by evolutionary dynamics. Reproduced from Furusawa & Kaneko (2018).

After altering the nutrient conditions, the abundances of all the components are changed. The average change of these abundances is denoted by $\delta X_j^{Env} \equiv \langle X_j(1) \rangle - \langle X_j(0) \rangle = \log \frac{\langle N_j(1) \rangle}{\langle N_j(0) \rangle}$, where generation 1 refers to the time point immediately following the environmental change, and the generation 0 denotes the generation right before this nutrient change. Similarly, the response by genetic evolution after m generations is given by $\delta X_j^{Gen}(m) = \langle X_j(m) \rangle - \langle X_j(0) \rangle$. Figure 9.11(a–c) shows the plot of δX_j^{Env} versus $\delta X_j^{Gen}(m)$ for $m = 5$, 10, and 50. The proportionality is observed between the environmental and genetic responses over all components.

Let us now define this proportion coefficient $r(m)$ for $\frac{\delta X_j^{Gen}(m)}{\delta X_j^{Env}}$ across components j. According to Eq. (9.8), this agrees with the growth rate change given by the ratio of $\delta \mu^{Gen}(m) = \mu(m) - \mu(0)(\leq 0)$ to $\delta \mu^{Env} = \mu(1) - \mu(0)(< 0)$ at each generation m. As shown in Fig. 9.11(d), the proportion coefficient $r(m)$ is plotted against this growth rate recovery $\delta \mu^{Gen}(m)/\delta \mu^{Env}$. The agreement between the two is discernible. This proportion coefficient $r(m)$ is initially close to 1 (i.e., $m \sim 1$); with increasing generation m, the value decreases toward zero, mirroring the recovery of the growth rate $\delta \mu^{Gen}(m)/\delta \mu^{Env}$. Thus, as stated in Le Chatelier's principle mentioned in Section 9.5.1, evolution shows a common tendency to reduce changes in components introduced by environmental change.

In this evolution to novel environment, we adopted the networks (individuals) that evolved in the old environment, and switched the condition. By comparing this evolution with those for the random networks, we further note the following two points.

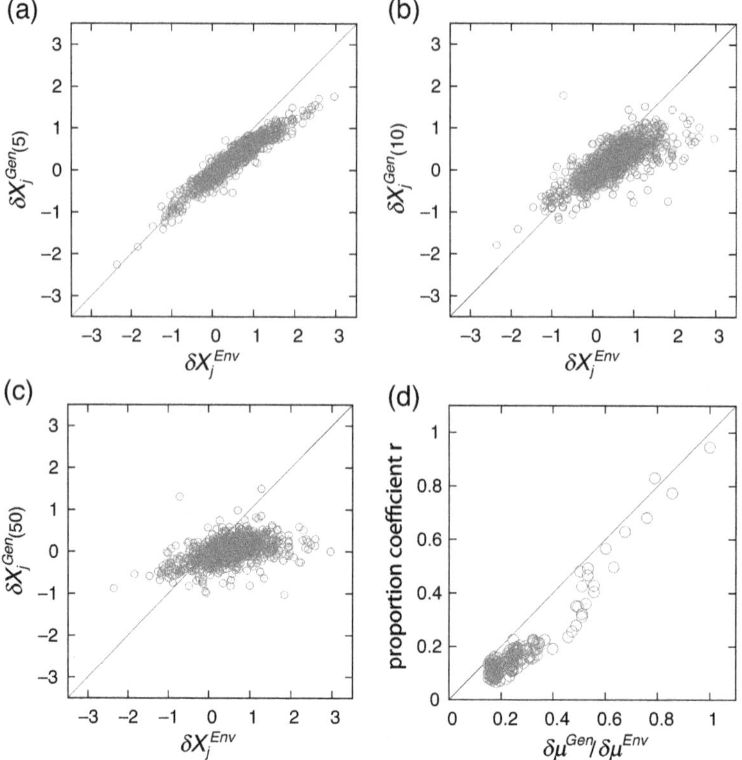

Figure 9.11 Response to environmental change versus response by evolution. Relationship between the environmental response δX_j^{Env} and genetic response $\delta X_j^{Gen}(m)$. (a), (b), and (c) show the plots for $m = 5$, 10, and 50, respectively. The black solid lines are $y = x$ for reference. (d) Relationship between the growth recovery rate $\delta \mu^{Gen}(m)/\delta \mu^{Env}$ and the proportion coefficient $r(m)$. The proportion coefficient $r(m)$ was obtained by using the least-squares method for the relationship of δX_j^{Env} and $\delta X_j^{Gen}(m)$ for $m = 1 \sim 200$. The black solid line is $y = x$ for reference. Reproduced from Furusawa & Kaneko (2015).

(i) The evolution process is generally faster when started from the adapted network. This might look a bit strange, as we started from those fitted to the old environment. The possible reason is that the evolution in this case takes advantage of the dimensional reduction, and then the optimization process to novel environment is faster under the restricted space than that from high-dimensional space. In fact, the evolutionary course in this case follows such constrained direction, as shown in Fig. 9.12(b), as compared from the evolution from random network in Fig. 9.12(a).

(ii) As the evolution occurs within a restricted space, the evolution in the phenotypic space is deterministic rather than stochastic. In Fig. 9.12, ten samples of evolutionary course are plotted, where the genetic change (i.e., the mutation

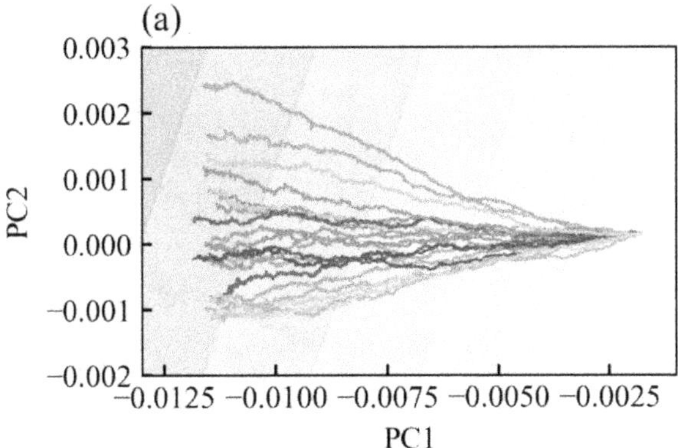

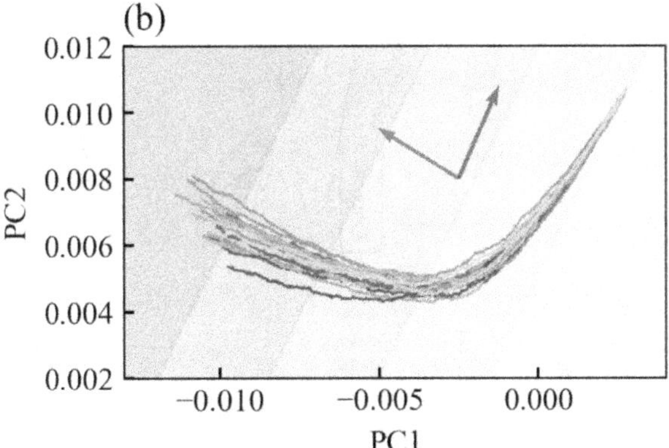

Figure 9.12 Evolutionary pathways plotted in the plane of PC1 and PC2: (a) evolution under a new environmental condition starting from a random network and (b) evolution under the new condition for genotypes that already evolved under the old condition. The vector pointing to the upper right in (b) is the projection of the eigenvector corresponding to the eigen value closest to zero, of the Jacobi matrix, for the evolved network under the old condition, whereas the one pointing to the upper left is that at the end of the evolution to new condition. The gradation in both figures represents the fitness value, with darker area corresponding to higher value as plotted in the PC plane. The PCs are calculated according to the phenotypes of the fittest cells in the population over generations for the 20 different strains. Reproduced from Sato & Kaneko (2020).

point in the network) differs by each. Still, when we trace the phenotypic change, they follow along the same course in phenotypic space, as shown in Fig. 9.12(b). In contrast, the evolution from the random network diversifies in phenotypic space as shown in Fig. 9.12(a).

These two salient features are also confirmed experimentally, as shown in Section 9.5.3.

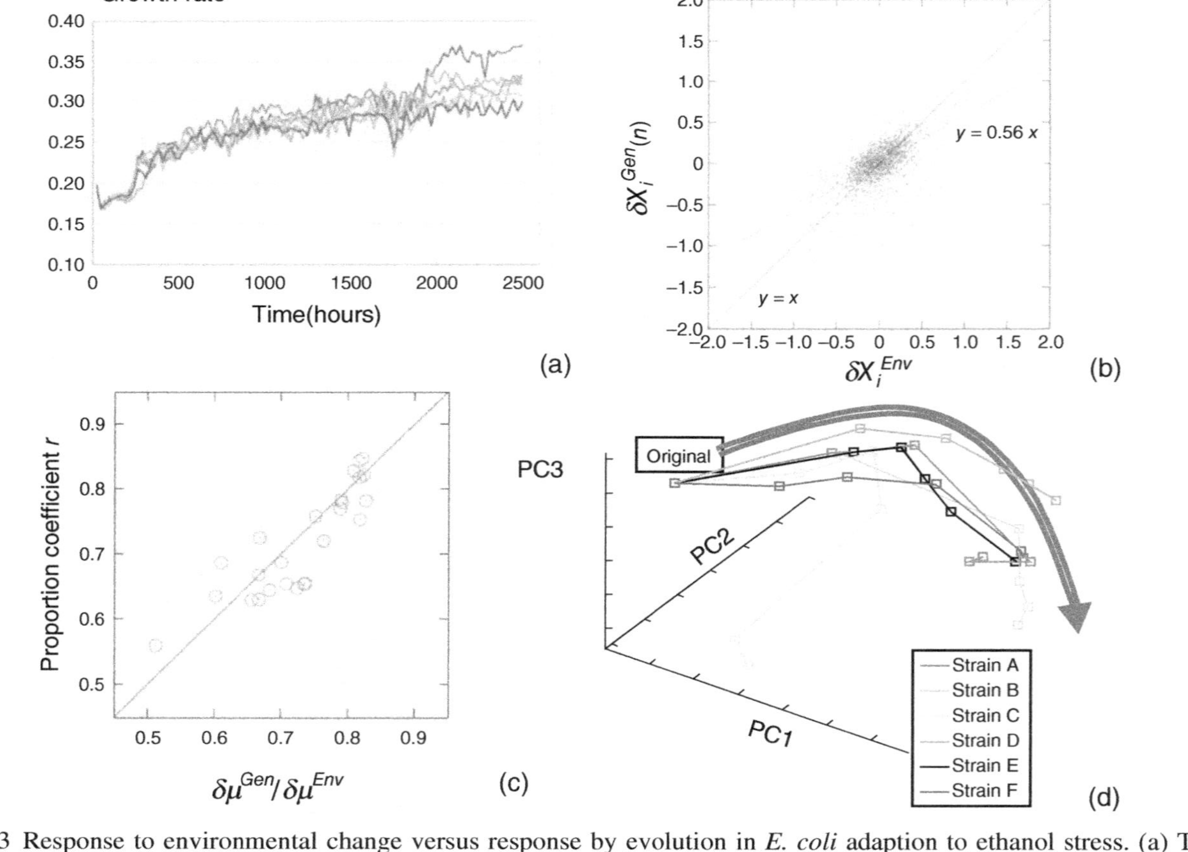

Figure 9.13 Response to environmental change versus response by evolution in *E. coli* adaption to ethanol stress. (a) The time course of the growth rate recovery through evolution, overlaid for six evolutionary samples leading to strains with different mutations. (b) Relationship between environmental response δX_j^{Env} and genetic response $\delta X_j^{Gen}(n)$ for $n = 104$ (days) as a representative example. δX_j^{Env} and $\delta X_j^{Gen}(n)$ were calculated by the log-transformed expression ratio between before and 1-day after exposure to ethanol stress. The plotted line $y = 0.56x$ was obtained by least-squares fitting, whereas the diagonal line is plotted for reference.

9.5.3 Experimental Confirmation by Laboratory Evolution

To verify the relationship given by Eq. (9.10), Furusawa et al. analyzed time-series transcriptome data obtained in an experimental evolution study of *E. coli* under conditions of ethanol stress (Furusawa & Kaneko, 2015; Horinouchi et al., 2015). In this experiment, after cultivation of approximately 1,000 generations (2,500 h) under 5 percent ethanol stress, 6 independent ethanol-tolerant strains were obtained, which exhibited an approximately 2-fold increase in specific growth rates compared with the ancestor (Fig. 9.13(a)). For all independent culture series, mRNA samples were extracted from approximately 10^8 cells at 6 different time points, and the absolute expression levels were quantified by microarray analysis. All mRNA samples were obtained from cells in an exponential growth phase, (see [Horinouchi et al., 2015] for details of Materials and methods).

Using the expression data taken at several generations through adaptive evolution, we analyze the common proportionality in expression changes. The environmental response of the j-th gene δX_j^{Env} is defined by the log-transformed ratio of the expression level of the j-th mRNA. Similarly, the evolutionary response at n days after the exposure to stress $\delta X_j^{Gen}(n)$ is defined by the log-transformed ratio of the expression level at the generation to that of the non-stress condition. Common trend between the environmental and genetic responses over all genes is demonstrated, as in Fig. 9.13(b).

Furthermore, as shown in Fig. 9.13(c), the agreement between $r(n)$ and the growth recovery ratio $\delta\mu^{Gen}(n)/\delta\mu^{Env}$, as predicted by Eq. (9.8), is discernible, where $\delta\mu^{Gen}(n)$ and $\delta\mu^{Env}$ are the growth rate differences at n days and 1 day after the exposure to stress, respectively. These results demonstrate that the evolutionary dynamics with growth recovery are accompanied by gene expression changes, which are reduced from those introduced by the new environment.

How does $\{X_j\}$ change in the state space of a few thousand dimensions? As we cannot display a high-dimensional state space, we use the first, second, and third principal components determined from the data for each generation of *E. coli* gene expression. (Approximately 31% of the change in each data set can be explained

---

Figure 9.13 (cont.) (c) Relationship between growth recovery rate $\delta\mu^{Gen}(n)/\delta\mu^{Env}$ and the proportion coefficient $r(n)$, obtained by the least-square fit for the ratio $\delta X_j^{Gen}(n)$ to δX_j^{Env}. Plotted for $n = 16, 31, 51, 76,$ and 104 days, over 5 samples. The diagonal line is plotted for reference. (d) Evolutionary orbits in PCA scores. Starting from the parent strain, changes in the expression profiles during the evolution are plotted in the three-dimensional PC plane. Among the six independent culture lines, the results of five lines follow the similar course. One example deviates at the later stage, due to the genome duplication that happened to occur. Reproduced from Horinouchi et al. (2015) and Furusawa & Kaneko (2018).

by the 1st component, whereas 15% is explained by the 2nd component). When the data points are plotted in the space of each principal component P^i-axis, they are distributed mainly along the P^1-axis direction, whereas the spread in the P^2 and P^3 axes is limited. Furthermore, the value of this first component is approximately proportional to the growth rate. This is consistent with the finding that the growth rate is a major factor in determining the change in the expression of each gene.

Figure 9.13(d) shows the evolutionary trajectory of the cell-state in the three-dimensional PC space. Here, six independent data are superimposed, which were obtained by repeating the same experiment. Mutations occurred at different sites by each experiment; each of the six strains has a different genetic sequence. Nevertheless, all experimental samples followed the same path. In other words, the way in which the phenotype changes to achieve high growth rates is largely deterministic as compared to random changes in genetic sequences. Shaping the relevant phenotypic change is a priority in evolution, whereas several possibilities exist to achieve such changes genetically.

9.6 Proportionality between Phenotypic Variance by Noise and by Genetic Changes across Components (C)

In Chapter 8, we have demonstrated the proportionality between the phenotypic variances by noise and by genetic change, through the course of evolution. Following the proportionality in phenotypic responses by external environment and by genetic change across components obtained in the present chapter, we may expect the proportionality of phenotypic variances across components. Here we show the proportionality of the two phenotypic variances X_i by noise and by genetic variation across components i. Let us represent the variance of the former by $V_{ip}(i)$ and that of the latter by $V_g(i)$.

As shown in Fig. 9.7(b), as well as in the hypothesis in Section 9.4, changes due to noise and genetic variations lie on the same major axis. Thus, phenotypic variances due to the former $V_{ip}(i) \equiv \langle \delta X_i^2 \rangle_{noise}$ and those due to the latter $V_g(i) \equiv \langle \delta X_i^2 \rangle_{mutation}$ are expected to be proportional across all components, where $\langle \cdots \rangle_{noise}$ and $\langle \cdots \rangle_{mutation}$ are the average over the distribution induced by phenotypic noise or mutations, respectively.

In fact, assuming that the relaxation of phenotypic changes is much slower along the dominant mode $\mathbf{W}$, phenotypic fluctuations due to noise are nearly confined to this axis. Thus, the variance of each component X_i is given by means of the variance in the variable W of the dominant mode (along the major axis) due to noise,

$$V_{ip}(i) = ((\mathbf{w_0})_i)^2 \langle \delta W^2 \rangle_{noise}. \tag{9.10}$$

Similarly, the variation due to the genetic change of each component is mostly constrained along the axis, so that

$$V_g(i) = ((\mathbf{w_0})_i)^2 \langle \delta W^2 \rangle_{mutation},\tag{9.11}$$

where $(\mathbf{w_0})_i$ is the ith component of $\mathbf{w_0}$. Hence,

$$V_{ip}(i) \propto V_g(i)\tag{9.12}$$

holds across the components i.

Note that in the derivation of Eq. (9.14), the growth rate term is not involved. Only the dominance of the mode along the major axis $\mathbf{w_0}$ must be considered in the expression changes. Major variance in the high-dimensional expression change occurs along the single dimension $\mathbf{w_0}$, irrespectively of the source of perturbations, for example, environmental change, noise, or genetic variation. Furthermore, as the evolution progresses along the manifold W, it is natural that V_{ip} and V_g are proportional through the evolutionary time course, which reconfirms the result of Chapter 8.

In fact, an alternative derivation of the above relationship is possible, based on the robustness argument to avoid error catastrophe in Section 8.3.1. We introduce a multivariate distribution function with regards to X_i and the genotype a, by extending the argument therein. (Here this peak position is set to be taken to be 0 by a suitable transformation of variables). By adopting Gaussian distribution form, it is given by

$$P(x_i, a) = N_0 exp\left(-\frac{x_i^2}{2\alpha_i} + C_i x_i a - \frac{a^2}{2\mu}\right),\tag{9.13}$$

with N_0, a normalization constant so that $\int P(x,a)dx = 1$. Here again, $\alpha_i \equiv V_{ip}(i)$ is the variance of the gene expression level, while μ is the mutation rate that determines the variance in the genotypes. This equation is rewritten as in the Eq. (9.13)

$$P(x_i, a) = N_0 exp\left(-\frac{(x_i - C_i a\alpha_i)^2}{2\alpha_i} - \frac{1}{2}\left(\frac{1}{\mu} - C_i^2\alpha_i\right)a^2\right).\tag{9.14}$$

As in the discussion of Section 8.3.1.2, let us assume the stability of the distribution $P(x_i, a)$, that is, a single-peak condition: It implies $1/\mu - C_i^2\alpha_i > 0$, that is, $\mu < \mu_{max}^i \equiv \frac{1}{C_i^2\alpha_i}$ is postulated. If a developmental dynamical system is chosen arbitrarily, this threshold mutation rate μ_{max}^i for the error catastrophe can generally take different values for each component i. However, when the biological system has achieved robustness to noise and mutation through evolution, some restriction on μ_{max}^i may be introduced across different i. To have higher robustness, the error catastrophe for the fitness should be postponed to a higher mutation rate. If each of the component change is independent of each other, this would influence on the fitness separately, and the dynamics leading to the fittest phenotype might be

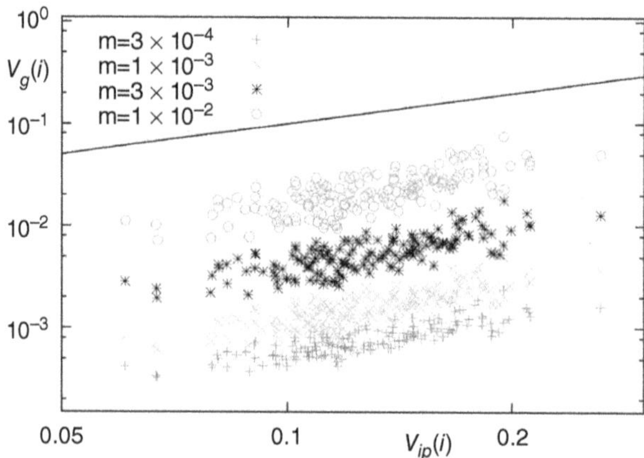

Figure 9.14 Relationship between $V_{ip}(i)$ and $V_g(i)$. $V_{ip}(i)$ and $V_g(i)$ were calculated based on the simulation results of randomly generated 10^5 networks with various mutation rates m (number of randomly replaced reaction paths divided by the total number of paths). The solid line is the $y = x$ line for reference. Reproduced from Furusawa & Kaneko (2015).

destroyed. When higher robustness is achieved, the error catastrophe at each gene will be suppressed so that once an error catastrophe for one component occurs, it can be propagated to change the other components. Hence, for a robust network having a higher threshold mutation rate for the error catastrophe, the error catastrophe in one component is propagated to others.[6] Accordingly, many components i are expected to share a common threshold mutation rate for the error catastrophe. If this sketchy argument is accepted, μ^i_{max} over many i take approximately a same value, when high robustness is evolved. Then, it is expected that

$$\mu^i_{max} = (C_i^2 \alpha_i)^{-1} = independent \ of \ i. \tag{9.15}$$

Since $V_g(i) = <(\delta x_i)^2> = C_i^2 \alpha_i^2 <(\delta a)^2>$. and $V_{ip}(i) = \alpha_i$, it follows that

$$V_g(i)/V_{ip}(i) = C_i^2 \alpha_i <(\delta a)^2>, \tag{9.16}$$

which is independent of gene i.

Now we examine the relationship by using the catalytic-reaction network model. As shown in Fig. 9.14, the numerical results revealed the relation $V_{ip}(i) \propto V_g(i)$ across most components i after evolution proceeded to achieve robustness (Furusawa & Kaneko, 2015). This proportionality is also observed in the evolution of the gene-regulation network model (Kaneko, 2011b, 2012b), whereas experimental supports will be presented in Section 9.8.

---

[6] This may correspond to the dimensional reduction in the present chapter.

9.7 Evolution of Slow Mode (D)

In summary, we demonstrated the 2-by-2 global proportionality of phenotypic changes occurring between responses and fluctuations and between the perturbations due to environmental (noise) and genetic changes, as already shown in Fig. 9.1. This proportionality is explained by evolutionary dimensional reduction, which states that phenotypic changes due to environmental changes and genetic variation are constrained along a unique low-dimensional manifold, as observed in bacterial and numerical evolution experiments. As a corollary of this global proportionality, all expression changes induced by environmental stress are reduced through evolution to restore the growth, which corresponds to the Le Chatelier's principle in thermodynamics mentions in Chapter 1.

We demonstrated in numerical evolution that high-dimensional phenotypic changes are mainly constrained along the mode $\mathbf{w}^0$, the eigenvector corresponding to the eigenvalue of the relaxation dynamics closest to zero. The change in the phenotypic state is larger along the direction of $\mathbf{w}^0$, and the variable W along this direction slowly returns to the steady-state value. The time scale of this mode is distinctively longer than others, as confirmed directly by evolution simulation of the catalytic reaction in Figs. 9.6 and 9.9. (It is also confirmed in the evolution of gene regulation network [Sato & Kaneko, 2023]).

This separation of the timescale of the slowest mode from others is theoretically expected, in order to make the evolutionary plasticity and robustness of the fitted state compatible. Because the dominant mode W is highly variable against environmental and genetic changes, the phenotype is plastic in the direction of W. In contrast, in the evolved state, the change in fitness (or growth rate) with respect to the environment and genetic changes would be small because of robustness, if the state is optimized against changes in the environment E and genetic evolution G. Recalling that the variations in W upon environmental and evolutionary changes are large, phenotypic changes upon environmental and genetic changes are buffered to the change in W, and then the change in the fitness F is drastically reduced, in the mapping from W to F, due to the robustness. This suggests that robustness and plasticity are compatible. Robustness in fitness is given by the insensitivity of the fitness along the mode W (see also the approximate neutrality shown Fig. 8.11), whereas the variability along the mode W implies the plasticity.

Formation of one (or few) slow mode as given by W separated from other modes is significant in evolutionary biology. It may be possible that this type of mode is straightforwardly given by the expression of some specific gene that changes more slowly than those of other genes. However, it may be more natural that this W is expressed as a collective change in the expressions of several genes rather than a single factor. Because the slow mode is expressed by the first principal component, determination of genes whose expression levels contribute more to the first principal component will improve the understanding of how plasticity and robustness are compatible in a cell.

The slow, dominant mode W emerges from evolution, but accelerates evolution. When faster and slower variables coexist and interact, the slower mode generally functions as a control parameter for the faster variables. Accordingly, if the slower mode is modified by a genetic change, most faster variables will be influenced simultaneously. Furthermore, because the mode W can influence the fitness F, the phenotypic evolution will be feasible simply by the change in this slow mode W (see also Appendix).

In contrast, if many variables change in a similar time scale, the genetic changes introduced to each will interfere and tend to cancel each other, which makes directional phenotypic changes harder. This situation is reminiscent of the proverb "Too many cooks spoil the broth" or its Japanese version, "Too many captains wreck the ship." The emergence of slow modes governing the others has also been observed in the evolution of pattern formation, as will be discussed in Section 10.4.

The correlation between evolutionary and environmental responses raises a question regarding how the two processes with quite different time scales are correlated. The presence of the slow mode W suggests a possible answer to this question. Adaptive dynamics, which originally show a much faster timescale than the evolutionary change, will be slowed along the mode W, whereas the evolutionary change, which originally has a much slower time scale is fastest, along the direction of the mode W. Thus, along the dominant mode W, the timescales of phenotypic adaptation and evolution can approach each other. This allows for the genetic assimilation of faster phenotypic changes.

9.8 Universality of Dimensional Reduction (C and E)

9.8.1 Theoretical Argument

To recapitulate the discussion in this chapter, we posit the following *evolutionary-dimensional reduction hypothesis*: Over the course of evolution, phenotypic states tend to exhibit robustness, yet they display plasticity in specific dimensions crucial for evolutionary adaptation. Consequently, a limited number of modes, characterized by eigenvalues close to zero (termed as slow modes), become isolated from the rest. In its simplest manifestation, this dimensionality reduces to one. However, in scenarios where multiple survival conditions must be met or when distinct environmental factors exert influence throughout evolution, the dimensionality may not be one. It may encompass multiple dimensions if various expression patterns correspond to different external conditions (Sato & Kaneko, 2023).

As most variations arising from noise, environmental alterations, and genetic mutations are constrained primarily along this slow manifold *W*, these changes from the different sources exhibit correlations, as proposed by the evolutionary-fluctuation response relationship. This phenomenon engenders both predictions and constraints in the evolution of phenotypes. Moreover, the separation of slow modes fosters enhanced evolvability. Few slow modes are separated, and most faster modes become influenced by these slow modes, akin to adiabatic elimination of fast variables. Consequently, mutational changes induce collective alterations in other variables, thereby amplifying their influence and accelerating the evolution speed.

Of course, additional studies are needed to establish a phenomenological theory for phenotypic evolution. The generality of the evolutionary dimensional reduction and resultant constraint in phenotypic evolution must be explored. The condition required for the emergence of dimensional reduction should also be determined. The models we studied satisfy the following conditions: (i) phenotypes with higher fitness are shaped by complex high-dimensional dynamical systems and (ii) the fraction of such fit states is rare in the state space and in the genetic-rule space. These two features are also consistent with our theoretical argument.

9.8.2 Universality in Theoretical Models

(I) Reaction-network and gene-regulation network models: The evolutionary dimensional reduction has also been observed in some other models. First, even when the fitness for selection does not affect the growth rate but rather some other quantity (such as the concentration of a component), the phenotypic change is mainly constrained along a one-dimensional manifold. Second, even when the environmental condition (e.g., concentrations of external nutrients) is not fixed but rather fluctuates over generations, restriction within the one-dimensional manifold is observed (Sato & Kaneko, 2019). Third, by using gene expression dynamics model as adopted in Chapter 8, and by imposing the fitness condition depending on the expression pattern of target genes, the dimensional reduction is observed.

(II) Evolution of protein structure: Through a combination of data analysis, elastic-network models, and theoretical analysis, the relationship between V_g and V_{ip} as well as the dimensional reduction of protein configuration dynamics have been confirmed. Estimated configuration changes resulting from noise and genetic mutations are constrained within a common low-dimensional space, typically around five dimensions, a stark reduction from the hundreds of residues giving rise to the high-dimensional state space (Tang

& Kaneko, 2021). Theoretical explanation is proposed therein, whereas consequence of few slow modes is discussed earlier (Togashi & Mikhailov, 2007), and some model studies also suggest such dimensional reduction (Tlusty et al., 2017).

(III) Evolving-interacting-spin model analogous to a spin-glass model: As mentioned in Section 8.3.2.3, the numerical evolution of spin dynamics S_i with the interaction matrix J_{ij} (where energy is given by $H = - \sum_j J_{ij}S_iS_j$) has shown the evolutionary fluctuation response relationship, for an intermediate range of temperature (corresponding to the replica symmetric phase) (Sakata et al., 2009; Pham & Kaneko, 2023). Within this temperature regime, the evolutionary dimensional reduction and V_g-V_{ip} relationship over spins are also numerically demonstrated (Sakata & Kaneko, 2020).

9.8.3 Experimental Confirmation

Besides the evolution of bacteria discussed here, recent studies of the bacteria exposed to a hundred antibiotics by Furusawa's group has provided further support. Transcriptome analysis of thousands of mRNA expression data indicates that changes in bacterial responses can be predicted by approximately seven principal components (Suzuki et al., 2014). In addition, the cross-fitness[7] after evolution under different types of antibiotics is shown to be correlated with the response prior to evolution, as is consistent with the theory of the present chapter (Sato et al., 2023). Further supports for such dimensional reduction are obtained from the Raman analysis of cellular components (Kamei et al., 2025).

The V_g-V_{ip} relationship, that is, the proportionality between the genetic and isogenic phenotypic variances across genes, is also consistent with global quantitative gene expression measurements from yeast. In budding yeast, global measurements have been made on the levels of isogenic expression fluctuation (expression "noise" (Landry et al., 2007)) and the variance induced by mutation. Instead of $V_g(i)$, the authors measured mutational variance that is the rate of spread of each gene expression level when mutations are added (diffusion constant by mutational change). Although this "mutational variance" is not equal to $V_g(i)$ exactly, the two would be highly correlated. According to their data, there is a correlation between the isogenic phenotypic fluctuation and this mutational variance, across all genes in the yeast (see also Fig. 3 of [Lehner & Kaneko, 2011]), which suggest the V_g-V_{ip}

---

[7] When bacteria evolved tolerance to antibiotics A, they can be either tolerant or vulnerable to antibiotics B. The cross-fitness gives a measure for such correlation.

relationship. Also, support on the relationship is recently obtained in the quantitative analysis in the gene expression pattern in Japanese *medaka* (fish) (Uchida et al., 2023).

Our argument on the proportionality between the variances is not restricted to gene expressions, but can be applied to other phenotypic traits. In a series of experiments Stearns et al. (1995) measured the isogenic fluctuation of five life history traits in Drosophila melanogaster. They also measured the genetic variance in the same five traits between different genetic lines. Strikingly, they observed the proportionality between the isogenic variance of each trait and its genetic variance (the figure 2 of [Stearns et al., 1995]), as in our theory. Further supports for the V_g-V_{ip} proportionality can be found in the wing morphology of *Drosophila* (fly) (Saito et al., 2024; Rohner & Berger, 2023).

Note that the proportionality relationship provides the possibility of predicting which component i is likely to evolve prior to evolution: changes due to noise itself are not inherited. Recall, however, that genetic variation is naturally transmitted to offspring, and the resulting phenotypic variance $V_g(i)$ is proportional to the evolutionary rate of that component (Fisher, 1930). Since $V_{ip}(i)$ is proportional to $V_g(i)$, the evolutionary rate of component i is proportional to $V_{ip}(i)$. In other words, the greater the phenotypic variance due to noise or environmental changes (prior to genetic change), the more feasible it is for the component (phenotype) i to evolve. In other words, the direction in which it is likely to evolve in phenotypic space can be predicted in advance, even before genetic variation (Kaneko & Furusawa, 2018).

The dimensional reduction has also gathered significant attention in other fields of biology. For instance, it is explored in development of *C. elegans* (Jordan & Miska, 2023), and ecological evolution (Frentz et al., 2015). Besides the issue in biophysics, dimensional reduction has attracted much attention in learning in the brain and neural networks (Schreier et al., 2017). In deep learning, the possible network configurations that can support a desired output are quite redundant, and reduction to the appropriate output space is achieved. This is similar to the phenotypic evolution in the present chapter. The genetic space (e.g., the network configuration) is high dimensional, so that there can be many ways to achieve the appropriate phenotype. Even though the final selected states are constrained in low-dimensional space, the use of high-dimensional space is essential for both machine learning and evolution.

Also, as a possible relationship, the *sloppy parameter hypothesis* by Sethna (Daniels et al., 2008) is proposed, which suggests that many parameters employed in biological models are irrelevant. Further studies are necessary to explain the universality of such evolutionary dimension reduction.

9.9 Toward Statistical Physics Theory for Evolutionary Dimensional Reduction

While the argument for dimensional reduction grounded in evolutionary robustness and plasticity appears plausible, it remains somewhat speculative in the absence of a rigorous theoretical foundation.[8] Theory to establish the macro-micro consistency is missing at present. Drawing inspiration from established theories in statistical physics, several avenues of investigation are expected:

(i) **Projection to dominant collective mode from high-dimensional phenotypic space**: In statistical physics, the projection method to collective mode is established by Mori and Zwanzig (Mori, 1965; Zwanzig, 1973), to be consistent with thermodynamics. In contrast, in the context of evolutionary dimensional reduction, the space onto which projection occurs emerges as a consequence of evolution itself. Consequently, we need a self-consistent approach to determine the space onto which the projection is made. One promising direction for it is the integration of dynamical mean field theory (Sompolinsky et al., 1988; Opper & Diederich, 1992) with evolutionary dynamics (Pham & Kaneko, 2024), although the formulation of the dimensional reduction process remains an ongoing endeavor.

(ii) **Renormalization group**: Through evolutionary process to achieve robustness, microscopic perturbations are smeared out, leaving behind only the low-dimensional aspects of plasticity. This process bears resemblance to the elimination of irrelevant variables in the renormalization group (Kadanoff, 1966, 2000; Wilson, 1983; Goldenfeld, 1992; Oono, 2012). Devising a methodology to integrate renormalization group concepts with evolutionary processes remains a challenging frontier.

(iii) **Slow-fast symmetry breaking**: As uncovered in the analysis of eigenvalue spectrum, few slow modes are separated from many other degrees of freedom, as represented by few outliers (whose eigenvalues are close to zero). As discussed, this separation of few slow modes is expected to facilitate the evolution. This phenomenon bears similarity to the origin of the central dogma, wherein symmetry breaking leads to the separation of information molecules and functional molecules exhibiting faster changes, as discussed in Chapter 3 (Takeuchi & Kaneko, 2019). A similar separation phenomenon also emerges in neural systems, where separation of area with slower neural activities is relevant to Bayesian inference (Ichikawa & Kaneko, 2024). Theoretical framework of slow-fast symmetry breaking for evolution will be required.

---

[8] This section is intended to those in theoretical physics.

(iv) **Avoidance of replica symmetry breaking**: In the evolving-interacting-spin model we mentioned in Section 9.8, robustness to noise and dimensional reduction is achieved at the replica symmetric phase (Sakata et al., 2009; Sakata & Kaneko, 2021), whereas the replica symmetry breaking (RSB) will undermine robustness. In the case of evolving interaction, the replica theory needs to be extended to both spins and couplings (Pham & Kaneko, 2023). On the other hand, when the evolution under multiple fitness against different environmental conditions is postulated, the region of fitted replica-symmetric phase shrinks toward RSB side (Sakata & Kaneko, 2023), which may suggest that the state achieving both robustness and plasticity to multiple conditions will approach "the edge of glass." This intriguing phenomenon warrants further investigation.

(v) **Macroscopic Potential theory**: If one assumes dimensional reduction, it raises the possibility of a macroscopic potential theory akin to thermodynamics. Analogous to thermodynamics where free energy potentials emerge from dimensional reduction, one might envision a similar description of biological systems employing a limited set of macroscopic variables. This could entail the formulation of a macroscopic potential, such as the growth rate or fitness, expressed as a function of environmental and genetic changes. This framework has the potential to yield linear relationships between evolutionary responses and phenotypic fluctuations caused by noise, as corroborated by bacterial evolution experiments, as have been discussed. In fact, landscape picture akin to potential was proposed by Waddington, with qualitative discussions on correlations between genetic and environmental responses, as is referred to *genetic assimilation* (1957), a concept that resonates with the V_g-V_{ip} relationship presented in this paper. With the formalization of the macroscopic potential, these discussions can advance to the level of quantitative theory.

The dimensional reduction we have observed suggests that the biological adaptation and evolution can be characterized by few variables, as thermodynamics have established. The observed Le Chatelier-type relationship, whereby evolution cancels out changes in component quantities caused by environmental change, also suggests such macroscopic description. In the Appendix, we introduce a potential-like function for the growth rate (or fitness) given as a function of the variables E and G representing environmental and genetic changes, respectively.

9.10 Collapse of Consistency in Dormant State? (F)

The result in this chapter is based on a robust cellular state with a steady growth. Then, can the state of a dormant cell that has stopped growing, as investigated in

Chapter 3, also be expressed in this way with a few degrees of freedom? So far, no definite answer has been obtained for this.

In the model mentioned in Section 3.4.2, the abundances of components are concentrated on few components in the autocatalytic subnetwork, in the exponentially growing phase, leading to the dimensional reduction. However, when the nutrient supply is reduced, the stagnation in reaction pathways within the autocatalytic network leads to increase the fraction of flux flowing to other components. Then, the increase in diversity of chemical components is observed as the cell enters a nongrowing quiescent state. This increase in diversity suggests a breakdown of dimensional reduction out of the steady growth (Yamagishi & Kaneko, 2024). In a dormant state, the dimensional reduction achieved for a robust state with consistency between macro and micro levels may collapse, leading to diversity in components.

To put it metaphorically, as in the beginning of Tolstoy's novel Anna Karenina, "Happy families are all alike; every unhappy family is unhappy in its own way," the consistent state of exponential growth can be described universally by few degrees of freedom, but the dormant "unhappy" states cannot be described by a few numbers of degrees of freedom.

9.11 Summary: Universal Laws Discussed in this Chapter

- For different types of environmental conditions (E, E') the changes in the log-concentration X_j are proportional across almost all components, over a wide range of perturbations. The proportionality coefficient is given by the ratio of the change in the cell growth rate μ. In other words. $\frac{\delta X_j(E)}{\delta X_j(E')} = \frac{\delta \mu(E)}{\delta \mu(E')}$ holds across most components j. (A, B, C, E).

- The above relationship (*deep linearity*) also holds between environmental change and evolutionary changes. Apply an environmental stress E, and let the changes in the log-concentration and the growth-rate by it be $\delta X_j(E)$ and $\delta \mu(E)$, respectively. Then, after evolution over some generations, let the change in the log concentration be $\delta X_j(G)$ and the growth-rate change be $\delta \mu(G)$. (These changes are measured from the original state before the environmental change). Then $\frac{\delta X_j(G)}{\delta X_j(E)} = \frac{\delta \mu(G)}{\delta \mu(E)}$ holds. With the evolution $|\delta \mu(G)|/|\delta \mu(E)| < 1$ gradually moves toward 0 over generations, so that the common proportionality coefficient $\delta X_i(G)/\delta X_i(E)$ also moves toward 0. In other words, the change in the concentrations induced by the environmental change decreases: Each expression level shows strong homeostasis, returning to its original state where no environmental change was applied. Subsequently, changes at the evolutionary scale occur. This is a process akin to the Le Chatelier's principle of thermodynamics, discussed in Chapter 1. (A, B, C, E).

- The relationships described above emerge, because phenotypic changes by environmental perturbations, noise, and genetic changes are constrained along a common low-dimensional manifold, as a result of evolution. This dimensional reduction is resulted, because (i) the fitted state after evolution should be stable against a variety of perturbations, so that strong attraction to the state against most directions in the phenotypic space should evolve whereas (ii) phenotypes keep plasticity to change to have evolvability, so that along the direction of evolution, the phenotypes can keep changeability. Hence, phenotypic dynamics change slowly along this low-dimensional direction, along which evolution proceeds (C, D, E).

- To achieve the above-mentioned dimensional reduction, there emerges separation of few slow modes in the dynamics for phenotypes, as represented by the separation of few eigenvalues in the relaxation dynamics. Such slow modes will control the dynamics, whose emergence facilitates the evolution (C, D, E).
- The variance of traits or phenotype i due to noise $V_{ip}(i)$ (or to environmental variations) and that due to genetic changes $V_g(i)$ are proportional (or highly correlated) across i. The latter is proportional to the speed of the evolution of each trait, so that whether a trait is feasible to evolve or not is given by the degree of fluctuation due to environmental variation or noise, prior to genetic variation. In other words, phenotypic evolution has a direction and is predictable (A, B, C, D, E).

9.12 Appendix: Potential Theory

Assume that evolution proceeds in a constant environment and the growth rate (or degree of adaptation) μ is optimized. Write it as $\mu(E, G)$ as a function of the environment E and genetic change G, and try to treat it like a "thermodynamic potential" and formulate the Le Chatelier's principle.

Thus, by considering the first-order change of $\delta\mu$ against the change in ΔE and ΔG: $\Delta\mu = \frac{\partial\mu}{\partial E}(\Delta E) + \frac{\partial\mu}{\partial G}(\Delta G)$,

$$\frac{\partial\mu}{\partial E} = 0; \qquad \frac{\partial\mu}{\partial G} = 0, \tag{9.17}$$

holds, as long as the reference state is optimized against the environmental and genetic change.

Next, as for the second order derivative, the following stability condition needs to be satisfied: By using the second-order expansion $\Delta\mu = \frac{\partial^2\mu}{2\partial E^2}(\Delta E)^2 + \frac{\partial^2\mu}{\partial G\partial E}(\Delta E)(\Delta G) + \frac{\partial^2\mu}{2\partial G^2}(\Delta G)^2$ is obtained. Thus, the maximality condition implies

$$\frac{\partial^2\mu}{\partial E^2} < 0; \qquad \frac{\partial^2\mu}{\partial G^2} < 0; \qquad \left(\frac{\partial^2\mu}{\partial G\partial E}\right)^2 - \frac{\partial^2\mu}{\partial E^2}\frac{\partial^2\mu}{\partial G^2} < 0, \tag{9.18}$$

Here E and G do not directly determine μ, but rather they affect the phenotypes X_i, through which μ is changed. Now, recall that all the expression changes X_i are highly constrained along the slow manifold. Let us denote this slow variable by X, which is the variable along the axis $\mathbf{w^0}$. Then the growth rate is a function of X, as $\mu(X)$, while X is a function of E, G, so that $\mu(X(E,G))$. Then, the change in $\mu(X(E,G))$ is represented as

$$\Delta\mu = \frac{\partial\mu}{\partial X}\frac{\partial X}{\partial E}(\Delta E) + \frac{\partial\mu}{\partial X}\frac{\partial X}{\partial G}(\Delta G), \tag{9.19}$$

Now, as X will change with E,G, it is unlikely that both $\frac{\partial X}{\partial E} = 0$ and $\frac{\partial X}{\partial G} = 0$ are satisfied, so that the optimization condition implies

$$\frac{\partial\mu}{\partial X} = 0. \tag{9.20}$$

Next, the 2nd order variation of μ is written as

$$\frac{\partial^2\mu}{2\partial X^2}\left(\frac{\partial X}{\partial E}\right)^2(\Delta E)^2 + \frac{\partial^2\mu}{\partial X^2}\frac{\partial X}{\partial E}\frac{\partial X}{\partial G}\Delta E\Delta G + \frac{\partial^2\mu}{2\partial X^2}\left(\frac{\partial X}{\partial G}\right)^2(\Delta G)^2. \tag{9.21}$$

Now consider the genetic adaptation after environmental change. Then, we consider that the environment is first changed to $E + \Delta E$. This leads to the change in $\delta\mu = \frac{\partial^2\mu}{2\partial X^2}(\frac{\partial X}{\partial E})^2(\Delta E)^2$. Through the evolution after this change, genetic change from G to $G + \Delta G$ follows to optimize μ, resulting in the change from $\delta\mu^E$ to $\delta\mu^{EG}$. According to the second-order calculation, it is given by

$$\Delta\mu_{EG} - \Delta\mu_E = \frac{\partial\mu}{\partial X}\left(\frac{\partial X}{\partial E}\right)_{G+\Delta G}(\Delta E) + \frac{\partial^2\mu}{2\partial X^2}\left(\frac{\partial X}{\partial G}\right)^2(\Delta G)^2$$

$$= \frac{\partial^2\mu}{2\partial X^2}(\Delta E)^2\left(\frac{z^2}{2}\left(\frac{\partial X}{\partial G}\right)^2 + z\frac{\partial X}{\partial E}\frac{\partial X}{\partial G}\right), \tag{9.22}$$

where we have introduced $z \equiv \Delta G/\Delta E$. Then by assuming that μ is optimized by genetic evolution, ΔG is determined by maximizing Eq. (9.22), so that $z_{max} \equiv -\frac{\frac{\partial X}{\partial E}}{\frac{\partial X}{\partial G}}$.

If the genetic adaptation is not completed, ΔG is not reached that maximal, so that $0 < z/z_{max} < 1$. By writing $z(= \Delta G/\Delta E) = cz_{max}$ with $0 < c < 1$,

$$\frac{\Delta X_G}{\Delta X_E} = \frac{\frac{\partial X}{\partial G}\Delta G}{\frac{\partial X}{\partial E}\Delta E} = \frac{z}{z_{max}} = c. \tag{9.23}$$

Thus,

$$0 \leq \frac{\Delta X_G}{\Delta X_E} \leq 1, \tag{9.24}$$

follows. Accordingly, the genetic change progresses to reduce the phenotypic change induced by environmental change. Thus, the Le Chatelier-type response observed both in evolution experiments of bacteria and numerical evolution of the present cell model follows.

Furthermore, as $\Delta\mu_{EG} = \frac{\partial^2\mu}{2\partial X^2}(\Delta E)^2(1-c)^2$, we get

$$\Delta\mu_{EG}/\Delta\mu_E = (1-c)^2, \tag{9.25}$$

which is between 0 and 1, representing that the growth rate is partially recovered.

Here we assumed that most expression changes are constrained by the slow manifold X. As X is the eigenvector for the smallest eigenvalue of the Jacobi matrix, the variation along this vector δX is large. Biologically speaking, this δX can be the concentration change of a specific component, but more naturally it represents a collective variable generated from concentrations (expression levels) of many components. In other words, we assumed that the dynamic changes of high-dimensional cellular state are constrained along a slow (collective) mode X. Then, we note

(1) Variation in X due to environmental or evolutionary change is rather large.
(2) Still, the growth rate (fitness) after evolution is rather robust, that is, $\partial\mu/\partial E$ *and* $\partial\mu/\partial G$ is rather small, which, in the present formulation, is achieved by $d\mu/dX \sim 0$.
(3) In other words, the environmental or genetic change is buffered to X, but from the change in X, the fitness (growth-rate) change is highly contracted.

10

Summary and Future Issues

10.1 Introduction

Throughout the present volume, we have investigated universal properties and laws of living systems, from the viewpoint of consistency between different scales as the key concept. When looking at the different levels of the hierarchy – molecule-cell or cell-multicellular individual – the lower-level (microscopic side) system consists of large degrees of freedom with a variety of components. On the other hand, the upper hierarchy (macroscopic side) has to be stable despite underlying microscopic, heterogeneous degrees of freedom. In order to have such macroscopic stability, the dynamics of the microscopic and macroscopic sides must keep a consistent relationship. With this consistency between hierarchies as the guiding principle, the basic properties of reproduction, adaptation, memory, differentiation, and evolution have been examined, with robustness and plasticity as underlying common concepts.

The lower levels of the hierarchy, for example, molecules in the molecule-cell hierarchy, or cells in cell-multicellular-organisms hierarchy generally consist of diverse components. In other words, the system is represented by a high-dimensional state space. However, the requirement to be consistent with the robustness at the upper level places strong constraints on the large-degree-of-freedom system at the lower level. As a result, the change in large-degree-of-freedom system is constrained into a low-dimensional space (manifold), which introduces few "macroscopic" variables, whereby general laws on the macroscopic response and statistical distribution and fluctuations for microscopic variables are extracted. These give a quantitative perspective on the robustness and plasticity of the organism. This is a standpoint that explores laws as a consequence of consistency.

Besides the consequence, understanding how consistency is brought about is also a step that universal biology should take. In this book, this is discussed mainly from the standpoint of dynamical systems and statistical physics. Indeed, the collective dynamics of a system of many components, which has recently gathered much attention in statistical physics, can be one of the fundamental concepts linking different levels in a hierarchy. In this sense, there exists a "collective variable" (or "mean field" in statistical mechanics). For instance, the growth rate of a cell is determined as a result of many components, whereas it affects the concentrations of all components through concentration dilution. This is an example how "collective variable" from many components influences all the components, which connects different levels in a hierarchy.

Another factor of inter-hierarchical consistency will be the interference of processes from different origins, such as chemical reactions, gene expression, x and $-$ intercellular dynamics. Such interplay will result in robustness of cells and organisms consisting of cells. We have also seen "consistency across time scales," where fast time scale phenomena are embedded into slower processes that stabilize the former, and sometimes are fixed as memories.

On the other hand, in order to relate the present macro-micro consistency approach with conventional molecular biology, it would be worthwhile to explore the molecular mechanisms that may be responsible for such consistency. For example, there may be "hub molecules" in a reaction network that affect many components. The ratio of ATP to ADP concentrations can work as a kind of "collective variable" molecules as they influence on most of reactions, whereas protein molecules such as chaperones also affect many components. The presence of such hub molecules, as global regulators, could synchronize multi-component changes and result in low-dimensional constrained changes. How these molecules contribute to correlated macroscopic changes needs to be explored further. However, there is a remark: while conventional molecular biology would pursue to conclude that such hub molecules are the *cause* of consistency, the present book emphasizes the consistency to achieve biological robustness, rather than the pursuit of specific causal molecular mechanisms; the view that processes mutually reinforce each other for macro-micro consistency. It is likely that the factors of consistency do not necessarily have to be specific molecules, but have been adopted and fixed through evolution (by chance) from among several possibilities.[1]

To recap, we have noted the following consequences of macro-micro consistency for robust biological systems:

- Dimensional reduction
- Autonomous control by slow mode(s)

---

[1] As a related example, recall that while phenotypic evolution was low-dimensional and deterministic, the genetic possibility that stabilized it was quite diverse (Chapter 9).

- Fluctuation–response relationship
- Le Chatelier Principle

Although not explored in depth in this book, it will be important to further investigate the processes and consequences of the breakdown of consistency. This could be, on the one hand, result in a state of arrested growth for a unicellular organism, or a cancer cells in multicellular organisms, but, on the other hand, it may bring about the restoration of plasticity that leads to adaptations to new environments and evolution of new species. Here, however, we should note that the consequence of the breakdown might lead to the deviation from universal characteristics. As mentioned in Chapter 9, "Happy states are all alike; every unhappy state is unhappy in its own way," as in the first sentence of Tolstoy's *Anna Karenina*. Further exploration of such *Anna Karenina* phenomena will be necessary in future.

In the following sections we will address some of the issues that are not fully addressed in this book but should be studied from the perspective of universal biology, for future.

The first issue is the origin of the hierarchy itself, such as molecule-cell or cell-multicellular organisms, with respect to the consistency of the assumption of hierarchy. For example, Maynard-Smith & Szathmary (1995) discussed the origin of several hierarchies in living systems as Major Transitions. Here, from our viewpoint, we will first discuss the origins of such hierarchies, followed by future issues, such as symbiosis, ecology, developmental processes, and neural systems.

10.2 Origin Problems

10.2.1 Origin of Life

As already mentioned, how a recursively reproductive set of molecules has emerged from a mere collection of molecules is precisely the question of the origin of life. The key question here is how it is possible for a diverse group of molecules to continue to produce something almost identical as a whole, that is, for molecular replication and (primitive) cell growth and division to continue in a consistent manner, and what the conditions are for this to happen.

First of all, as mentioned in Chapter 1, unless nutrition is provided without limitation, the replication system will soon face the problem of nutrient deficiency. Under such conditions, diversity will be resulted: diverse molecules catalyze each other to replicate under the condition of diverse environments with resource-limitation (Kamimura & Kaneko, 2015, 2016). These molecules will come together to form the primordial cell. Here, as mentioned in Section 3.6, while they are working as catalysts to help the replication of other molecules, they are not replicated. Hence, as for the evolution at a single-molecule level, molecules tend to lose their catalytic activity. Parasitic molecules that are replicated with the aid of others but do not help the replication of others will spread (Takeuchi et al., 2016). In this situation, as discussed in Chapter 3, differentiation into two types of molecules emerges those that retain catalytic function and lose replication information and those that lose catalytic function but act as replication information molecules (Takeuchi & Kaneko, 2019). In other words, the separation of molecules carrying information and function, also known as the central dogma of molecular biology, is an inevitable consequence of hierarchical replication systems. Furthermore, at this time, the number of molecules with function will be larger, whereas that of molecules carrying information will be smaller, realizing the situation of minority control leading to the acquisition of evolutionary potential (Kaneko & Yomo, 2002). As the minority molecules, replicate and mutate more slowly than the majority molecules, the minority control is also consistent with the achievement of evolvability by slow-mode control discussed in Chapter 9. Once this minority control is established. The system can sustain exponential growth (Kamimura & Kaneko, 2018). Furthermore, cell-growth-division events and replication of information molecules are synchronized (Kamimura & Kaneko, 2010).

The above theoretical studies will be experimentally verified. In fact, protocells with oil droplet including RNA replications, have been synthesized, in which the role of minority molecules or parasitic molecules to evolution has been discussed (Matsuura et al., 2011; Ichihashi et al., 2010, 2013; Mizuuchi & Ichihashi, 2018). The origin of function/information separation and diversification will be explored in such protocell systems. Here, it should be remarked that the above theoretical studies are also relevant to the hierarchy at a higher level. For instance, the separation between functional units with majority and information units with minority, as studied by the symmetry breaking theory of the central dogma, will be relevant to understand the separation between germ cells and somatic cells in multicellular organisms, or queens and workers in the social insects (see Table 10.1).

Table 10.1 *Division of labor into function and information units in a different level of hierarchy in a biological system.*

level	functional unit	information unit
(characteristic)	fast, majority	slow, minority
cell	protein	DNA
multicellular organisms	somatic cell	germ cell
colony of social insects	worker	queen

10.2.2 Endogeneous Symbiosis

As organisms proliferate, their density increases, resulting in strong interactions with each other. Individual interactions can include predation or parasitism in which one utilizes the other unidirectionally, but often these relationships are eventually replaced by symbiotic relationships by evolution, in which both organisms mutually benefit. Although these are exogeneous symbiotic relationships between different individuals, there also exists intra-individual symbiosis, in which different organisms live within and replicate as a group, synchronized with the host individual (Margulis, 1981).

For instance, multicellular organisms often have bacteria living within, which are useful and essential for the survival of the host. A wide variety of bacteria can live in our gut, which is often essential to our survival. In some insects, bacteria are strongly involved in the developmental process of individuals, without which development into adult individuals cannot proceed. For example, in some *Pentatomoidea* (stink bugs), bacteria are required for the formation of the gastrointestinal tract, and the mother transmits the bacteria to offspring by licking the eggs (Hosokawa et al., 2010). In such instances, it is impossible for the individual to survive and continue replicating, without the internal symbiont. In other words, a hierarchical replication system is shaped, involving internal bacteria. This is also true for humans having bacteria within.

A further well-known example of endosymbiosis is intracellular symbiosis. Eukaryotic cells have intracellular organelles, some of which have their own genome and thus are internal replicators. Mitochondria, essential for respiration, and chloroplasts, essential for photosynthesis, are examples of such endosymbionts constituting a hierarchical replication system (Margulis, 1981).

From the standpoint of this book, the first question to be addressed here is how this hierarchical replication system can maintain its consistency. In other words, the growth rates of the endosymbiont and the host must be aligned. If the symbiont multiplies faster than the host, the number of endosymbionts will increase as the host repeatedly divides, leading to the collapse of the host. If the opposite is the case, the endosymbiont will be lost.

In fact, in *Paramecium bursaria*, an organism with Chlorella inside (but unlike the *Euglena*, this organism can live without Chlorella and can move these in and out, so that it has not yet reached full endosymbiosis), the number of Chlorella increases in proportion to the *Paramecium* growth. In other words, the density of Chlorella (i.e., its number divided by the volume of the host) is constant. One interpretation of this is that the density of the internal symbionts increases within the host, so that it has reached the transition point from the exponential growth phase to the dormant state (stationary phase), and the number growth of symbionts stops (see Chapter 3). On the other hand, as the host grows, the density decreases so that the symbionts regain the growth by going out of the stationary phase. Thus, the density of symbionts stays at the condition close to the transition to the stationary phase, so that the growth of symbionts is synchronized with the host.

The next question is how endosymbionts with their own genes evolve when they compete within the host. If a mutation increases the replication speed of the endosymbiont it may grow faster than the host, thus breaking the symbiotic relationship. How is consistency maintained in spite of this? Mutations on the host side may also occur, and if the growth rate increases further, consistency will be restored. However, this may lead to a kind of arm's race, and it may be difficult to keep the race.

On the other hand, the endosymbiont would be able to replicate faster if it had fewer internal genomes, so it is conceivable that the number of internal genomes would decrease as a result of their competition. If this were to progress to the point where symbionts completely lose their genes and are created as an organ of the host, they would no longer be independent replicators, and no more units of selection that compete with others. In other words, there will be a paradox that it is more advantageous to lose the genome, but if it is lost, it ceases to be a competing replicator. When internal genomes are maintained under this paradox is a question that needs to be answered.[2]

In considering the evolution of endosymbiosis, the optimal number of endosymbionts needs to be investigated. In a symbiotic state, it may be beneficial for the host to have a large number of symbionts, but the costs of maintaining a large number may outweigh the merit. Further, if there are too many symbionts, mutations to the symbiont's genome may be averaged out to lose their evolutionary potential, and eventually the genome itself may be lost. Conversely, if the number of symbionts is not so large, a bias in the number by the division may occur. In such a case, it will be necessary to consider some mechanism to make an equal split.[3]

It will also be necessary to understand the conditions under which either endo- or exo-symbiosis is favored. In the former, the symbionts are on the same boat of host, whereas in the latter, the hosts and symbionts exist outside and help each other. For example, if they are inside, they are always constrained to share the same chemical components they need with each other, which would be more advantageous than the case they live separately, where the chemicals are diluted. Still, when they are on the same boat, the flexibility to relax the mutual dependence will be lost.

---

[2] It is also thought that in the case of mitochondria, they are maintained because they can respond quickly to change at that location (see for example, Lane (2015).

[3] The distribution mechanism of this has been actively discussed (Adachi et al., 2006; Surovtsev & Jacobs-Wagner, 2018; Gerdes et al., 2010; Sugawara & Kaneko, 2011).

Finally, noting that the normal eukaryotic cells grow satisfying the consistency with the growth of endosymbionts, it would also be interesting to study the breakdown of the consistency. This will be related with the dormant state with arrested growth and other pathological states from this perspective.

10.2.3 Origin of Multicellular Organism

In Chapter 7 we investigated how cells differentiate from stem cells to multiple cell types that coexist stably. How, then, did such a multicellular system arise in the first place? This requires that many different types of cells with different phenotypes arise from one cell, and with the differentiation, cells grow faster or survive more than a homogeneous population of a single cell type.

In general, as the number of cells increases, they turn to be crowded and compete for limited resources. In other words, there arises a situation of "strong interactions with limited resources." In this case, depending on the reaction rules in the cells (i.e., the genotypes that determine them), cell differentiation as studied in Chapter 7 follows where cells with different phenotypes coexist (Kaneko & Yomo, 1997, 1999; Furusawa & Kaneko, 1999, 2002; Yamagishi et al., 2016). Of course, whether the phenotype differentiates in this way depends on the reaction network (gene regulatory network) within the cell. Still, when appropriate networks are chosen, differentiation occurs when the cells are crowded and strongly interacting with each other under lack of nutrition.

Is it advantageous, then, for differentiation to occur? If differentiated cell types help each other by division of labor, they will grow faster as a result. For example, Yamagishi et al. (2016) considered the catalytic reaction system in Fig. 10.1 and found that this system (1) differentiates into cell types with high concentrations of x_1, x_3 in network parts 1 and 3 (on the left side of the figure) and those with high concentrations of x_2, x_4 in the opposite parts 2 and 4, when the cell density is high and nutrition is low, and then (2) the growth rate increases when differentiated cells coexist.

The reason for the increased growth rate here is that concentrating the concentration on some components makes it easier for the reaction to proceed.[4] Indeed, in Fig. 10.1, the cell types concentrated on the two components on the left and on the right. Then, each cell has a higher growth rate than the cell type that

---

[4] For example, the rate of the catalytic reaction $x_i + x_j^2 \rightarrow x_k + x_j^2$ proceeds as the cube of the concentration. If there are N molecules in the cell and they are allocated to k components with equal amounts, then each reaction rate is proportional to k^{-3}. Even if the number of reaction types increases in proportion to k if there are k types, the synthesis rate of each component by the reaction is proportional to k^{-2}, so the synthesis rate decreases as the number of components increases. Therefore, the synthesis is faster if the molecular weight in the cell is concentrated on the necessary few components. In fact, this acceleration has been confirmed by numerical calculations.

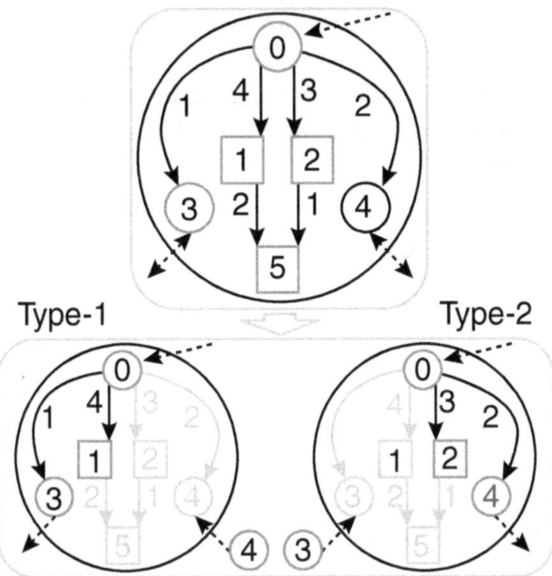

Figure 10.1 Cellular model of symbiotic growth with phenotypic differentiation. Components 1, 2, 3, 4, and 5 are produced from nutrient 0 by the reaction pathway shown in the diagram. Here, the symbol 1 on the arrow of $0 \rightarrow 3$ means that the reaction from 0 to 3 is catalyzed by the component 1. The concentration of each component of this cell model oscillates when a single cell is isolated. Then this cell grows with the reaction to convert the nutrient to internal components 1–5, and is divided to increase the number. Here components 3 and 4 are assumed to permeate the membrane and are exchanged between cells by diffusion. When this model is simulated and the number of cells increases, the cell population differentiates into cells having dominantly components 0, 1, 3, and 5 as shown on the left in the bottom row, and those having dominantly components 0, 2, 4, and 5 as shown on the right. Here, the cell on the left needs 4 to produce component 1, which is transported by diffusion from the cell on the right. Similarly, the right cell receives the necessary component 3 from the left cell Yamagishi et al. (2016). (Courtesy of Jumpei Yamagishi).

keeps and synthesizes all components. Here, if a cell has only the left partial network alone, it does not synthesize the catalytic component 4 that is necessary for the reaction to proceed. Indeed, the component is synthesized in the right partial network. Similarly, the reaction on the right side alone does not produce the necessary component 3. Now this necessary component is supplied by the opposite type of cell, which leaks the component 4 or 3, respectively. In other words, the cells specializing in each of these types help each other grow by exchanging the components produced within. This corresponds to the situation in economics, whereby each country concentrates on production of specific goods and exchanges the necessary goods with each other through trade. In this way, "symbiotic cellular differentiation" is achieved. (Yamagishi et al., 2016).

When a variety of multi-component, complex reaction networks are examined by using the catalytic reaction network discussed in Chapter 5, sufficient fraction of them exhibit this symbiotic differentiation. Hence such symbiotic differentiation is expected to be found as the cells are crowded by their growth in number and resources turn to be poor.

The emergence of the system with many different cell types that cooperate mutually for their growth, however, is not sufficient for the origin of multicellular organism. Such multicellular system must reproduce itself. Here, in many multicellular organisms, the organism of the next generation is not produced from all the cells of their mother, but from a small proportion of them, called as germ cells. A small number of germ cells and a large number of somatic cells are separated at an early stage of development (Fig. 10.2(a)). This differentiation is distinct from the symbiotic division of labor just described (Fig. 10.2(b)). Rather, cells are separated to those carrying information transmission and function (Furusawa & Kaneko, 2000b, 2002). As mentioned in Section 10.1, this structure, in which only one or few cells carry information and is transmitted to the next generation, is common to the differentiation between information-carrying units (DNA) and function-carrying units (see Table 10.1), and will be a universal property of hierarchical reproduction systems.

Indeed, if the next generation is produced from a cluster of cells, the next generation is formed from various cells that may be genetically different due to mutations. In this way, mutations are accumulated on the average. The proportion

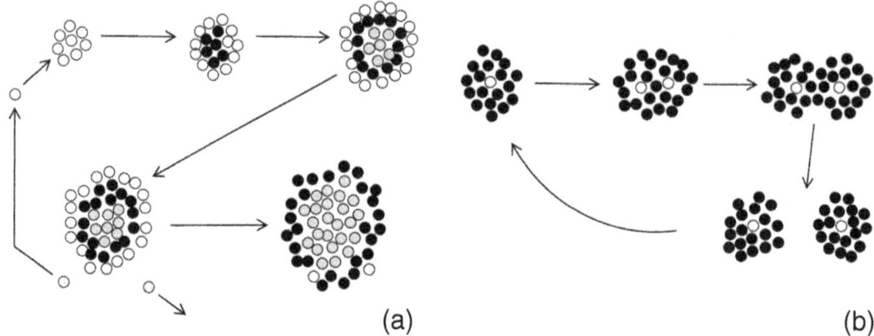

(a)   (b)

Figure 10.2 Schematic representation of the recursive production of a multicellular colony. (a) After cell differentiation with an increase in cell number, a cell (dotted one) separates from the original colony by losing its adhesion to form a novel colony of the next generation (A demonstration of such example by a cell model in Chapter 7. with inclusion of spatial structure and cell–cell adhesion is given in (Furusawa & Kaneko, 2002). (b) Schematic representation of the recursive production of a multicellular colony by the division of a minority cell at the center. The type A cell (black) and type B cell (white) form a mutually symbiotic relationship.

of mutations that increase function is generally small, so that on average the cells gradually decline their function. In contrast, if the next generation is formed from only one or few cells, there is a possibility that each mutation will be transferred without being averaged out. Of course, mutational changes leading to fitter phenotypes are rare, but once such rare mutations occurs, this will be selected for the next generation. In this way, *minority control* mechanism (Chapter 3) is at work, and this multicellular system has evolutionary potential. This is probably why for most multicellular organisms, few germ cells and somatic cells are separated, and only the former is transmitted to the next generation (see Table 10.1).

To sum up, the following scenario is proposed for the origin of multicellular organisms, based on the consistency between growth of an ensemble of cells and of a single cell incorporating intracellular reaction dynamics:

(1) As the cell number increases because of cell division, cells become crowded, and resources available for growth become limited. With resource limitation and cell–cell interactions, cellular states are diversified. At this stage, cells are diversified in phenotype, but genetic mutations can also occur.

(2) Cell–cell interaction leads to differentiation of roles for symbiotic growth: Under these strong interactions, cellular states are no longer rigidly determined and change over time. From a single genotype, diverse phenotypes can be generated. For some genotypes, different cell types attain a symbiotic relationship by achieving a division of labor. Indeed, there are several examples of intracellular reaction dynamics that support this *isologous diversification*. Even though genotypes giving rise to such reaction dynamics may be rare, colonies of cells without such reaction dynamics cease to grow. Thus, such genotypes that produce isologous diversification and symbiotic relationships, albeit rare, will remain.

(3) By taking advantage of state-dependent cell adhesion, a multicellular colony is recursively produced via a detached cell of a specific type: These cells form a cell colony by adhering with each other. With differentiated cell types, some cell types may lose their adhesion with neighboring cells. Thus, one or few cell(s) will be detached from the original colony, and a new colony of the next generation is generated from the detached cell(s). Then, the next generation of multicellular colony will be produced.

(4) This primitive form of germ-soma separation allows for evolution. Cell differentiation is established from a cell type with an oscillatory expression state, whereas pluripotency is lost with the loss of oscillation. Once this prototypical multicellular organism is achieved, genetic evolution will fix the cell differentiation scheme, as a system that reproduces itself from a minority component within has greater evolvability in general. Genetic changes in minority cells

producing the next generation are a major influence for selection. Thus, germ-somatic differentiation is established.

(5) Differentiation mediated by cell–cell interactions is established and results in regulation in the proportion of each cell type: With the interplay between intracellular reaction and cell–cell interaction, both the phenotype of each cell type and the number ratio of cells of each type is robust to developmental perturbation. Evolution further enhances developmental robustness to noise and mutation, whereas cancer cells may appear, which are originated with the loss (or decrease) of the robustness.

Note that this evolutionary process is a result of multilevel dynamics, a combination of intra- and intercellular dynamics. As these developmental dynamics are highly noisy, evolution occurs to enhance robustness to noise. Diverse, distinct cell types emerge from a single cellular state, containing diverse components, and have nonstationary dynamics. These cell types achieve differentiation of roles, and they sustain their phenotype and grow collectively. Evolution stabilizes this differentiation process. Thus, a developmental process from a stem cell with pluripotency is established, where differentiated cell types take rigid and fixed states with a loss of pluripotency. Now, the present form of cell differentiation is achieved.

Following the above viewpoint, it may be possible to propose a research project to construct prototypes of multicellular organisms by putting unicellular organisms in a moderately crowded situation, since multicellularity may be a generic process under limited resources and crowded cell conditions. Indeed, multicellular aggregates of the snowflake type are found to be generated, by using yeast (Ratcliff et al., 2012), whereas differentiation into adherent and swimming types that mutually sustain the growth are found in bacteria (*Pseudomonas fluorescens*) (Rainey & Rainey, 2003; Hammerschmidt et al., 2014). Constructive biology for multicellular organisms has just been started, and in the future, it will be interesting to examine the validity of our theoretical scenario in such constructive experiments, or to propose an alternative multilevel dynamics scenario that may match with experimental observations.

On the other hand, our multilevel dynamics provides an integrated picture of intracellular dynamics, cell–cell interactions, differentiation, morphogenesis, and growth. It provides experimentally verifiable predictions, including a possible relationship between pluripotency and oscillatory gene-expression dynamics, autonomous regulation of cell-type ratio, and robust pattern formation. Also, the hypothesis on the non-robust cancer attractor and the accumulation of mutations as discussed in Chapter 7 should be examined in the future.

Of course, several questions remain on the origin and evolution of multicellular organisms, which have to be answered in terms of consistency in multilevel dynamics. They include the following: (1) Which form of multicellular states is possible (logically and physio-chemically) as a multilevel stable state allowing for growth? (2) Which condition allows for isogenic multicellularity rather than heterogenic colonies? (3) Is there a reciprocal relationship between robustness at a cell ensemble and plasticity at a single cell level to achieve differentiation? (4) Which type of development from single (stem) cells enables the achievement recursive production of multicellular organisms? (5) How can robust developmental process (*homeorhesis*) be achieved? (6) Is there evolution–development congruence, as a result of robust multilevel dynamics? (The issues (5–6) will be discussed in Section 10.5). (7) Is cancer an unavoidable occurrence in multicellular systems beyond some level of complexity?

10.3 Ecosystem

10.3.1 Diversity

Higher up on the hierarchy of living systems is, of course, the ecosystem. Individuals of many species co-exist, influence each other, maintain their populations, and sometimes evolve. The first question in this regard is how the strongest species will not dominate to monopolize resources and many species can coexist.

To put it in layman's terms, why does not the strongest organism exist? If organisms have evolved that can grow faster than others in a certain range of given environmental conditions, then the coexistence of multiple species will not occur. The answer to this question would be that environmental conditions change depending on the existence of the organisms. In fact, the major part of the environment for an individual organism will be the surrounding organisms of the same or different species. If an organism is adapted to a certain environment and the population of the species increase, the nutrient in the environmental they use will be depleted. Each microorganism takes chemicals from the environment, and synthesizes cellular components from nutrients, and releases some of them to the outside world. As the number of organisms increases, the components that are secreted out accumulate in the environment, whereas the uptake of resources would be reduced to suppress the growth. On the other hand, the components accumulated in the environment may facilitate the increase of different species that utilize the components, and conversely, the presence of those species reduces their accumulation in the environment, allowing the two to coexist (Yamagishi et al., 2020).

In Chapter 1, we discussed why diverse components coexist within a cell, and noted that when the growth is suppressed due to the lack of nutrients, multiple components coexist. By applying the same logic to interspecies interactions, we may expect multiple species coexist when populations increase and available resources are limited. In short, the deficiency of nutritional resources due to population growth and strong interactions between individuals will lead to coexistence of multiple species. In fact, evolution to coexistence of multiple types that support each other by leakage of necessary components has been demonstrated (Yamagishi et al., 2021).

Of course, for the achievement of multispecies coexistence through evolution, species must differentiate. This requires the emergence of individuals with different genotypes and phenotypes. As mentioned at the end of Chapter 8, sympatric speciation can proceed along the following route: stronger interaction → phenotypic differentiation → genetic differentiation (Kaneko & Yomo, 2000; Kaneko, 2002a). Based on the discussion in this section, it is then expected that the following positive feedback loop: Resource depletion due to increased population leads to strong interaction → phenotypic plasticity → species diversification → increase in interaction. With this loop, species diversity will be increased. Although the appropriate modeling and theoretical formulation have not yet been completed, it is expected that when the positive feedback loop starts to work, the diversification transition will follow.

10.3.2 Resilience

Coexistence of diverse replicating units within a system is common to a cell and an ecosystem. Could we then expect a relationship similar to that found in Chapter 9? Instead of dimensional reduction on the chemical changes in a cell, can we expect similar reduction on the changes in the abundances of each species? Let us note here the differences and similarities between cells and ecosystems. The difference between ecosystems and cells or individuals is that the ecosystems themselves do not replicate in contrast to cells. Although they can spread spatially, ecosystems do not usually replicate as a whole. In this respect, it may seem difficult at first glance to obtain laws such as those found in Chapter 9.

On the other hand, both the ecosystem and cells share the postulate of robustness. Ecosystems that sustain a diversity of species can often remain stable over the long term. In a long-term span, under environmental changes, the emergence of new species as well as the extinction of some species can occur, whereas many species remain to coexist over a long time span. There will be then strong constraint to maintaining stability of diversity in this context. Stability allows for keeping population of each species, while in adapting and evolving in response to environmental change, plasticity will be needed. This situation is similar to that discussed in Chapter 9, where dimensional reduction follows to make plasticity and robustness compatible. Then, a strong constraint on the possible phenotypes and the population change of each species is expected, and a relationship analogous to Eq. (9.3) may be expected.

In fact, Sanchez and his collaborators (Goldford et al., 2018) recently investigated the nature of coexisting bacteria in leaves. The coexisting bacterial species are different by leaves, in terms of species (genes). However, at the family level, five families coexist commonly over leaves, with a certain range of population densities. Each family represents a group of a different functional phenotype. Hence, genetically, the coexisting species are different, but as for functional phenotypes the coexisting types and their distribution are identical for each leaf. This echoes Chapter 9, in which genetic evolution differs by strains, but phenotypic evolution is deterministic. Theory for such constraints in phenotypes in ecosystems need to be explored (see also Moran & Tikhonov, 2022).

On the other hand, if the number of individuals of each species in an ecosystem increases and saturates to a certain level. The rate of increase in the number of individuals of each species should be almost zero overall, and they should all be balanced (equal growth and death), in order to keep the coexistence. In other words, in a stable ecosystem, the population change of each species is neutral, as none of the species dominates over others. Indeed, by requesting this neutrality, Hubbell (2001) succeeded in explaining the log-normal distribution of each population observed in nature. This neutral theory at an ecological level is in a different hierarchy than the neutral theory in molecular evolution (Kimura, 1983). In the ecological system, neutrality emerges as a consequence of coexistence, even though each mutation is not neutral. In any case, the fact that an ecosystem is stable to many perturbations and neutral in the direction of possible change will be important to derive general laws therein.

10.4 Evolution–Development Relationship

Another macro-micro consistency emphasized in this book concerns different time scales. Indeed, we have shown the correspondence between phenotypic changes as an environmental adaptation within one generation and those across an evolutionary scale. This is concerning with the variations of the evolved phenotypic *state* (Chapters 8 and 9). Then, is there correspondence between the ontogenetic process within one generation and evolutionary change over a long period of generations, with regards to the temporal *path* to shape the phenotypes, rather than the final *state*?

Indeed, the relationship between developmental processes and evolution is one of the classic conundrums of biology. As a pioneering study, Haeckel (1866) proposed recapitulation theory, in which developmental processes follow a similar path with the phylogenic change, that is, with the evolutionary branching sequence: Its simplest and most naive picture saying for instance that mammals share a common developmental stage with fish at an earlier stage, and later with reptiles, and so forth, has been rejected. Nevertheless, the idea that there may be some relationship between evolutionary and developmental changes have attracted the attention of biologists. Furthermore, the field of Evo-Devo (the field to discuss the relationship between evolution and development) has recently been growing, which renews the interest in the possible relationship between the two (Raff, 1996; Hall, 2012; Kuratani, 2017). However, limited genetic and developmental data on past organisms in the evolutionary lineage have prevented clear conclusions from being drawn.

To understand this evolution-development relationship, Kohsokabe et al. performed evolutionary simulations of the developmental process (Kohsokabe & Kaneko, 2016). Specifically, they considered cells with gene expression changes controlled by gene regulatory network as described in Chapter 7. These cells are placed in a one-dimensional line, where some chemical products are assumed to be exchanged with neighboring cells by diffusion. In addition, external chemical components (morphogens) are supplied with a spatial gradient, providing input to the gene expression system. Under these conditions, the dynamics for the protein expression level in each cell are studied. In the model, gene (protein) expression levels reach a steady state, which may be position dependent, leading to pattern formation. Starting from the initial homogeneous state without gene (protein) expression, a spatial pattern for expressions is generated by the above gene expression dynamics with diffusion.

Let us first assume that fitness, the degree of adaptation, differs depending on the pattern generated after development, and that the fitness is higher if the pattern for output proteins generated after development is closer to a specific, predetermined, target spatial pattern. As a specific example, fitness is given as the similarity (inverse of distance) between the final resulting output expression pattern and the pre-assigned target pattern, so that evolution proceeds to reduce the distance (increase fitness): the gene regulatory network paths and parameters are slightly modified by mutation each generation, and those individuals with higher fitness are selected for the next generation.

After a few hundred generations, the evolution of the networks reaches the fittest, that is, the target pattern is generated after development from the homogeneous non-expressed initial state. Now, during the development of these individuals, several stripes are formed as (developmental) time progresses from an initial uniform pattern to the target pattern. On the other hand, if we follow the final pattern formed in each generation from an evolutionary point of view, changes from a uniform pattern to the target pattern progress. The question now is whether there is a similarity between the developmental process of pattern formation and the evolutionary process. As can be seen in the example in Fig. 10.3, in both cases stripe formation occurs in a stepwise manner, and the way the stripes are branched or created, as well as their order, is well matched between the developmental

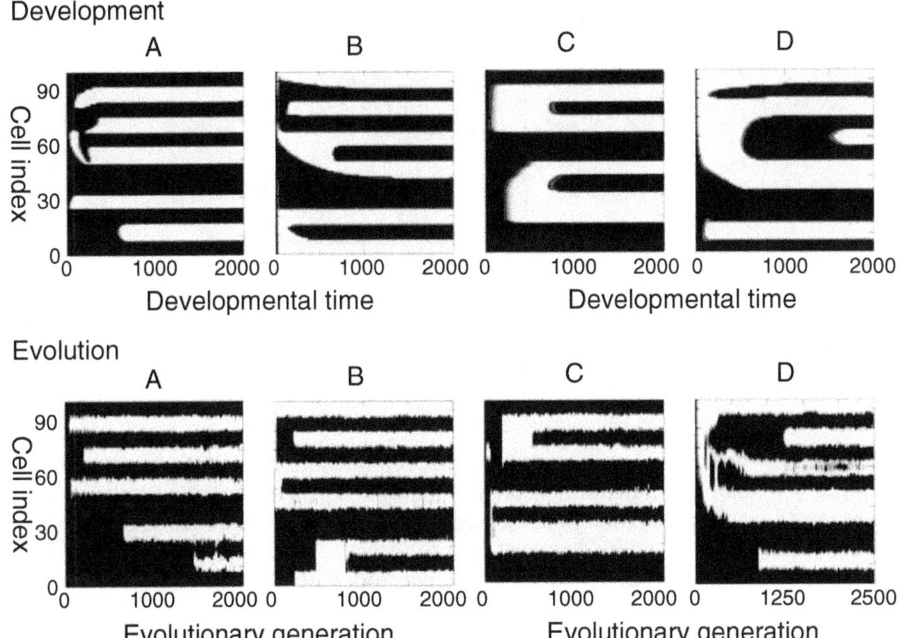

Figure 10.3 Evolution–development correspondence. The vertical axis represents the indices of cells arranged in one dimension, while the horizontal axis represents the developmental time (upper) or evolutionary generation (lower). The density in the grayscale represents the expression level of the target gene in the cell at the spatial location. In most cases (95 percent), the two correspond, as in A, B, and C. In rare cases, the correspondence is broken, as in D. Adapted from Kohsokabe & Kaneko (2016).

and evolutionary paths. Comparison of many examples of pattern formation in developmental and evolutionary processes using quantitative indicators confirms this congruence. In this sense, a relationship is found that can be described as a modern expression of recapitulation theory.[5]

In considering this correspondence, it is important to note that both developmental and evolutionary processes consist of periods of quiescence without much change and epochs of abrupt change. In the case of evolution, such stepwise changes are observed because relevant mutations must be accumulated to change the stripe pattern to increase fitness. After the accumulation of neutral mutations, the pattern changes, which is known as punctuated equilibrium (Gould & Eldredge, 1977). Why, then, do developmental processes also proceed in such a stepwise manner? It turns out that during development, the expression of certain genes (proteins) changes slowly, controlling the expression of the output genes.

---

[5] This correspondence is not identical to Haeckel's comparison between phylogeny and ontogeny, in which the comparison is made for currently existing species. In contrast, the comparison above is with ancestral species. This comparison may not be so easy in nature, as the available fossil data is often limited.

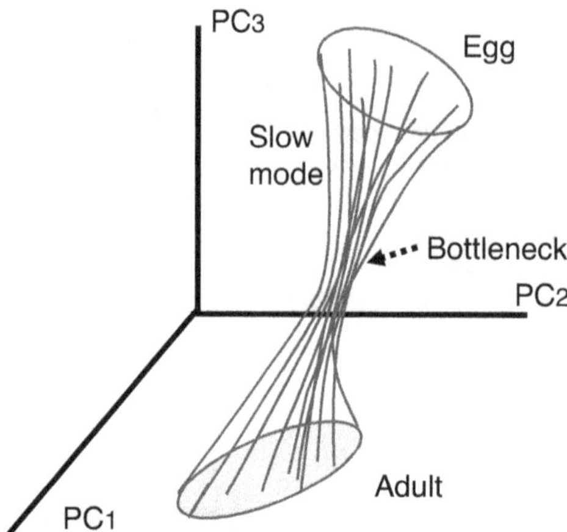

Figure 10.4 Schematic representation of developmental hourglass, robustness in developmental orbits, and dimensional reduction: The variation over initial variance or by genetic variation is decreased up to the mid-developmental stage, and then the orbits are diversified toward the adult stage. The developmental orbits are constrained in the low-dimensional space, controlled by slow modes.

This slow change in gene expression changes the pattern. Stepwise development is a result of this slowly changing component that controls patterning. Both evolutionary and developmental step changes are expressed as bifurcations, where the attractor (state) of the dynamical system changes significantly in response to a slowly changing parameter. The fact that this bifurcation coincides in evolution and development explains the correspondence between the two.

From the point of view of gene regulatory networks, the correspondence can be understood as follows: When a new pattern is generated in evolution, a new (feedforward or feedback) subnetwork is added downstream of the ancestral network. In the evolved individuals, in contrast, the entire gene regulatory networks are present from the beginning of the developmental process. However, the inputs that make the downstream networks work are missing in the early stages of development, so that they do not work effectively. Only when the slow gene expression starts to work do the downstream networks start to work. Thus, the working networks succeed in the same order as in evolution, resulting in a correspondence between the two.

For a detailed analysis, see the original paper (Kohsokabe & Kaneko, 2016, 2022). In any case, it is interesting to note that in contrast to the correlation between adaptive response and evolutionary change in the *final phenotype (state)* identified in Chapters 8 and 9, the results here reveal correspondence between the *processes* to shape the final pattern, whereas the existence of one or a few modes that change more slowly than others (protein expression levels) is consistent with the results of Chapter 9. As shown there, such a slow mode that controls others emerges evolutionarily, which also facilitates evolution. In the present case, such a slow mode (slowly changing expression) emerges through evolution, which provides the evolution–development congruence mentioned above. These results suggest that "control by separation of time scales" is one of the basic principles in evolution, which allows for evolution–development congruence: Indeed, one might wonder how the two can correspond with such different time scales. Probably the slowest mode that controls development can correspond to the fastest mode in evolution as a common bifurcation process, thus allowing the connection between the two with such distant time scales.

In most examples of numerical simulations, evolution and development correspond each other as described, but there are rare cases where the correspondence is broken. This occurs when a gene regulatory network added downstream feeds back upstream and influences the upstream, to change the dynamics at the upstream. In other words, this is a case of a breakdown in the correspondence between developmental and evolutionary processes, in which case a novel pattern-forming mechanism emerges. In this sense, collapse of consistency between development and evolution, leads to evolutionary novelty.

Studies on the evolution–development correspondence have remained at a qualitative level until recently, but it is now taking on a quantitative dimension because it has been possible to measure the developmental dynamics quantitatively by gene expression and to analyze the data on genome changes over different lineages. Among these, the developmental hourglass hypothesis has attracted much attention in recent years. According to it, the difference among different species is large in the egg stage and then shrinks in the middle stage of development and then diversifies again to form a variety of different forms. Following the decrease in the differences in gene expressions or shapes at this mid-stage, it is termed as the hourglass hypothesis (see Fig. 10.4). When it was first proposed (Duboule, 1994), it was difficult to determine quantitatively, but recently it has become possible to compare gene expression at each stage, and quantitative arguments have been made, and data that support the hypothesis have been accumulating (Kalinka et al., 2010; Irie & Kuratani, 2011; Hu et al., 2017).[6]

To test the validity of the developmental hourglass hypothesis, Kohsokabe et al. (Kohsokabe et al., 2024) performed evolutionary simulations using the

---

[6] This view is not contradictory with the evolutionary developmental correspondence described above. Once the differences have narrowed in the mid-stage, evolution and development can correspond in the later stages where the developmental processes of each species diverge.

above model. Again, they considered organisms consisting of cells aligned on a one-dimensional line containing identical gene networks controlling morphogenesis. Instead of the fitness mentioned above, selection pressure was imposed to induce more cell types, since organisms need to evolve to different spatial patterns rather than toward a unique specific pattern, in order to compare development across species. Again, after a few hundred generations, the networks have evolved to reach the fittest.

Now, by computing the similarity between the spatial patterns of gene expression of two species that evolved from a common ancestor, a developmental hourglass is indeed observed, that is, the difference in gene expression patterns is minimal in the mid-developmental stage, or in other words, the similarity between the expression patterns of two different species takes a maximum in the middle stage of development. The similarity peak in the middle developmental stage is prominent if the species being compared are phylogenetically close, while it becomes vague and eventually disappears as the species diverge further in phylogeny.

In addition to this observation, other experimentally reported characteristics in the developmental hourglass is confirmed in the model simulations. In particular, the variances of gene expression patterns across clones are minimal at the developmental bottleneck stage: Here, the phenotypic variance due to initial or developmental noise is computed with the developmental time t, as $V_{ip}(t)$, in contrast to V_{ip} of the steady state in Chapter 9. This $V_{ip}(t)$ decreases from the egg stage to the hourglass bottleneck stage, followed by an increase that leads to diversification in later developmental processes.

Behind the emergence of the developmental hourglass in the model, there exists the acquisition of genes that govern a slow scheduling process in development and induce developmental robustness. Such a slow mode controls the expression dynamics of other genes, which also facilitates evolution. In fact, these slow expression dynamics were conserved among species within a certain phylogenetic range sharing the hourglass in the model.

Thus, the developmental hourglass can be related to the robustness and dimensional reduction by a slow mode as discussed in Chapter 9: The phenotypic diversity inherited from the parents at the egg stage must be reduced at some point, in order for the developmental process to proceed stably. This leads to the low-dimensional constraint on the developmental path of phenotype, leading to the decrease in variance $V_{ip}(t)$ at mid-developmental stage, the timing of which is given by this slowly changing gene expression. By presuming that it is correlated with the variance over mutants $V_g(t)$ through the developmental time. the developmental bottleneck over species will be expected as $V_g(t)$ also takes a minimal value therein (Kohsokabe & Kaneko, 2025). In any case, correlation between evolution and development will be an issue to be explored in future, in relation to the robustness and consistency with processes of different time scales.

10.5 Memory, Cognition, and Neural Dynamics

The concepts based on consistency between hierarchical levels described in this volume, especially the interference and consistency of dynamics on different time scales, will provide fresh perspectives on memory, learning, and cognition in neural systems.

For example, the idea of consistency between phenomena on different time scales, such as genetic and phenotypic changes, discussed in evolution, may also be useful in understanding learning processes. The firing of neurons is a result of activation and inhibition of pulses from other neurons, and the firing activity is responsible for the information processing in brain. This activation-inhibition is determined by synaptic connections between neurons, and the dynamics of the overall neural firing activity is governed by the neural network, while synaptic connections change slowly during the learning process. In this respect, the dynamics of the neural activities have much in common with that of gene expression dynamics. Both have networks consisting of excitatory and inhibitory interactions (corresponding to the matrix J_{ij} in the model used in Chapters 5 and 7), and almost the same form of models with the matrices is adopted in both:[7] Both have on-off type dynamics, where expression or firing is turned on when the input to it goes beyond a threshold. In addition, the network itself changes on a slower time scale, through evolution or learning. In both cases, dynamical system on the state of the units (neural activity or gene expression patterns) depends on the network, and such network giving rise to higher function is selected accordingly.

Fluctuation–response

In the case of evolution, the rate of phenotypic evolution is correlated with the variance of fluctuations caused by noise (without genetic changes). Now, by reinterpreting evolution by learning, the learnability of the input-output relationship is expected to be correlated (proportional) to the fluctuations in neural firing before learning occurs. The direction that can be learned is expected to be constrained in the space where the variances are larger, following the result of Chapter 9. Indeed, experimental studies that can be related to those of Chapter 9 have been pursued, where high-dimensional neural firing dynamics are constrained to a low-dimensional manifold, within which the learnability is higher (Sadtler et al., 2014). In addition, an explicit formula linking the variance of neural activity to the speed

---

[7] In fact, the neural network model was proposed much earlier.

of learning has recently been obtained theoretically (Kurikawa & Kaneko, 2024). This suggests a possible link between learnability and spontaneous variances in neural firing dynamics.

In the case of evolution, behind this fluctuation–response relationship is a correlation between the robustness to noise and to genetic variation in gene expression dynamics. Similarly, in the case of neural systems, there is noise in neuronal firing dynamics on the fast timescale and noise in synaptic changes on the slow timescale. Since our memory and cognition should have robustness to these two types of noise, the correlation between the two types of robustness may be essential to study the plasticity of learning or the constraints on it, where the consistency between robustness to perturbations on these different time scales can be expected, as in Chapter 8.

Memory recall as bifurcation and dimensional reduction in spontaneous neural dynamics

The next point to be addressed in neural dynamics is analogy with differentiation from a plastic stem cell state. In the theory of Chapter 7, highly plastic stem cell states are represented by a time-varying (oscillatory) state, from which cells are differentiated into distinct cell types with fixed states, in response to inputs from other cells. The original state is an attractor without inputs, and can bifurcate into many different states depending on inputs. In this case, the initial single-cell dynamics does not necessarily have many attractors (multi-stability). Can such dynamical-systems view of cell differentiation be applied to memory and recall in the brain?

As already mentioned, dynamics of neural networks have similar characteristics to those by gene regulatory networks. In particular, the dynamical-systems models of neural networks have been established, that assign memories to each of multiple attractors (Amari, 1978; Hopfield, 1982). This is similar to the view of cell types as attractors corresponding to valleys in Waddington's landscape in Chapter 7. In contrast, we emphasized the position linking memory to dynamic processes in Chapter 7 (see also [Tsuda, 1992; Tsuda, 2002, 2016]). Indeed, memory retrieval is the result of the dynamics of neural activity in the presence of external inputs (or those from other areas of the brain) to this network. Therefore, even if a dynamical system without inputs does not have multiple attractors (stable states), multiple memories can be embedded and recalled, if bifurcation to distinct output states occurs in response to each input.

In fact, Kurikawa et al. have designed such neural network by introducing a simple synaptic connection and learning rule leading to it (Kurikawa & Kaneko, 2012, 2013, 2015; Kurikawa et al., 2019). First, by designing the neural connectivity matrix J_{ij} depending on the input and output patterns, it was shown that when an input pattern η_i^α is injected to each neuron i, the neural activity generates the corresponding output pattern ξ_i^α. This coupling is explicitly represented by a combination of inputs η_i^μ and the corresponding outputs ξ_i^μ, where $\mu = 1, \cdots, k$ is the index of stored memories.[8] Up to a certain memory capacity k_c, the corresponding output ξ_i^μ is recalled as by bifurcation upon the input η_i^μ, where k_c is about 30 percent of the total number of neurons.

Next, it is shown that such a coupling form J_{ij} can be generated by a simple learning rule to change the couplings under input. This coupling change is given by a combination of only two rules: the Hebbian rule (Hebb, 1949), which strengthens the coupling if firings of neuron i and target j are synchronized, and weakens it otherwise, and the anti-Hebbian rule, which weakens the connection between ith and jth neurons if their firings are synchronized. This type of rule for changing the coupling is consistent with neural experiments, since the strength of the coupling changes depending on the correlation of the firings. Without introducing complex mechanisms as is often done in machine learning, memories are stored by embedding fast-scale changes in firing into slow scale changes in synaptic coupling.

In this case, there are not necessarily multiple attractors corresponding to the stored patterns, but rather the state of neural activity is changed by the bifurcation upon the input to form the corresponding output (see Fig 10.5). What is interesting in this case is the neural state without inputs. As learning progresses and multiple input-output patterns are stored as embedded in J_{ij}, the neural activity in the absence of inputs is no longer constant, but varies over time (with chaotic oscillations). This dynamic activity approaches and departs from each output pattern (which must be recalled upon input). In other words, spontaneous neural activity in the absence of input can be viewed as rehearsing the output for the input. Indeed, such spontaneous activity has been found experimentally and correlated with activity under input (Kenet et al., 2003; Luczak et al., 2009).

One can see that this spontaneous activity is similar to the oscillatory state of gene expression in a pluripotent cell as discussed in Chapter 7. In both cases, there is spontaneous oscillation in the absence of input, and there appears a transition from this oscillatory state to a stable steady state in response to the external (or intercellular) input. In this respect, the cell differentiation process and the memory recall process in the brain have something in common in terms of dynamical systems.

---

[8] The simplest form for it is given by $J_{ij} = \sum_\mu (\xi_i^\mu + \eta_i^\mu)(\xi_j^\mu - \eta_j^\mu)$ (Kurikawa & Kaneko, 2012, 2013), if there is no correlation among ξ^μ's and η^μ's whereas the explicit form including the correlation is also obtained (Kurikawa & Kaneko, 2023).

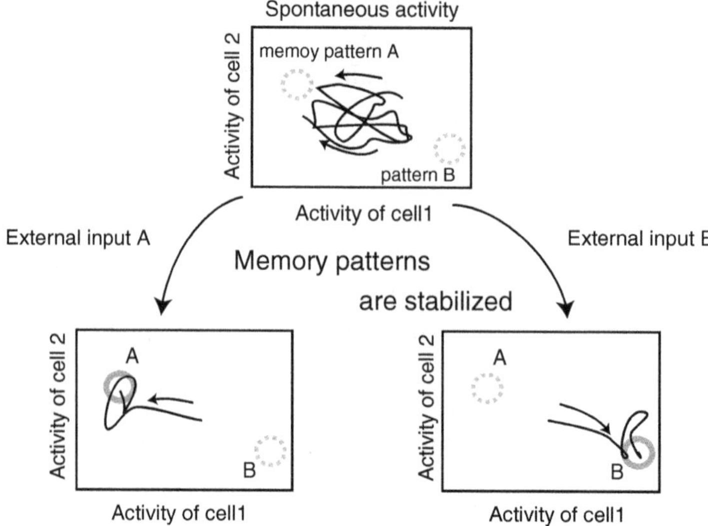

Figure 10.5 Spontaneous brain activity and memory. Memory of input-output patterns results in spontaneous activity that exhibits chaotic oscillations in the state space of neural activity. Depending on the input, the neural state diverges so that it exhibits a corresponding constant output pattern. Schematic diagram based on simulation results. (Courtesy of Tomoki Kurikawa).

In both theoretical models and actual experiments, spontaneous neural activity is not random, but is represented by an attractor in which neural output patterns change in time and wander around within a certain region in the state space. Here, changes in neural activity patterns are restricted to lower-dimensional regions within a high-dimensional state space. Such dimensional reduction is also a common feature with the cellular state in Chapters 7 and 9.

Macroscopic robustness in cognition and dimensional reduction
Although the microscopic neural dynamics are quite high-dimensional given the huge number of neurons, the macroscopic collective neural activity, is **robust**. Our cognition is not much affected by changes in individual neural activity fluctuations or by external or internal noise. On the other hand, our cognition changes in response to external inputs (on a fast timescale) and also by learning (on a slower timescale). Therefore, the neural dynamics must be **plastic**. This compatibility between robustness and plasticity is also discussed in Chapter 9 for phenotypic evolution. Following the discussion therein, the change in neural states is expected to be constrained in a low-dimensional space. As long as the response in neural activity to external inputs is not orthogonal to the constrained low-dimensional space, there will be an appropriate response. Learning can enhance the response

efficiency (Kurikawa & Kaneko, 2024). Indeed, there are some reports suggesting a drastic reduction in the dimensionality of the response of neural activity and facilitation of learning to inputs along the constrained dimension (Stadtler et al., 2024). It will be important to develop a theory of neural learning along the lines of Chapter 9.

10.6 Toward "Universal Theory" for Universal Biology

The theory of life phenomena is far from established compared to the theory of physics and still needs to be developed. At the end of this volume, a few remarks will be made on future directions. (Note that this part is quite theoretical, so you can skip it unless you are particularly interested in that direction.)

Diversity limit

In this volume, we have considered a living system as a robust system composed of diverse components, and discussed universal properties of adaptation, memory, differentiation, development, and evolution. So how *many* components are needed to consider a system "diverse"? It would make sense to consider the number that can be considered diverse even when considering the conditions for the origin of life.

As the number of components, or degrees of freedom N, increases, the combinatorial diversity over such units increases drastically. For example, the number of possible combinations of units increases with the factorial $(N-1)!$. On the other hand, the volume in the state space of an N-dimensional dynamical system increases exponentially if one considers the expressed or non-expressed (on/off) states, that is, with 2^N. In general, the combinatorial variety exceeds the exponential increase. Roughly speaking, when N exceeds $5 \sim 10$, the combinatorial variety exceeds the on/off variety. In fact, it has been found that the state of globally coupled dynamical systems and layered neural networks can no longer exhibit on-off behaviors when N exceeds $\sim 7 \pm 2$ (and this is also consistent with the number of short-term memories known in psychology) (Kaneko, 2002b; Ishihara & Kaneko, 2005). The estimation of the minimum number to achieve sufficient diversity needs to be pursued in general.

Although this is an attempt to explore the minimum number for diversity, it is also interesting if one can extract some property mathematically by taking the ideal limit for this diversity to infinity. In fact, in statistical mechanics, a rigorous formulation is possible by taking the *thermodynamic limit*, where the number of particles is infinite. With this limit, one can rigorously define the phase-transition (critical) point as a point where an order parameter becomes nonzero or the susceptibility diverges. Of course, the number of particles is not infinite in nature, so the true divergence does not occur, but if the number is reasonably large (even 10 in practice), the theory is applicable with sufficient precision. In fact, even if the number of particles is finite and there is no divergence to infinity at the phase transition point, its shadow can be seen sufficiently (*in the sense of Plato*). In the same way, in a living system, it may be possible to first consider a hypothetical limit of the infinite number of component species in order to make a clear mathematical distinction between life/nonlife, and then to understand the real cells as its shadow of the ideal limit. In fact, it might not be easy to define the point at which an ensemble of chemicals turns to be alive, whereas such a point for life might be defined as the "singular point" by taking the ideal limit of infinite diversity.

Possibility classes

Universal Biology intends to clarify possible forms of a living system, not limited to the Earth. It will then be relevant to such forms as a possible class in mathematical, physical, or chemical systems. Mathematically speaking, the life system we are concerned with is a state that has the possibility to grow and reproduce almost identical states within a high-dimensional dynamical system. Life systems are then considered as a possible class of mathematical systems that satisfy the postulates – robust systems that are consistent across hierarchies in large-degree-of-freedom dynamical systems.

This is a view of life as a mathematically possible class of universality. Next, physics requires that such a state can exist in the class without violating the thermodynamics and statistical physics of many components. So far, this is life as a physically possible class. Next is whether this state can be realized in reacting macromolecules that exist in nature, which is a possible universality class as chemistry. So far, these possible classes should be satisfied not only on Earth, but also in other places in the cosmos. Next, life that has evolved on Earth must be within a possible class of systems that can be realized under Earth's environment. The successive classification of possible systems will be important to relate universal biology to astrobiology or constructive biology.

Stability

Since the number of chemical species is quite large, while the total number of molecules in a cell is limited, the molecule numbers of some (or several) components must be small. Then the fluctuations in the concentration of such components will be quite large. In order for a cell to continue to reproduce, it must be robust to such large fluctuations in the concentration of such molecules. Such stability can be expressed in terms of dynamical systems having a strong attraction to attractors or having a large volume of basins. The former corresponds to a deep valley and the latter to a wide valley in the potential landscape. The two are generally uncorrelated, and how each of the stability is achieved needs to be explored in high-dimensional dynamical systems (Kaneko, 1997a, 1998b).

On the other hand, living systems also possess evolutionary stability, that is, the robustness of attractors against genetic changes. The genetic changes slightly alter the rule of dynamical systems, such as adding or deleting terms, or slightly changing the parameters in the equations of the dynamical system. Thus, stability against slight changes in the dynamical system itself is postulated. In the mathematics of dynamical systems there is a concept of **structural stability** as stability of attractors against changes in the system itself. It states that the topological properties of the attractor remain unchanged against arbitrarily small perturbations of the equations themselves. However, the concept of structural stability may be too general and difficult to apply to problems in the natural and life sciences, because it postulates stability against any arbitrary changes, however small. It would be important to loosen the "arbitrary" part and express the notion of evolutionary stability. A mathematical expression of robustness and plasticity based on such a concept needs to be developed in the future.

Evolution and cognition as selection of dynamical systems by dynamical systems

When considering living systems, it is necessary to consider seriously what dynamical systems have been and will be selected. This is different from many mathematical-physical models that investigate time evolution and steady states, given the rule for time evolution (by dynamical systems or stochastic processes).[9] If the evolutionary process of the phenotypes of each individual can be represented by a dynamical system, the population can be represented by a distribution of slightly different dynamical systems. The result is a distribution of phenotypes that evolves over time. The stability of this distribution provides evolutionary stability. It will be important to consider the distribution of dynamical systems itself when considering the problem of evolutionary development.

Dimensional reduction from large degrees of freedom

Although the state of living systems originally contains a very large number of degrees of freedom and is represented in a high-dimensional space, the regions in which the state can vary in response to environmental, stochastic, or evolutionary changes are constrained to a much lower-dimensional space as a consequence of robustness, as discussed in Chapters 8 and 9. Similarly, the equilibrium state in thermodynamics, which is macroscopically stable, is represented by a set of a few

---

[9] Probably cosmology could also be a discipline in which the equation itself can be selected by evolution.

variables, although microscopically it contains many degrees of freedom. Here, in statistical mechanics, projection onto collective variables (Mori, 1965) is often used to extract a small number of macroscopic degrees of freedom from a system with many degrees of freedom. The renormalization group method (Kadanoff, 1966; Wilson, 1983; Goldenfeld, 1992; Oono, 2009) is also useful for coarsening the microscopic degrees of freedom and extracting relevant modes. Extending these methods to extract a few dominant modes from many degrees of freedom in biological systems would be important for establishing the mathematical foundations of universal biology, as also discussed in Chapter 9.

Extension of the dynamical-systems framework

In this volume, large-degree-of-freedom dynamical systems and stochastic processes are used as mathematical representations of living systems. Dynamical systems are relatively new areas in the long history of mathematics, and there is still much room for further development.

Dynamical systems (and stochastic processes) follow the evolution of states in time, given three sets of "state variables, rules of evolution, and initial conditions" in exogenously given time. However, when considering biological systems, in particular development, learning, and evolution, it will be necessary to extend these three sets themselves so that they are not fixed, but change autonomously as a result of the dynamical system itself (Kaneko, 2014).

For example, consider N cells represented by state variables with k degrees of freedom that interact through the developmental process. Then the whole system is represented by a dynamical system with Nk degrees of freedom. If the cells divide and increase in number, as we saw in Chapter 7, N is not fixed but increases with time. If the state of each cell remains identical, the whole system can be described by the k degrees of freedom of a single cell, but if the cell state is differentiated due to cell–cell interactions, more degrees of freedom are needed. In this case, the number of degrees of freedom depends on the current state of the cell, that is, the result of the dynamical system, and is not given independently as an external parameter of the dynamical system (Kaneko, 1994, 1997b).

In considering a dynamical system of a single cell after cell division, its initial state is given by the state of the mother cell. In the developmental process of multicellular organisms, the initial condition of the "egg" is provided by the mother. In considering the external flux to a cell, the boundary conditions are also provided by the mother, which can change with the developmental process. In contrast to standard dynamical systems where the initial and boundary conditions are given externally, independently of the state variables, here they change as a result of changes in the cellular state during the developmental process. The formulation of dynamical systems in which the initial or boundary conditions are variable needs to be pursued.

In addition, we have discussed the interference and consistency between processes on different time scales (fast/slow processes) and the view of memory as the embedding of fast phenomena in slow processes. In learning processes, the synaptic connections between neurons change slowly, which determines the temporal evolution of fast neural activity, while the connections change as a function of fast neural activity. Epigenetic modifications are responsible for the dynamics of gene expression levels, while the former change slowly as a function of the latter, as discussed in Chapters 6 and 7. It will be necessary to provide a general mathematical formulation of such a fast-to-slow embedding process within the framework of dynamical systems.

Furthermore, in living systems, the time scale of variable changes can also vary depending on the state, as seen, for example, in the change of reaction rate regulated by the abundance of enzymes, as discussed in Chapter 6. This perspective of "time-scale machines" and the theory of dynamical systems with plastic time scales will be essential for understanding memory and dynamic information processing, as well as how life shapes its own internal time.

Self-referential dynamics

In such slow-fast dynamical systems, the slower process acts as a parameter or rule for dynamical systems of many faster variables, while the feedback from fast to slow variables leads to a change in the rule for the dynamics of the state, depending on the state variables themselves. Thus, this feedback could be viewed as a dynamical rule system for the dynamical system itself. The importance of such autonomous rule change by the state is sometimes noted as an essential character of living systems. In this respect, a dynamical-systems framework that separates the rule governing the dynamics from the states whose change is driven by the rule may have limitations. Indeed, Rosen, who wrote a pioneering textbook on the dynamical systems approach to life (Rosen, 1970), in his later years mentioned the limitations of the dynamical systems approach and sought a different path (Rosen, 2000).

Ultimately, it should be important to develop a mathematical theory that incorporates the autonomous dynamics of the rule governing the dynamics. This direction requires the study of self-referential dynamics, where the rule applied to itself changes the rule in time. Here we must pursue a system in which the rule for the dynamics and the variables driven by the rule are not separated. For this, the rule must operate on itself in order to change itself. For this direction, dynamical systems (map) of a function were introduced earlier, in which the dynamics of the function $f(x)$ contains the self-operation term $f \circ f = f(f)$. With this term, the function f can also be a state variable that is changed by it. This term $f \circ f$ leads to a self-reference, because the evolution of the function $f(x)$ is controlled by the function $f(x)$ itself. For example, the system

$$f_{n+1}(x) = (1 - \epsilon)f_n(x) + \epsilon f_n(f_n(x)),$$

with n as a time step, and ϵ as a coupling parameter is investigated (Kataoka & Kaneko, 2000, 2001, 2003; Takahashi et al., 2001).

Function dynamics studies the dynamics of rule formation as well as the mutual interference between rule and state. Unfortunately, however, the study of function dynamics remains in the realm of abstract mathematics, and so far there are no applications of it to evolution or learning. It should be important to further develop function dynamics in order to apply it to evolution, adaptation, or cognitive processes.

A possible limitation here is that in the model of function dynamics the timescale for rule and state changes is of the same order, whereas in most biological phenomena, they are separated. From the perspective of this volume, the first step is to develop an extension of dynamical systems to include the interference between slow and fast processes, the evolution of the separation of timescales and hierarchies, and the changes in degrees of freedom. These studies are intended to extend standard dynamical systems from within. It will be a challenge for the future to see how functional dynamics can emerge from or be embedded in this framework.

References

Adachi, S., Hori, K., and Hiraga, S. (2006). Subcellular positioning of F plasmid mediated by dynamic localization of SopA and SopB. *Journal of Molecular Biology*, 356(4), 850–863.

Alberts, B., Bray, D., Lewis, J., Raff, M., Roberts, K., and Watson, J. D. (1983, 1989, 1994, 2002). *The Molecular Biology of the Cell*. Garland Science.

Altmeyer, S., and McCaskill, J. S. (2001). Error threshold for spatially resolved evolution in the quasispecies model. *Physical Review Letters*, 86, 5819–5822.

Alon, U. (2006). *An Introduction to Systems Biology: Design Principles of Biological Circuits*. CRC Press.

Alon, U., Surette, M. G., Barkai, N., and Leibler, S. (1999). Robustness in bacterial chemotaxis. *Nature*, 397, 168–171.

Alon, U., Barkai, N., Notterman, D. A., et al. (1999). Broad patterns of gene expression revealed by clustering analysis of tumor and normal colon tissues probed by oligonucleotide arrays. *Proceedings of the National Academy of Sciences of the United States of America*, 96, 6745–6750.

Amari, S. (1978). *Mathematics of Neural Networks* (in Japanese). Sangyo Tosho.

Ancel, L. W., and Fontana, W. (2002). Plasticity, evolvability, and modularity in RNA. *Journal of Experimental Zoology*, 288, 242–283.

Aoki, H., and Kaneko, K. (2013). Slow stochastic switching by collective chaos of fast elements. *Physical Review Letters*, 111(14), 144102.

Ariizumi, T., and Asashima, M. (2001). In vitro induction systems for analyses of amphibian organogenesis and body patterning. *The International Journal of Developmental Biology*, 45, 273–279.

Asakura, S., and Honda, H. (1984). Two-state model for bacterial chemoreceptor proteins: The role of multiple methylation. *Journal of Molecular Biology*, 176(3), 349–367.

Awazu, A., and Kaneko, K. (2007). Discreteness-induced transition in catalytic reaction networks. *Physical Review E*, 76(4), 041915.

Balaban, N. Q., Merrin, J., Chait, R., Kowalik, L., and Leibler, S. (2004). Bacterial persistence as a phenotypic switch. *Science*, 305(5690), 1622–1625.

Bar-Even, A., Paulsson, J., Maheshri, N., et al. (2006). Noise in protein expression scales with natural protein abundance. *Nature Genetics*, 38(6), 636.

Barkai, N., and Leibler, S. (1997). Robustness in simple biochemical networks. *Nature*, 387(6636), 913.

Beisson, J., and Sonneborn, T. M. (1965). Cytoplasmic inheritance of the organization of the cell cortex in Paramecium aurelia. *Proceedings of the National Academy of Sciences of the United States of America*, 53(2), 275–282.

Blelloch, R. H., Hochedlinger, K., Yamada, Y., et al. (2004). Nuclear cloning of embryonal carcinoma cells. *Proceedings of the National Academy of Sciences of the United States of America*, 101, 13985–13990.

Benner, S. A., and Sismour, A. M. (2005). Synthetic biology. *Nature Reviews Genetics*, 6(7), 533–543.

Berg, H. C. (1975). Chemotaxis in bacteria. *Annual Review of Biophysics and Bioengineering*, 4(1), 119–136.

Berthier, L., and Biroli, G. (2011). Theoretical perspective on the glass transition and amorphous materials. *Reviews of Modern Physics*, 83(2), 587.

Biancalani, T., Rogers, T., and McKane, A. J. (2012). Noise-induced metastability in biochemical networks. *Physical Review E*, 86(1), 010106.

Boerlijst, M., and Hogeweg, P. (1991). Spiral wave structure in pre-biotic evolution – hypercycles stable against parasites. *Physica*, 48D, 17.

Braun, E. (2015). The unforeseen challenge: from genotype-to-phenotype in cell populations. *Reports on Progress in Physics*, 78(3), 036602.

Bremer, H. (1981). Simulated bacterial growth. *Journal of Theoretical Biology*, 92, 23–38.

Callahan, H. S., Pigliucci, M., and Schlichting, C. D. (1997), Developmental phenotypic plasticity: where ecology and evolution meet molecular biology. *Bioessays*, 19, 519–525.

Campbell, A. (1957). Synchronization of cell division. *Bacteriological Reviews*, 21(4), 263–272.

Campbell, K. H. S., McWhir, J., Ritchie, W. A., and Wilmut, I. (1996). Sheep cloned by transfer from a cultured cell line. *Nature*, 380, 64–66.

Cannon, W. B. (1932). *Wisdom of the Body*. W.W. Norton and Company Inc.

Carathéodory, C. (1976). Investigations into the foundations of thermodynamics. *The Second Law of Thermodynamics*, 5, 229–256.

Cavagna, A. (2009). Supercooled liquids for pedestrians. *Physics Reports*, 476(4–6), 51–124.

Chambers, I., Silva, J., Colby, D., et al. (2007). Nanog safeguards pluripotency and mediates germline development. *Nature*, 450(7173), 1230.

Chang, H. H., Hemberg, M., Barahona, M., Ingber, D. E., and Huang, S. (2008). Transcriptome-wide noise controls lineage choice in mammalian progenitor cells. *Nature*, 453(7194), 544–547.

Chow, M., Yao, A., and Rubin, H. (1994). Cellular epigenesis: Topochronology of progressive spontaneous transformation of cells under growth constraint. *Proceedings of the National Academy of Sciences*, 91, 599–603.

Ciliberti, S., Martin, O. C., and Wagner, A. (2007). Robustness can evolve gradually in complex regulatory gene networks with varying topology. *PLOS Computational Biology*, 3, e15.

Daniels, B. C. Chen, Y. J. Sethna, J. P. Gutenkunst, R. N., and Myers, C. R. (2008). Sloppiness, robustness, and evolvability in systems biology. *Current Opinion in Biotechnology*, 19(4), 389–395.

Darwin, C. (1859). *On the Origin of Species by Means of Natural Selection or the Preservation of Favored Races in the Struggle for Life* (Murray, London).

Debenedetti, P. G., and Stillinger, F. H. (2001). Supercooled liquids and the glass transition. *Nature*, 410(6825), 259.

Derrida, B., and Pomeau, Y. (1986). Random networks of automata: a simple annealed approximation. *Europhysics Letters*, 1(2), 45.

Dodd, I. B. Micheelsen, M. A. Sneppen, K., and Thon, G. (2007). Theoretical analysis of epigenetic cell memory by nucleosome modification. *Cell*, 129, 8132.

Duboule, D. (1994). Temporal colinearity and the phylotypic progression: a basis for the stability of a vertebrate Bauplan and the evolution of morphologies through heterochrony. *Development*, (Supplement), 135–142.

Dunn, S. J., Martello, G., Yordanov, B., Emmott, S., and Smith, A. G. (2014). Defining an essential transcription factor program for naive pluripotency. *Science*, 344(6188), 1156–1160.

Dyson, F. (1985). *Origins of Life*. Cambridge University Press.

Edwards, A. W. F. (2000). *Foundations of Mathematical Genetics*, Cambridge University Press.

Eigen, M. (1992). *Steps Towards Life*. Oxford University Press.

Eigen, M., and Schuster, P. (1979). *The Hypercycle* (Springer).

Einstein, A. (1905). Über die von der molekularkinetischen Theorie der Wärme geforderte Bewegung von in ruhenden Flüssigkeiten suspendierten Teilchen. *Annalen der Physik*, 17, 549–560.

Einstein, A. (1906). Zur Theorie der Brownschen Bewegung. *Annalen der Physik*, 19, 371–381.

Elena, S. F., and Lenski, R. E. (2003). Evolution experiments with microorganisms: the dynamics and genetic bases of adaptation. *Nature Reviews Genetics*, 4(6), 457.

Elowitz, M. B., and Leibler, S. (2000). A synthetic oscillatory network of transcriptional regulators. *Nature*, 403, 335–338.

Elowitz, M. B., Levine, A. J., Siggia, E. D., and Swain, P. S. (2002). Stochastic gene expression in a single cell. *Science*, 297, 1183–1186.

Frentz, Z., Kuehn, S., and Leibler, S. (2015). Strongly deterministic population dynamics in closed microbial communities. *Physical Review X*, 5(4), 041014.

Fisher, R. A. (1930;1958). *The Genetical Theory of Natural Selection*. Oxford University Press.

Forgacs, G., and Newman, S. A. (2005). *Biological Physics of the Developing Embryo*. Cambridge University Press.

Fujimoto, K., and Kaneko, K. (2003). How fast elements can affect slow dynamics. *Physica D: Nonlinear Phenomena*, 180(1–2), 1–16.

Furusawa, C., and Kaneko, K. (1998a). Emergence of rules in cell society: differentiation, hierarchy, and stability. *Bulletin of Mathematical Biology*, 60(4), 659–687.

Furusawa, C., and Kaneko, K. (1998b). Emergence of multicellular organisms with dynamic differentiation and spatial pattern. *Artificial Life*, 4, 78–89.

Furusawa, C., and Kaneko, K. (2000a). Origin of complexity in multicellular organisms. *Physical Review Letters*, 84, 6130–6133.

Furusawa, C., and Kaneko, K. (2000b). Complex organization in multicullarity as a necessity in evolution. *Artificial Life*, 6, 265–281.

Furusawa, C., and Kaneko, K. (2001). Theory of robustness of irreversible differentiation in a stem cell system: Chaos hypothesis. *Journal of Theoretical Biology*, 209(4), 395–416.

Furusawa, C., and Kaneko, K. (2002). Origin of multicellular organisms as an inevitable consequence of dynamical systems. *Anatomical Record*, 268, 327–342.

Furusawa, C., and Kaneko, K. (2003a). Zipf's Law in gene expression. *Physical Review Letters*, 90, 088102.

Furusawa, C., and Kaneko, K. (2003b). Robust development as a consequence of generated positional information. *Journal of Theoretical Biology*, 224, 413–435.

Furusawa, C., and Kaneko, K. (2008). A generic mechanism for adaptive growth rate regulation. *PLoS Computational Biology*, 4(1), e3.

Furusawa, C., and Kaneko, K. (2009). Chaotic expression dynamics implies pluripotency: when theory and experiment meet. *Biology Direct*, 4(1), 17.

Furusawa, C., and Kaneko, K. (2012a). A dynamical-systems view of stem cell biology. *Science*, 338(6104), 215–217.

Furusawa, C., and Kaneko, K. (2012b). Adaptation to optimal cell growth through self-organized criticality. *Physical Review Letters*, 108(20), 208103.

Furusawa, C., and Kaneko, K. (2013). Epigenetic feedback regulation accelerates adaptation and evolution. *PloS One*, 8(5), e61251.

Furusawa, C., and Kaneko, K. (2015). Global relationships in fluctuation and response in adaptive evolution. *Journal of The Royal Society Interface*, 12(109), 20150482.

Furusawa, C., and Kaneko, K. (2018). Formation of dominant mode by evolution in biological systems. *Physical Review E*, 97(4), 042410.

Furusawa, C., Suzuki, T., Kashiwagi, A., Yomo, T., and Kaneko, K. (2005). Ubiquity of log-normal distributions in intra-cellular reaction dynamics. *Biophysics*, 1, 25.

Futuyma, D. J. (1986). *Evolutionary Biology Second Edition*. Sinauer Associates Inc., Sunderland, Mass.

Ganti, T. (1975). Organization of chemical reactions into dividing and metabolizing units: The chemotons. *Biosystems*, 7, 189.

Gefen, O., Fridman, O., Ronin, I., and Balaban, N. Q. (2014). Direct observation of single stationary-phase bacteria reveals a surprisingly long period of constant protein production activity. *Proceedings of the National Academy of Sciences*, 111(1), 556–561.

Gehring, W. (1998). *Master Control Genes in Development and Evolution: The Homeobox Story*. Yale University Press.

Gerdes, K., Howard, M., and Szardenings, F. (2010). Pushing and pulling in prokaryotic DNA segregation. *Cell*, 141(6), 927–942.

Glass, L., and Kauffman, S. (1973). The logical analysis of continuous nonlinear biochemical control networks. *Journal of Theoretical Biology*, 39, 103–129.

Goodsell, D. S. (1998). *The machinery of Life*. Springer.

Goentoro, L., Shoval, O., Kirschner, M. W., and Alon, U. (2009). The incoherent feedforward loop can provide fold-change detection in gene regulation. *Molecular Cell*, 36(5), 894–899.

Goldenfeld, N. (1992). *Lectures on phase transitions and the renormalization group*. CRC Press.

Goldford, J. E., Lu, N., Bajić, D., et al. (2018). Emergent simplicity in microbial community assembly. *Science*, 361(6401), 469–474.

Goodwin, B. (1963). *Temporal Organization in Cells*. Academic Press, London.

Goto, Y., and Kaneko, K. (2013). Minimal model for stem-cell differentiation. *Physical Review E*, 88(3), 032718.

Gould, S. J., and Eldredge, N. (1977). Punctuated equilibria: the tempo and mode of evolution reconsidered. *Paleobiology*, 3, 115–151 (1977).

Grainger, R. M., and Gurdon, J. B. (1989). Loss of competence in amphibian induction can take place in single nondividing cells. *Proceedings of the National Academy of Sciences of the USA*, 86, 1900904.

Grey, D., Hutson, V., and Szathmary, E. (1995). A Re-examination of the stochastic corrector model. *Proceedings of the Royal Society of London. Series B*, 262, 29–35.

Gurdon, J. B., Laskey, R. A., and Reeves, O. R. (1975). The developmental capacity of nuclei transplanted from keratinized skin cells of adult frogs. *Journal of Embryology and Experimental Morphology*, 34, 93–112.

Haeckel, E. (1866). *Generelle morphologie der organismen: Allgemeine grundzuge der organischen formen – wissenschaft, mechanisch begrundet durch die von Charles Darwin reformirte Descendenz – Theorie*. Georg Reimer.

Haken, H. (1979). *Synergetics*. Springer.

Haldane, J. B. S. (1949). Suggestion as to quantitative measurement of the rate of evolution. *Evolution*, 3, 51–56.

Hall, B. K. (2012). *Evolutionary Developmental Biology*. Springer Science and Business Media.

Hammerschmidt, K., Rose, C. J., Kerr, B., and Rainey, P. B. (2014). Life cycles, fitness decoupling and the evolution of multicellularity. *Nature*, 515(7525), 75.

Hartl, D. L., and Clark, A. G. (2007). *Principles of Population Genetics*, 4th ed. Sinauer Associates, MA, USA.

Hashimoto, M., Nozoe, T., Nakaoka, et al. (2016). Noise-driven growth rate gain in clonal cellular populations. *Proceedings of the National Academy of Sciences*, 113(12), 3251–3256.

Hasty, J., Pradines, J., Dolnik, M., and Collins, J. J. (2000). Noise-based switches and amplifiers for gene expression. *Proceedings of the National Academy of Sciences USA*, 97, 2075–2080.

Hatakeyama, T. S., and Kaneko, K. (2012). Generic temperature compensation of biological clocks by autonomous regulation of catalyst concentration. *Proceedings of the National Academy of Sciences*, 109(21), 8109–8114.

Hatakeyama, T. S., and Kaneko, K. (2014a). Kinetic memory based on the enzyme-limited competition. *PLoS Computational Biology*, 10(8), e1003784.

Hatakeyama, T. S., and Kaneko, K. (2014b). Homeostasis of the period of post-translational biochemical oscillators. *FEBS Letters*, 588(14), 2282–2287.

Hatakeyama, T. S., and Kaneko, K. (2015). Reciprocity between robustness of period and plasticity of phase in biological clocks. *Physical Review Letters*, 115(21), 218101.

Hatakeyama, T. S., and Kaneko, K. (2017). Robustness of spatial patterns in buffered reaction-diffusion systems and its reciprocity with phase plasticity. *Physical Review E*, 95(3), 030201.

Hatakeyama, T. S., and Kaneko, K. (2020). Transition in relaxation paths in allosteric molecules: Enzymatic kinetically constrained model. *Physical Review Research*, 2(1), 012005.

Hatakeyama, T. S., and Kaneko, K. (2023). A linear reciprocal relationship between robustness and plasticity in homeostatic biological networks. *Plos One*, 18(1), e0277181.

Hayashi, K., de Sousa Lopes, S. M. C., Tang, F., and Surani, M. A. (2008). Dynamic equilibrium and heterogeneity of mouse pluripotent stem cells with distinct functional and epigenetic states. *Cell Stem cell*, 3(4), 391–401.

Hebb, D. O. (1949). *The organization of behavior*. A neuropsychological theory. John Wiley & Sons.

Heppner, G. H. (1984). Tumor heterogeneity. *Cancer Research*, 44(6), 2259–2265.

Himeoka, Y., and Kaneko, K. (2014). Entropy production of a steady-growth cell with catalytic reactions. *Physical Review E*, 90(4), 042714.

Himeoka, Y., and Kaneko, K. (2017). Theory for transitions between exponential and stationary phases: universal laws for lag time. *Physical Review X*, 7(2), 021049.

Himeoka, Y., and Kaneko, K. (2020). Epigenetic ratchet: Spontaneous adaptation via stochastic gene expression. *Scientific Reports*, 10(1), 459.

Hirsch, M. W., Smale, S., and Devaney, R. L. (2003). *Differential Equations, Dynamical Systems, and an Introduction to Chaos*. Academic Press.

Hochedlinger, K., Blelloch, R., Brennan, C., et al. (2004). Reprogramming of a melanoma genome by nuclear transplantation. *Genes & Development*, 18(15), 1875–1885.

Hogeweg, P. (1994). Multilevel evolution: replicators and the evolution of diversity. *Physica 75 D*, 275–291.

Hopfield, J. J. (1974). Kinetic proofreading: a new mechanism for reducing errors in biosynthetic processes requiring high specificity. *Proceedings of the National Academy of Sciences*, 71(10), 4135–4139.

Hopfield, J. J. (1982). Neural networks and physical systems with emergent collective computational abilities. *Proceedings of the National Academy of Sciences*, 79(8), 2554–2558.

Horinouchi, T., Suzuki, S., Hirasawa, T., et al. (2015). Phenotypic convergence in bacterial adaptive evolution to ethanol stress. *BMC Evolutionary Biology*, 15(1), 180.

Hoshino, E., and Kaneko, K. (2007). Unpublished note.

Hosokawa, T., Koga, R., Kikuchi, Y., Meng, X. Y., and Fukatsu, T. (2010). Wolbachia as a bacteriocyte-associated nutritional mutualist. *Proceedings of the National Academy of Sciences*, 107(2), 769–774.

Hu, H., Uesaka, M., Guo, S., et al. (2017). Constrained vertebrate evolution by pleiotropic genes. *Nature Ecology & Evolution*, 1(11), 1722.

Huang, S., and Ingber, D. E. (2006). A non-genetic basis for cancer progression and metastasis: self-organizing attractors in cell regulatory networks. *Breast Diseases*, 26, 27–54.

Huang, S. (2009). Reprogramming cell fates: reconciling rarity with robustness. *Bioessays*, 31(5), 546–560.

Huang, S., Ernberg, I., and Kauffman, S. (2009). Cancer attractors: a systems view of tumors from a gene network dynamics and developmental perspective. *Seminars in Cell and Developmental Biology*, 20, 869–876.

Hubbell, S. P. (2001). *The Unified Neutral Theory of Biodiversity and Biogeography*. Princeton University Press.

Ichihashi, N., Matsuura, T., Kita, H., Sunami, T., Suzuki, H., and Yomo, T. (2010). Constructing partial models of cells. *Cold Spring Harbor Perspectives in Biology*, a004945.

Ichihashi, N., Usui, K., Kazuta, Y., Sunami, T., Matsuura, T., and Yomo, T. (2013). Darwinian evolution in a translation-coupled RNA replication system within a cell-like compartment. *Nature Communications*, 4, 2494.

Ichikawa, K., and Kaneko, K. (2024). Bayesian inference is facilitated by modular neural networks with different time scales. *PLOS Computational Biology*, 20(3), e1011897.

Inoue, M., and Kaneko, K. (2011). Weber's law for biological responses in autocatalytic networks of chemical reactions. *Physical Review Letters*, 107(4), 048301.

Inoue, M., and Kaneko, K. (2013). Cooperative adaptive responses in gene regulatory networks with many degrees of freedom. *PLoS Computational Biology*, 9(4), e1003001.

Irie, N., and Kuratani, S. (2011). Comparative transcriptome analysis reveals vertebrate phylotypic period during organogenesis. *Nature Communications*, 2, 248.

Ishihara, S., and Kaneko, K. (2005). Magic number 7 ± 2 in networks of threshold dynamics. *Physical Review Letters*, 94, 058102.

Ito, Y., Toyota, H., Kaneko, K., and Yomo, T. (2009). How selection affects phenotypic fluctuation. *Molecular Systems Biology*, 5(1), 264.

Ito-Miwa, K., Furuike, Y., Akiyama, S., and Kondo, T. (2020). Tuning the circadian period of cyanobacteria up to 6.6 days by the single amino acid substitutions in KaiC. *Proceedings of the National Academy of Sciences*, 117(34), 20926–20931.

Jennings, H. S. (1906). *Behavior of the Lower Organisms*. The Colombia University press, New York.

Jones, P. A., and Baylin, S. B. (2002). The fundamental role of epigenetic events in cancer. *Nature Reviews Genetics*, 3, 415–426.

Jordan, D. J., and Miska, E. A. (2023). Canalisation and plasticity on the developmental manifold of Caenorhabditis Belegans. *Molecular Systems Biology*, 19 e11835.

Kadanoff, L. P. (1966). Scaling laws for Ising models near Tc. *Physics Physique Fizika*, 2(6), 263.

Kadanoff, L. P. (2000). *Statistical Physics: Statics, Dynamics and Renormalization* (World Scientific).

Kaern, M., Elston, T. C., Blake, W. J., and Collins, J. J. (2005). Stochasticity in gene expression: from theories to phenotypes. *Nature Reviews Genetics*, 6, 451–464.

Kalinka, A. T., Varga, K. M., Gerrard, D. T., et al. (2010). Gene expression divergence recapitulates the developmental hourglass model. *Nature*, 468(7325), 811.

Ken-ichiro, F. K., Kobayashi-Kirschvink, K. J., Nozoe, T., Nakaoka, H., Umetani, M., and Wakamoto, Y. (2025). Revealing global stoichiometry conservation architecture in cells from Raman spectral patterns. *eLife*, 14.

Kamimura, A., and Kaneko, K. (2010). Reproduction of a protocell by replication of a minority molecule in a catalytic reaction network. *Physical Review Letters*, 105(26), 268103.

Kamimura, A., and Kaneko, K. (2015). Transition to diversification by competition for multiple resources in catalytic reaction networks. *Journal of Systems Chemistry*, 6(1), 5.

Kamimura, A., and Kaneko, K. (2016). Negative scaling relationship between molecular diversity and resource abundances. *Physical Review E*, 93(6), 062419.

Kamimura, A., and Kaneko, K. (2018). Exponential growth for self-reproduction in a catalytic reaction network: relevance of a minority molecular species and crowdedness. *New Journal of Physics*, 20(3), 035001.

van Kampen, N. G. (1971). The case against linear response theory. Phys. Norveg. 5, 279.

van Kampen, N. G. (1992). *Stochastic Processes in Physics and Chemistry*. Elsevier.

Kaneko, K. (1990). Clustering, coding, switching, hierarchical ordering, and control in network of chaotic elements. *Physica D: Nonlinear Phenomena*, 137–172.

Kaneko, K. (1994a). Relevance of clustering to biological networks. *Physica 75D*, 55.

Kaneko, K. (1997a). Dominance of milnor attractors and noise-induced selection in a multi-attractor system. *Physical Review Letters*, 78, 2736–2739.

Kaneko, K. (1997b). Coupled maps with growth and death: An approach to cell differentiation. *Physica D: Nonlinear Phenomena*, 103, 505–527.

Kaneko, K. (1998a). Life as complex systems: Viewpoint from intra-inter dynamics. *Complexity*, 3, 53–60.

Kaneko, K. (1998b). On the strength of attractors in a high-dimensional system: Milnor attractor network, robust global attraction, and noise-induced selection. *Physica D*, 124, 322–344.

Kaneko, K. (2002a). Symbiotic sympatric speciation: Compliance with Interaction-driven phenotype differentiation from a single genotype. *Population Ecology*, 44, 71–85.

Kaneko, K. (2002b). Dominance of milnor attractors in globally coupled dynamical systems with more than 7 ± degrees of freedom. *Physical Review E.* 66, 055201(R).

Kaneko, K. (2005). On recursive production and evolvabilty of cells: Catalytic reaction network approach. *Advances in Chemical Physics*, 130, 543–598.

Kaneko, K. (2006). *Life: An Introduction to Complex Systems Biology*. Springer.

Kaneko, K. (2007). Evolution of robustness to noise and mutation in gene expression dynamics. *PLoS One*, 2(5), e434.

Kaneko, K. (2009). Relationship among phenotypic plasticity, phenotypic fluctuations, robustness, and evolvability; Waddington's legacy revisited under the spirit of Einstein. *Journal of Biosciences*, 34(4), 529.

Kaneko, K. (2011a). Characterization of stem cells and cancer cells on the basis of gene expression profile stability, plasticity, and robustness. *Bioessays*, 33(6), 403–413.

Kaneko, K. (2011b). Proportionality between variances in gene expression induced by noise and mutation: consequence of evolutionary robustness. *BMC Evolutionary Biology*, 11(1), 27.

Kaneko, K. (2012a). Phenotypic plasticity and robustness: evolutionary stability theory, gene expression dynamics model, and laboratory experiments in Evolutionary systems biology (pp. 249–278). Springer New York.

Kaneko, K. (2012b). Evolution of robustness and plasticity under environmental fluctuation: Formulation in terms of phenotypic variances. *Journal of Statistical Physics*, 148(4), 687–705.

Kaneko, K. (2014). Dynamical systems++ for a theory of biological system in *Chaos, Information Processing and Paradoxical Games: The Legacy of John S Nicolis* (ed.), G. Nicolis and B. Vasileois, World Scientific Publishing Co. Pte. Ltd.

Kaneko, K. (2016). *Scenario for the Origin of Multicellular Organisms: Perspective from Multilevel Consistency Dynamics in Multicellularity: Origins and evolution.* MIT Press., Niklas, K. J., Newman, S. A., eds.

Kaneko, K. (2024). Constructing universal phenomenology for biological cellular systems: an idiosyncratic review on evolutionary dimensional reduction. *Journal of Statistical Mechanics: Theory and Experiment*, 2024(2), 024002.

Kaneko, K., and Furusawa, C. (2006). An evolutionary relationship between genetic variation and phenotypic fluctuation. *Journal of Theoretical Biology*, 240(1), 78–86.

Kaneko, K., and Furusawa, C. (2018). Macroscopic theory for evolving biological systems akin to thermodynamics. *Annual Review of Biophysics*, 47, 273–290.

Kaneko, K., Furusawa, C., and Yomo, T. (2015). Universal relationship in gene-expression changes for cells in steady-growth state. *Physical Review X*, 5(1), 011014.

Kaneko, K., Sato, K., Michiue, T., et al. (2008). Developmental potential for morphogenesis in vivo and in vitro. *Journal of Experimental Zoology Part B: Molecular and Developmental Evolution*, 310(6), 492–503.

Kaneko, K., and Tsuda, I. (1994). Constructive complexity and artificial reality: an introduction. *Physica D: Nonlinear Phenomena*, 1–10.

Kaneko, K., and Tsuda, I. (2000). *Complex Systems: Chaos and Beyond ——A Constructive Approach with Applications in Life Sciences.* Springer.

Kaneko, K., and Tsuda, I. (2003). Chaotic itinerancy. *Chaos*, 13(3), 926.

Kaneko, K., and Yomo, T. (1994). Cell division, differentiation, and dynamic clustering. *Physica D: Nonlinear Phenomena*, 89–102.

Kaneko, K., and Yomo, T. (1997). Isologous diversification: a theory of cell differentiation. *Bulletin of Mathematical Biology*, 59, 139–196.

Kaneko, K., and Yomo, T. (1999). Isologous diversification for robust development of cell society. *Journal of Theoretical Biology*, 199, 243–256.

Kaneko, K., and Yomo, T. (2000). Symbiotic speciation from a single genotype. *Proceedings of the Royal Society B*, 267, 2367–2373.

Kaneko, K., and Yomo, T. (2002a). On a kinetic origin of heredity: minority control in replicating molecules. *Journal of Theoretical Biology*, 312, 563–576.

Kaneko, K., and Yomo, T. (2002b). Symbiotic sympatric speciation through interaction-driven phenotype differentiation. *Evolutionary Ecology Research*, 4, 317–350.

Kasemeier-Kulesa, J. C., Teddy, J. M., Postovit, L. M., et al. (2008). Reprogramming multipotent tumor cells with the embryonic neural crest microenvironment. *Developmental Dynamics*, 237, 2657–2666.

Kashiwagi, A., Urabe, I., Kaneko, K., and Yomo, T. (2006). Adaptive response of a gene network to environmental changes by fitness-induced attractor selection. *PloS One*, 1(1), e49.

Kataoka, N., and Kaneko, K. (2000). Functional dynamics. I: Articulation process. *Physica D: Nonlinear Phenomena*, 138(3–4), 225–250.

Kataoka, N., and Kaneko, K. (2001). Functional dynamics: II: Syntactic structure. *Physica D: Nonlinear Phenomena*, 149(3), 174–196.

Kataoka, N., and Kaneko, K. (2003). Dynamical networks in function dynamics. *Physica D: Nonlinear Phenomena*, 181(3–4), 235–251.

Kauffman, S. A. (1969). Metabolic stability and epigenesis in randomly constructed genetic nets. *Journal of Theoretical Biology*, 22, 437.

Kauffman, S. (1971). Differentiation of malignant to benign cells. *Journal of Theoretical Biology*, 31, 429–451.

Kauffman, S. A. (1993). *The Origin of Order*, Oxford University Press.

Kenet, T., Bibitchkov, D., Tsodyks, M., Grinvald, A., and Arieli, A. (2003). Spontaneously emerging cortical representations of visual attributes. *Nature*, 425(6961), 954.

Kimura, M. (1983). *The Neutral Theory of Molecular Evolution*. Cambridge University Press.

Kirkpatrick, S., Gelatt, C. D., and Vecchi, M. P. (1983). Optimization by simulated annealing. *Science*, 220(4598), 671–680.

Kobayashi, T., Mizuno, H., Imayoshi, I., Furusawa, C., Shirahige, K., and Kageyama, R. (2009). The cyclic gene Hes1 contributes to diverse differentiation responses of embryonic stem cells. *Genes & Development*, 23(16), 1870–1875.

Kohsokabe, T., and Kaneko, K. (2016). Evolution–development congruence in pattern formation dynamics: Bifurcations in gene expression and regulation of networks structures. *Journal of Experimental Zoology Part B: Molecular and Developmental Evolution*, 326(1), 61–84.

Kohsokabe, T., and Kaneko, K. (2022). Dynamical systems approach to evolution–development congruence: Revisiting Haeckel's recapitulation theory. *Journal of Experimental Zoology Part B: Molecular and Developmental Evolution*, 338(1–2), 62–75.

Kohsokabe, T., and Kaneko, K. (2025). in preparation.

Kohsokabe, T., Kuratanai, S., and Kaneko, K. (2024). Developmental hourglass: Verification by numerical evolution and elucidation by dynamical-systems theory. *PLOS Computational Biology*, 20(2), e1011867.

Komatsu, S. (1968). *Tsugu-no-wa dareka* (Who is the successor) in Japanese (Hayakawa Pub.)

Kondepudi, D., and Prigogine, I. (2014). *Modern Thermodynamics: From Heat Engines to Dissipative Structures*. John Wiley and Sons.

Kondo, Y., and Kaneko, K. (2011). Growth states of catalytic reaction networks exhibiting energy metabolism. *Physical Review E*, 84(1), 011927.

Koshland, D. E., Goldbeter, A., and Stock, J. B. (1982). Amplification and adaptation in regulatory and sensory systems. *Science*, 217, 220–225.

Kotte, O., Volkmer, B., Radzikowski, J. L., and Heinemann, M. (2014). Phenotypic bistability in Escherichia coli's central carbon metabolism. *Molecular Systems Biology*, 10(7), 736.

Kubo, R. (1957). Linear response theory and fluctuafion. Dissipation Theorem. *J. Phys. Soc. Japan*, 12, 570–586.

Kubo, R., Toda, M., and Hashitsume, N. (1985). *Statistical Physics II: Nonequilibrium Statistical Mechanics* (Vol. 31). Springer Science and Business Media.

Kuratani, S. (2017). *Dobutsu-Shinka-Keitaigaku*. University of Tokyo Press, in Japanese.

Kurikawa, T., and Kaneko, K. (2012). Associative memory model with spontaneous neural activity. *Europhysics Letters*, 98(4), 48002.

Kurikawa, T., and Kaneko, K. (2013). Embedding responses in spontaneous neural activity shaped through sequential learning. *PLoS Computational Biology*, 9(3), e1002943.

Kurikawa, T., and Kaneko, K. (2015). Memories as bifurcations: Realization by collective dynamics of spiking neurons under stochastic inputs. *Neural Networks*, 62, 25–31.

Kurikawa, T., and Kaneko, K. (2024). Fluctuation-learning relationship in neural networks. *arXiv preprint arXiv:2409.13597*.

Landry, C. R., Lemos, B., Rifkin, S. A., Dickinson, W. J., and Hartl, D. L. (2007). Genetic properties influencing the evolvability of gene expression. *Science*, 317(5834), 118–121.

Lane, N. (2015). *The Vital Question: Energy, Evolution, and the Origins of Complex Life*. WW Norton and Company.

Levy, S. F. Ziv, N., and Siegal, M. L. (2012). Bet Hedging in yeast by heterogeneous, age-correlated expression of a stress protectant. *PLoS Biol*, 10(5), e1001325.

Li, F., Long, T., Lu, Y., Ouyang, Q., and Tang, C. (2004). The yeast cell-cycle network is robustly designed. *Proceedings of the National Academy of Sciences of the United States of America*, 101(14), 4781–4786.

Li, L., Connelly, M. C. Wetmore, C., et al. (2003). Mouse embryos cloned from brain tumors. *Cancer Research*, 63, 2733–2736.

Lieb, E. H., and Yngvason, J. (1999). The physics and mathematics of the second law of thermodynamics. *Physics Reports*, 310(1), 1–96.

Lisman, J. E., and Goldring, M. A. (1988). Feasibility of long-term storage of graded information by the Ca2+/calmodulin-dependent protein kinase molecules of the postsynaptic density. *Proceedings of the National Academy of Sciences*, 85(14), 5320–5324.

Lisman, J., Schulman, H., and Cline, H. (2002). The molecular basis of CaMKII function in synaptic and behavioural memory. *Nature Reviews Neuroscience*, 3, 1750.

López-Maury, L., Marguerat, S., and Bähler, J. (2008). Tuning gene expression to changing environments: from rapid responses to evolutionary adaptation. *Nature Reviews Neuroscience*, 9(8), 583–593.

Luczak, A., Bartho, P., and Harris, K. D. (2009). Spontaneous events outline the realm of possible sensory responses in neocortical populations. *Neuron*, 62(3), 413–425.

MacArthur, R., and Wilson, E. O. (1967). *The Theory of Island Biogeography*. Princeton University Press.

Maeda, Y., Mizuuchi, R., Shigenobu, S., et al. (2022). Experimental evidence for the correlation between RNA structural fluctuations and the frequency of beneficial mutations. *RNA*, 28(12), 1659–1667.

Matsumoto, Y., Murakami, Y., Tsuru, S., Ying B-Y., and Yomo, T. (2013). Growth rate-coordinated transcriptome reorganization in bacteria. *BMC Genomics*, 14, 808.

Matsuura, T., Ichihashi, N., Sunami, T., Kita, H., Suzuki, H., and Yomo, T. (2011). Evolvability and Self-Replication of Genetic Information in Liposomes in *The Minimal Cell* (pp. 275–287). Springer Netherlands.

Matsuura, T., Yomo, T., Yamaguchi, M., et al. (2002). Importance of compartment formation for a self-encoding system. *Proceedings of the National Academy of Sciences USA*, 99, 7514–7517.

Margulis, L. (1981). *Symbiosis in Cell Evolution*. W.H. Freemand and Company.

Maynard-Smith, J. (1979). Hypercycles and the origin of life. *Nature*, 280, 445–446.

Maynard-Smith, J. (1989). *Evolutionary Genetics*. Oxford University Press.

Maynard-Smith, J., and Szathmary, E. (1995). *The Major Transitions in Evolution* (W.H. Freeman).

McCulloch, W. S., and Pitts, W. (1943). A logical calculus of the ideas immanent in nervous activity. *The Bulletin of Mathematical Biophysics*, 5(4), 115–133.

Mezard, M., Parisi, G., and Virasoro, M. A. eds. (1987). *Spin Glass Theory and Beyond*. World Scientific Publishing.

Mikhailov, A., and Hess, B. (1995). Fluctuations in living cells and intracellular traffic. *Journal of Theoretical Biology*, 176, 185–192.

Milo, R., and Phillips, R. (2015). *Cell Biology by the Numbers*. Garland Science.

Mintz, B., and Ilmensee, K. (1975). Normal genetically mosaic mice produced from malignant teratocarcinoma cells. *Proceedings of the National Academy of Sciences of the USA*, 72, 3585–3589.

Mizuuchi, R., and Ichihashi, N. (2018). Sustainable replication and coevolution of cooperative RNAs in an artificial cell-like system. *Nature Ecology & Evolution*, 2(10), 1654.

Mjolsness, E., Sharp, D. H., and Reinitz, J. (1991). A connectionist model of development. *Journal of Theoretical Biology*, 152, 429–453.

Monod, J. (1949). The growth of bacterial cultures. *Annual Reviews in Microbiology*, 3(1), 371–394.

Moran, J., and Tikhonov, M. (2022). Defining coarse-grainability in a model of structured microbial ecosystems. *Physical Review X*, 12(2), 021038.

Mori, H. (1965). Transport, collective motion, and Brownian motion. *Progress of Theoretical Physics*, 33(3), 423–455.

Nakajima, A., and Kaneko, K. (2008). Regulative differentiation as bifurcation of interacting cell population. *Journal of Theoretical Biology*, 253(4), 779–787.

Nakajima, M., Imai, K., Ito, H., et al. (2005). Reconstitution of circadian oscillation of cyanobacterial KaiC phosphorylation in vitro. *Science*, 308, 414–415.

Nakaoa, Y., Tokui, H., Gion, Y., Inoue, S., and Oosawa, F. (1982). Behavioral adaptation of Paramecium caudatum to environmental temperature. *Proceedings of the Japan Academy, Series B*, 58(7), 213–217.

Newman, S. A. (1994). Generic physical mechanisms of tissue morphogenesis: a common basis for development and evolution. *Journal of Evolutionary Biology*, 7, 467–488.

Newman, S. A., and Comper, W. D. (1990). Generic physical mechanisms of morphogenesis and pattern formation. *Development*, 110, 1–18.

Nicolis, G., and Prigogine, I. (1977). *Self-organization in Nonequilibrium Systems* (Wiley).

Ohno, S. (1970). *Evolution by Gene Duplication*. Springer Berlin.

Ohta, T. (1992). The nearly neutral theory of molecular evolution. *Annual Review of Ecology and Systematics*, 23(1), 263–286.

Ohta, T. (2011). Near-neutrality, robustness, and epigenetics. *Genome Biology and Evolution*, 3, 1034–1038.

Opper, M., and Diederich, S. (1992). Phase transition and 1/f noise in a game dynamical model. *Physical Review Letters*, 69, 1616.

Oono, Y. (2012). *The Nonlinear World: Conceptual Analysis and Phenomenology*. Springer Science and Business Media.

Pfeuty, B., and Kaneko, K. (2014). Reliable binary cell-fate decisions based on oscillations. *Physical Review E*, 89(2), 022707.

Pfeuty, B., and Kaneko, K. (2016). Requirements for efficient cell-type proportioning: regulatory timescales, stochasticity and lateral inhibition. *Physical Biology*, 13(2), 026007.

Pham, M. T., and Kaneko, K. (2023). Double-replica theory for evolution of genotype-phenotype interrelationship. *Physical Review Research*, 5, 023049.

Pham, T. M., and Kaneko, K. (2023). Theory for adaptive systems: collective robustness of genotype-phenotype evolution. *arXiv*:2306.01403.

Pirt, S. J. (1965). The maintenance energy of bacteria in growing cultures. *Proceedings of the Royal Society B*, 163(991), 224–231.

Price, G. R. (1970). Selection and covariance. *Nature*, 227, 520–521.

Price, G. R. (1972). Extension of covariance selection mathematics. *Annals of Human Genetics*, 35(4), 485–490.

Rashevsky, N. (1940). Physicomathematical aspects of some problems of organic form. *The Bulletin of Mathematical Biophysics*, 2(3), 109–121.

Rashevsky, N. (1960). *Mathematical biophysics: physico-mathematical foundations of biology*. Literary Licensing, LLC.

Raff, R. A. (1996). *The Shape of Life: Genes, Development, and the Evolution of Animal Form*. University of Chicago Press.

Rafols, I., Sawada, Y., Amagai, A., Maeda, Y., and MacWilliams, H. K. (2001). Cell type proportioning in Dictyostelium slugs: lack of regulation within a 2.5-fold tolerance range. *Differentiation*, 67(4–5), 107–116.

Rainey, P. B., and Rainey, K. (2003). Evolution of cooperation and conflict in experimental bacterial populations. *Nature*, 425(6953), 72.

Ratcliff, W. C., Denison, R. F., Borrello, M., and Travisano, M. (2012). Experimental evolution of multicellularity. *Proceedings of the National Academy of Sciences*, 109(5), 1595–1600.

Reya, T., Morrison, S. J., Clarke, M. F., and Weissman, I. L. (2001). Stem cells, cancer, and cancer stem cells. *Nature*, 414, 105–111.

Rice, S. H. (2004). *Evolutionary Theory: Mathematical and Conceptual Foundations*. Sunderland, MA: Sinauer Associates.

Ritort, F., and Sollich, P. (2003). Glassy dynamics of kinetically constrained models. *Advances in Physics*, 52(4), 219–342.

Rohner, P. T., and Berger, D. (2023). Developmental bias predicts 60 million years of wing shape evolution. *Proceedings of the National Academy of Sciences*, 120(19), e2211210120.

Rosen, R. (1970). *Dynamical System Theory in Biology*. Wiley and Sons.

Rosen, R. (2000). *Essays on Life Itself*. Columbia University Press.

Ronen, M., Rosenberg, R., Shraiman, B., and Alon, U. (2002). Assigning numbers to the arrows: parameterizing a gene regulation network by using accurate expression kinetics. *Proceedings of the National Academy of Sciences of the USA*, 99, 10555–10560 (2002).

Rubin, H. (1990). The significance of biological heterogeneity. *Cancer and Metastasis Reviews*, 9 (1990), 1–20.

Rubin, H. (1994a). Cellular epigenetics: control of the size, shape, and spatial distribution of transformed foci by interactions between the transformed and nontransformed cells. *Proceedings of the National Academy of Sciences*, 91, 1039–1043.

Rubin, H. (1994b). Experimental control of neoplastic progression in cell populations; Fould's rules revisited. *Proceedings of the National Academy of Sciences*, 91, 6619–6623.

Russell, J. B., and Baldwin, R. L. (1979). Comparison of maintenance energy expenditures and growth yields among several rumen bacteria grown on continuous culture. *Applied and Environmental Microbiology*, 37(3), 537–543.

Sadtler, P. T., Quick, K. M., Golub, M. D., et al. (2014). Neural constraints on learning. *Nature*, 512(7515), 423.

Saito, N., Sughiyama, Y., and Kaneko, K. (2016). Motif analysis for small-number effects in chemical reaction dynamics. *The Journal of Chemical Physics*, 145(9), 094111.

Saito, K., Tsuboi, M., and Takahashi, Y. (2024). Developmental noise and phenotypic plasticity are correlated in Drosophila simulans. *Evolution Letters*, 8(3), 397–405.

Sakata, A., Hukushima, K., and Kaneko, K. (2009). Funnel landscape and mutational robustness as a result of evolution under thermal noise. *Physical Review Letters*, 102(14), 148101.

Sakata, A., and Kaneko, K. (2020). Dimensional reduction in evolving spin-glass model: correlation of phenotypic responses to environmental and mutational changes. *Physical Review Letters*, 124(21), 218101.

Sakata, A., and Kaneko, K. (2023). Evolutionary shaping of low-dimensional path facilitates robust and plastic switching between phenotypes. *Physical Review Research*, 5(4), 043296.

Salazar-Ciudad, I. Newman, S. A., and Sole, R. V. (2001). Phenotypic and dynamical transitions in model genetic networks I. *Evolution and Development*, 3, 84.

Salazar-Ciudad, I., Sole, R. V., and Newman, S. A. (2001). Phenotypic and dynamical transitions in model genetic networks II. Application to the evolution of segmentation mechanisms. *Evolution & Development*, 3(2), 95–103.

Sato, K., Ito, Y., Yomo, T., and Kaneko, K. (2003). On the Relation between Fluctuation and Response in Biological Systems. *Proceedings of the National Academy of Sciences of the USA*, 100, 14086–14090.

Sato, K., and Kaneko, K. (2006). On the distribution of state values of reproducing cells. *Physical Biology*, 3(1), 74.

Sato, T., and Kaneko, K. (2020). Evolutionary dimension reduction in phenotypic space. *Physical Review Research*, 2, 013197.

Sato, T. U., Furusawa, C., & Kaneko, K. (2023). Prediction of cross-fitness for adaptive evolution to different environmental conditions: Consequence of phenotypic dimensional reduction. *Physical Review Research*, 5, 043222.

Schmidt, A., Kochanowski, K., Vedelaar, S., et al. (2016). The quantitative and condition-dependent *Escherichia coli* proteome. *Nature Biotechnology*, 34(1), 104.

Schreier, H. I. Soen, Y., and Brenner, N. (2017). Exploratory adaptation in large random networks. *Nature Communications*, 8, 14826.

Schrödinger, E. (1946). *What is Life*. Cambridge University Press.

Scott, M., Gunderson, C. W., Mateescu, E. M., Zhang, Z., and Hwa, T. (2010). Interdependence of cell growth and gene expression: origins and consequences. *Science*, 330(6007), 1099–1102.

Shimizu, Y., Kanamori, T., and Ueda, T. (2005). Protein synthesis by pure translation systems. *Methods*, 36(3), 299–304.

Shoval, O., Goentoro, L., Hart, Y., Mayo, A., Sontag, E., and Alon, U. (2010). Fold-change detection and scalar symmetry of sensory input fields. *Proceedings of the National Academy of Sciences*, 107(36), 15995–16000.

Silva, A. J. Kogan, J. H. Frankland, P. W., and Kida, S. (1998). CREB and memory. *Annual Review of Neuroscience*, 21, 1278.

Sompolinsky, H., Crisanti, A., and Sommers, H. J. (1988). Chaos in random neural networks. *Physical Review Letters*, 61, 259.

Sprinzak, D., and Elowitz, M. B. (2005). Reconstruction of genetic circuits. *Nature*, 438, 443–448.

Spudich, J. L., and Koshland, D. E. Jr. (1976). Non-genetic individuality: chance in the single cell. *Nature*, 262, 467–471.

Stearns, S. C., Kaiser, M., and Kawecki, T. J. (1995). The differential genetic and environmental canalization of fitness components in *Drosophila melanogaster*. *Journal of Evolutionary Biology*, 8, 539.

Stern, S., Dror, T., Stolovicki, E., Brenner, N., and Braun, E. (2007). Genome-wide transcriptional plasticity underlies cellular adaptation to novel challenge. *Molecular Systems Biology*, 3(1), 106.

Stern, S., Fridmann-Sirkis, Y., Braun, E., and Soen, Y. (2012). Epigenetically heritable alteration of fly development in response to toxic challenge. *Cell Reports*, 1(5), 528–542.

Steward, F. C., Mapes, M. O., and Mears, K. (1958). Growth and organized development of cultured cells. II. Organization in cultures from freely suspended cells. *American Journal of Botany*, 45, 705–708.

Strogatz, S. (2001). *Nonlinear Dynamics and Chaos: With Applications to Physics, Biology, Chemistry, and Engineering*. Perseus Books.

Sugawara, T., and Kaneko, K. (2011). Chemophoresis as a driving force for intracellular organization: theory and application to plasmid partitioning. *Biophysics*, 7, 77–88.

Surovtsev, I. V., and Jacobs-Wagner, C. (2018). Subcellular organization: a critical feature of bacterial cell replication. *Cell*, 172(6), 1271–1293.

Suzuki, N., Furusawa, C., and Kaneko, K. (2011). Oscillatory protein expression dynamics endows stem cells with robust differentiation potential. *PLoS One*, 6(11), e27232.

Suzuki, S., Horinouchi, T., and Furusawa, C. (2014). Prediction of antibiotic resistance by gene expression profiles. *Nature Communications*, 5, 5792.

Szathmary, E., and Demeter, L. (1987). Group selection of early replicators and the origin of life. *Journal of Theoretical Biology*, 128, 463–486.

Szathmary, E., and Maynard-Smith, J. (1997). From replicators to reproducers: the first major transitions leading to life. *Journal of Theoretical Biology*, 187, 555–571.

Takagi, H., and Kaneko, K. (2005). Dynamical Systems Basis of Metamorphosis: Diversity and Plasticity of Cellular States in Reaction Diffusion Network. *Journal of Theoretical Biology*, 234, 173–186.

Takagi, H., Furusawa, C., Sawai, S., and Kaneko, K. (2025). Theoretical Biology of the Cell – Dynamical Systems Perspective (Cambridge University Press).

Takahashi, K., and Yamanaka, S. (2006). Induction of pluripotent stem cells from mouse embryonic and adult fibroblast cultures by defined factors. *Cell*, 126(4), 663–676.

Takahashi, Y., Kataoka, N., Kaneko, K., and Namiki, T. (2001). Function dynamics. *Japan Journal of Industrial and Applied Mathematics*, 18(2), 405.

Takeuchi, N., Kaneko, K., and Hogeweg, P. (2016). Evolutionarily stable disequilibrium: endless dynamics of evolution in a stationary population in *Proceedings of the Royal Society B* (Vol. 283, No. 1830, p. 20153109). The Royal Society.

Takeuchi, N., Hogeweg, P., and Kaneko, K. (2017). The origin of a primordial genome through spontaneous symmetry breaking. *Nature Communications*, 8(1), 250.

Takeuchi, N., and Kaneko, K. (2019). The origin of the central dogma through conflicting multilevel selection. *Proceedings of the Royal Society B*, 286(1912), 20191359.

Takeuchi, N., Kaneko, K., and Koonin, E. V. (2014). Horizontal gene transfer can rescue prokaryotes from Mullers ratchet: benefit of DNA from dead cells and population subdivision. *G3: Genes, Genomes, Genetics*, 4(2), 325–339.

Takeuchi, N., Mitarai, N., and Kaneko, K. (2022). A scaling law of multilevel evolution: how the balance between within-and among-collective evolution is determined. *Genetics*, 220(2), iyab182.

Tan, C., Marguet, P., and You, L. (2009). Emergent bistability by a growth-modulating positive feedback circuit. *Nature Chemical Biology*, 5(11), 842–848.

Tang, Q. Y., and Kaneko, K. (2021). Dynamics-evolution correspondence in protein structures. *Physical Review Letters*, 127(9), 098103.

Taniguchi, Y., Choi, P. J., Li, G. W., et al. (2010). Quantifying *E. coli* proteome and transcriptome with single-molecule sensitivity in single cells. *Science*, 329(5991), 533–538.

Thom, R. (1975). *Structural Stability and Morphogenesis*. (Pergamon).

Tlusty, T., Libchaber, A., and Eckmann, J. P. (2017). Physical model of the genotype-to-phenotype map of proteins. *Physical Review X*, 7(2), 021037.

Togashi, Y., and Kaneko, K. (2001). Transitions induced by the discreteness of molecules in a small autocatalytic system. *Physical Review Letters*, 86, 2459.

Togashi, Y., and Kaneko, K. (2005). Discreteness-induced stochastic steady state in reaction diffusion systems: Self-consistent analysis and stochastic simulations. *Physica D*, 205, 87–99.

Togashi, Y., and Mikhailov, A. S. (2007). Nonlinear relaxation dynamics in elastic networks and design principles of molecular machines. *Proceedings of the National Academy of Sciences of the United States of America*, 104(21), 8697–8702.

Tsuda, I. (1991). Chaotic itinerancy as a dynamical basis of Hermeneutics in brain and mind. *World Futures*, 32, 167.

Tsuda, I. (1992). Dynamic link of memory–chaotic memory map in nonequilibrium Neural networks. *Neural Networks*, 5, 313.

Tsuru, S., Ichinose, J., Kashiwagi, A., Ying, B. W., Kaneko, K., and Yomo, T. (2009). Noisy cell growth rate leads to fluctuating protein concentration in bacteria. *Physical Biology*, 6(3), 036015.

Tsuru, S., Yasuda, N., Murakami, Y., et al. (2011). Adaptation by stochastic switching of a monostable genetic circuit in Escherichia coli. *Molecular Systems Biology*, 7(1), 493.

Turing, A. M. (1952). The chemical basis of morphogenesis. *Philosophical Transactions of the Royal Society B*, 237, 37–72.

Uchida, Y., Takeda, H., Furusawa, C., and Irie, N. (2023). Stability in gene expression and body-plan development leads to evolutionary conservation. *EvoDevo*, 14(4).

de Visser, J. A. G. M., Hermisson, J., Wagner, G. P., et al. (2003). Evolution and detection of genetic robustness. *Evolution*, 57, 1959–1972.

van Zon, J. S. Lubensky, D. K. Altena, P. R. H., and ten Wolde, P. R. (2007). An allosteric model of circadian KaiC phosphorylation. *Proceedings of the National Academy of Sciences of the United States of America*, 104, 7420–7425.

Veening, J. W., Hamoen, L. W., and Kuipers, O. P. (2005). Phosphatases modulate the bistable sporulation gene expression pattern in Bacillus subtilis. *Molecular Microbiology*, 56, 1481–1494.

Waddington, C. H. (1957). *The Strategy of the Genes*. George Allen and Unwin LTD., Bristol.

Wagner, A. (2007). *Robustness and Evolvability in Living Systems*. Princeton University Press.

Wakamoto, Y., Dhar, N., Chait, R., Schneider, K., Signorino-Gelo, F., Leibler, S., and McKinney, J. D. (2013). Dynamic persistence of antibiotic-stressed mycobacteria. *Science*, 339(6115), 91–95.

Wang, P., Robert, L., Pelletier, J., Dang, W. L., Taddei, F., Wright, A., and Jun, S. (2010). Robust growth of Escherichia coli. *Current Biology*, 20(12), 1099–1103.

West-Eberhard, M. J. (2003). *Developmental Plasticity and Evolution*, Oxford University Press.

Wilson, K. G. (1983). The renormalization group and critical phenomena. *Reviews of Modern Physics*, 55(3), 583.

Yamagishi, J. F., and Kaneko, K. (2024). Universal transitions between growth and dormancy via intermediate complex formation. *Physical Review Letters*, 132(11), 118401.

Yamagishi, J. F., Saito, N., and Kaneko, K. (2016). Symbiotic cell differentiation and cooperative growth in multicellular aggregates. *PLoS Computational Biology*, 12(10), e1005042.

Yamagishi, J. F., Saito, N., and Kaneko, K. (2020). Advantage of leakage of essential metabolites for cells. *Physical Review Letters*, 124(4), 048101.

Yamagishi, J. F., Saito, N., and Kaneko, K. (2021). Adaptation of metabolite leakiness leads to symbiotic chemical exchange and to a resilient microbial ecosystem. *PLOS Computational Biology*, 17(6), e1009143.

Yoshida, T., Murayama, Y., Ito, H., Kageyama, H., and Kondo, T. (2009). Nonparametric entrainment of the in vitro circadian phosphorylation rhythm of cyanobacterial KaiC by temperature cycle. *Proceedings of the National Academy of Sciences of the United States of America*, 106, 1648–1653.

Zhuangzi (Chuang Tzu) (around 2C BC), in Chinese; English translation, by Watson, B. (1968). *The Complete Works of Chuang Tzu*. New York: Columbia University Press.

Zipf, G. K. (1949). *Human Behavior and the Principle of Least Effort*. Addison-Wesley, Cambridge.

Zwanzig, R. (1973). Nonlinear generalized Langevin equations. *Journal of Statistical Physics*, 9, 215–220.

Index

For EU product safety concerns, contact us at Calle de José Abascal, 56–1°,
28003 Madrid, Spain or eugpsr@cambridge.org.